THE POETRY AND MUSIC OF SCIENCE

PRAISE FOR *THE POETRY AND MUSIC OF SCIENCE*

'By carefully engaging with historic and contemporary mathematicians, scientists, writers, artists, and composers, McLeish is able to illuminate the commonalities inherent in scientific and artistic endeavours—both the moments of creative inspiration and the diligent, methodical efforts that follow. The author provides the most sustained, learned, and insightful exploration of the mutual characteristics of music and mathematics that I know of. McLeish offers us an expansive perspective on the human yearning to discover and create. In doing so, he also provides a release from narrow views of the sciences and the arts and an opportunity to reconcile diverse academic specializations. As a musician and historian, I cherish this book for the insights it gives into both the compositions I perform and the academic work I engage in. The book is a delight to read and a profound encouragement for those of us who create—on stage, or in academic forums.'

Elaine Stratton Hild, University of Würzburg

'In this wide-ranging and engaging meditation on the parallel work of imagination, inspiration, and intuition found in both art and science, McLeish closes the illusory gap between the two cultures. McLeish, a physicist, heals the divisions in our modern communities of learning by drawing on his considerable knowledgeable of history, art, philosophy, music, and literature.'

Bernard Lightman, President of the History of Science Society and Distinguished Research Professor of the Humanities at York University

'Tom McLeish has written a book which shows as well as tells. It reinstates creativity and imagination as being as much part of the practice of science as close observation, experiment, and rigour. Reading it is a pleasure for all parts of the brain, and the heart as well as the mind. It offers insights to help everyone interested in the future conduct of science, and in the creation and sharing of wisdom and beauty in the twenty-first century.'

Claire Craig Provost of the Queen's College, Oxford

'Tom McLeish writes as a lifelong scientist with a rich appreciation of the arts and a deep historical knowledge. He makes a compelling case for seeing the scientific and the artistic imagination as intertwined and interdependent. McLeish shows us that science and the arts never have been, never could be, and never should be two divided cultures.'

John Holmes, Professor of Victorian Literature and Culture, University of Birmingham

'This kind of book is rarer than it should be, and all the more valuable for that. It dares to take seriously and probe deeply the interplay of the arts and the sciences, neither setting one above the other nor looking for glib connections but instead interrogating both as expressions of human creativity and imagination. In place of the tired and reductive notion of "Two Cultures", Tom McLeish reveals—passionately, persuasively, and with great scholarship—that so many meaningful points of contact exist between the sciences and music, literature and visual art that by the end of the book it would seem strange to put them back in separate boxes. May this be the beginning of a new and rich conversation!'

Philip Ball, Science Writer

'Where do creative ideas come from? There is an answer, and it is the same for ideas in art as it is for ideas in science. There is a hidden wellspring inside the human mind from which they arise continuously. Tom McLeish provides meticulous evidence by interrogating the greatest minds. The result is a brilliant kaleidoscopic view of the history of imagination.'

Uta Frith FBA FRS, UCL Institute of Cognitive Neuroscience

'Within the short compass of this subtle and elegant exposition, McLeish tackles one of the most disabling narratives of our time. Creativity is neither a luxury nor a disqualification in a world whose survival requires all our imaginative resources, and it infuses the arts and sciences in uncannily similar ways. The author has also created a rare and beautiful thing: few could embrace such a range of artistic and scientific endeavour with such an uplift.'

Marilyn Strathern DBE, Professor of Social Anthropology, Cambridge University

'Reading this book is like eavesdropping on a series of intimate and detailed conversations among a set of accomplished scientists and artists on the subject of imagination, play and creativity, conversations in which McLeish himself participates. What McLeish's book reveals is the astounding fact that these conversations are seamless, involving moments of transformation and insight that are not only familiar but frequent.'

Robert Crease, Chair of the Department of Philosophy, Stony Brook University

'An interdisciplinary tour-de-force of astonishing clarity and depth, this book blows apart the "Two Cultures" argument for the rigid standoff between the sciences and the rest of the arts and humanities. McLeish is a twenty-first century Renaissance man, and his words are just the tonic for anyone who loudly doubts that science involves creativity—as well as for those who quietly know from lived experience that it certainly does.'

Jennifer Rohn, Principle Research Associate at University College London and author of *Cat Zero*

'Many things have been said and written about the connections between maths and music but I have never encountered anything remotely as deep as this. McLeish exhibits mind-boggling inter-disciplinarity with depth as well as breadth, and his prose is as rich in poetry and music as the subject of the book.'

Eugenia Cheng, Scientist in Residence at the School of the Art Institute of Chicago

'*The Poetry and Music of Science* will be recognized for years to come as a triumph that bridged the sciences and the humanities in an effort to call out the creativity and imagination that feeds them both. This book on creativity is creatively written and a pleasure to read.'

Elaine Howard Ecklund, author of *Science vs. Religion*

'Anyone who believes that imagination, inspiration and creativity are the preserve of the arts should read this beautifully crafted ode to the enterprise of scientific discovery.'

Jim Al-Khalili OBE FRS, Professor of Theoretical Physics, University of Surrey

THE POETRY AND MUSIC OF SCIENCE

Comparing Creativity in Science and Art

Tom McLeish

Professor of Natural Philosophy, University of York, UK

OXFORD
UNIVERSITY PRESS

OXFORD
UNIVERSITY PRESS

Great Clarendon Street, Oxford, OX2 6DP,
United Kingdom

Oxford University Press is a department of the University of Oxford.
It furthers the University's objective of excellence in research, scholarship,
and education by publishing worldwide. Oxford is a registered trade mark of
Oxford University Press in the UK and in certain other countries

Impression: 1

Published in the United States of America by Oxford University Press
198 Madison Avenue, New York, NY 10016, United States of America

British Library Cataloguing in Publication Data
Data available

Library of Congress Control Number: 2022934657

ISBN 978–0–19–284537–5 (pbk)

DOI: 10.1093/oso/9780192845375.001.0001

Printed and bound by
CPI Group (UK) Ltd, Croydon, CR0 4YY

To Juanita Kalerghi Rothman,

who exemplified how loving and thinking of science, writing,
music, and art bring them together in the human soul.

Preface to first edition

This all began on a school visit—I believe in the North Yorkshire town of Harrogate. I was a visiting university speaker to a group of bright high-school pupils in support of their 'General Studies' activity. I can't be sure now of the topic—it might have been the history of science, or the science of plastics, or even the fascinating story of religion and science—but the message for me came during the discussion. The UK school system, as many readers will know, is uniquely specialized after the age of 16, by any international comparison. This meant that some of the 17- and 18-year-olds in the group were no longer studying any science or mathematics topics at all. From the level of their discussion, many of them were clearly extremely bright and could have pursued any subject they had wished, so I asked them why they had not chosen to continue with any further studies in science. 'Because I saw no room for any imagination, or my personal creativity, in science', was the common response. I still remember feeling pain, and knew then that something had gone very awry in the way we talk about science in education and in the media. My own experience as a practising scientist had been completely different. The school I attended, a generation before, had unusually allowed me to continue studies in French, alongside sciences and mathematics, before going to university, and there I had been helped to see that science cannot begin without the first creative step. I began to perceive that these two fortunate experiences were not unrelated. Scientists had failed somehow to communicate the creative core of science to the young people I was with, and clearly to many others besides.

The later privilege of working as a University Pro-Vice-Chancellor for Research (the post would be called Vice-President for Research in the USA) at Durham University gave me a wonderful daily insight into the thinking of imaginative academic colleagues from the humanities and social sciences, as well as of the university's great scientists. I had, of course, read C. P. Snow's *Two Cultures*, and recognized the failed intercultural communication that concerned him, but became increasingly convinced that our disciplines share far more than we recognize. Our schools and universities are not in general structured in support of interdisciplinary conversation, but whenever we created space to talk, some precious spark seemed to be struck. Some of those sparks took hold and became serious projects that are still growing.

The first thanks therefore go to the institutions and individuals at Durham who gave their time, energies, and encouragement to an interdisciplinary vision. Professor Veronica Strang, Executive Director of

Durham's Institute for Advanced Studies, is an interdisciplinary inspiration, and kindly hosted the introductory seminar that started the project that became this book. Prof. Giles Gasper is the visionary leader of the extraordinary *Ordered Universe* collaboration which brings the complementary perspectives of medieval scholars and scientists together to study the science of the thirteenth century. This project shows in a fascinating way that science possesses a much longer and deeper story than is usually told. It also illustrates in a tangible way that scientists and humanities scholars have much in common, while introducing me to the extraordinary thirteenth-century polymath, Robert Grosseteste, who will also assist us at several points in this book. Revd. Prof. David Wilkinson, Principal of St John's College Durham, has my eternal gratitude for his intellectual, faithful, and personal encouragement to frame theology and science in fresh ways. I am greatly indebted to Prof. Patricia Waugh for pointing me to the entanglements of science and the writing of novels, to Prof. Julian Horton for a shared love of Schumann and for his expert analysis of the *Konzertstück*, and to Prof. Martin Cann for a long and patient collaboration between biology and physics.

An invitation from the Notre Dame Institute for Advanced Study (Indiana) as Director's Fellow 2017–8 was instrumental, especially in the research for Chapter 4, but also in support of the book at a whole. NDIAS Director, Prof. Brad Gregory, inspired me with his vision for the 'unity of knowledge', and I am deeply grateful for the discussions, readings, and comments of other NDIAS fellows and ND academic staff, especially Profs. Laura Dassow-Walls, Celia Deane-Drummond, Steve Fallon, Margot Fassler, Patrick Griffin, Chris Kolda, David Bentley Hart, Elaine Stratton Hild, Bernard McGinn, Richard Taylor, Emily Dumler Winckler, Xinyu Dong, and my two wonderful undergraduate research assistants Sofia Carozza and Jeremy Cappello Lee. Without the creative support of Don Stelluto and Carolyn Sherman, none of that would have happened.

The third important setting for this book is my new academic home of the University of York. The imaginative decision to create an interdisciplinary Chair of Natural Philosophy, so regenerating a tradition dormant in England for about two centuries, has already proved a joy to work in. In a short time, new colleagues have already nourished the content of this book, including and especially Dr. Mary Garrison, Profs. Keith Allen, Kevin Killeen, Dr. Jeanne Nuechterlein, and Prof. Elizabeth Tyler.

Listening to artists, art curators, musicians, scientists, and educationalists talk about their work candidly and honestly has been an immense privilege and an essential source of wisdom. Thanks go especially to the *Ordered Universe* project's artists in residence and collaborators Alexandra

Carr, Colin Rennie, and Cate Watkinson, and to Vanessa Chamberlin, Janet Graham, Jeremy Mayall, the late Graeme Willson, Prof. Berry Billingsley, Prof. Michael Reiss, Prof. Stephen Blundell, Prof. Julie Kornfield, Katie Lewis, Prof. Ard Louis, Prof. Wilson Poon, Prof. Scott Milner, Prof. Christopher Southgate, and Prof. Hannah Smithson. I am grateful to Jill Cook of the British Museum for the inspiration of her *Living with the Gods* exhibition, and to Jennifer Thompson of the Philadelphia Museum of Art for access to and discussion of Monet's *Pines at Sunset*. Of great value was the *Imagination Institute* workshop held in Kings College, Cambridge, in 2016. For this, thanks are due to Ashley Zauderer, Philip Ball and to Profs. Robert Sternberg, James Kaufman, Martin Rees, Michael Berry, Michael Cates, Melissa Franklin, Naomi Leonard, Jon Keating, John Pendry, and Herbert Huppert.

Several colleagues, both in the humanities and sciences, have been particularly generous in working through details of their stories of creativity and discovery with me, greatly enriching the book. I am especially grateful to Prof. Henrike Lange (Berkeley) for sharing her ground-breaking work on Giotto and the Arena Chapel, to Prof. Julie Kornfield (Caltech) for the story of the jet-fuel additive, and to Prof. Jan Vermant (Zürich) for his tale of biophysical pattern-formation.

I am very grateful to friends and colleagues who have gone the extra mile by reading and commenting on drafts, often suggesting vital new discussions and sources that I would otherwise have missed, but at the same time kindly picking up a myriad of small errors. Any that remain sit very firmly at my door. Especial thanks go to Prof. Carl Gombrich, Dr. Mary Garrison, Profs. Michael Reiss, Andrew Steane, and Prof. Adrian Sutton for comprehensive overhauls, and to Prof. Mark Miller, Julie McLeish, Profs. Marilyn Strathern, Victoria Lorrimar, and Iain McGilchrist for particular comments. Thanks go to Rosie McLeish for very efficient typesetting of the musical quotations in Chapter 6 and to Brian Slater for the excellent photography of Graeme Willson's artwork. Graeme himself was inspirational throughout the writing of the book, and is sorely missed. Those on this list who share my surname have also supported the necessary labours with their patience and love more than they know. Juanita Kalerghi Rothman has been a lifelong inspiration to connect the scientific and the artistic, to find aesthetics in engineering, and contemplation in creativity of all kinds. To her, and to her memory, this book is gratefully dedicated.

The editorial team at OUP, especially Ania Wronski and Sonke Adlung, are a perpetual joy to work with, and are most certainly due thanks for their encouragement but especially their patience. It was fun.

Tom McLeish

York,

Feast of St Barnabas, the Son of Encouragement, 2018

Preface to the revised edition

I did not anticipate the extent to which *The Poetry and Music of Science* would touch such a deep nerve within educational, research, communication, and religious communities, when it first appeared. The two years since publication have been a wonderful learning experience for me, and I recall with immense gratitude the many discussions following public presentations of the book's ideas, in contexts from university interdisciplinary events and graduate training programmes, to cathedral public lecture series, literary festival events, summer schools on science and poetry, and webcast series. The 2019 *York Festival of Ideas* held an entire day's sequence on the theme of the book, with three panels of inspirational artists and scientists each engaging with one of the 'modes' of creativity I proposed. There is 'creativity talk' in the air—and surely no coincidence that Prof. Marcus du Sautoy's book on artificial creativity, *The Creativity Code*[1] appeared at the same time (we shared a fascinating evening's public discussion at the Royal Society for Arts just before the pandemic hit).

Three events in particular have catalysed this revised edition, edited throughout and with one entirely new chapter. The first was a launch event at Emmanuel College, Cambridge University, at the kind invitation of the Master, Dame Fiona Reynolds, who chaired a quite remarkable panel of discussants. Dame Ottoline Leyer FRS (now Chief Executive of UK Research and Innovation), poet, priest, and Coleridge scholar, Revd Dr Malcolm Guite, and theologian of science and former biologist, Dr Andrew Davison, together explored ways in which a renewed vision of public confidence and contemplation through science, an urgent environmental mandate, and how Coleridge's (and Shakespeare's) poetic theology could act as windows into the interdisciplinary roots of creativity.

The second was an invitation from the editor of the journal, *Interdisciplinary Science Reviews*, Prof. Willard McCarty, to engage with a special issue of the journal dedicated to multiple reviews of the book, together with a précis and response from me. The result[2] should henceforth be read as a critical companion volume to the present one. I am very grateful indeed to Profs. Michael Whitworth, Steve Fuller, Irmtraud Huber, Andrew Hugill, Bennett Zon, James Leach, and to poet John Barnie as well as to Willard

[1] Marcus du Sautoy (2019), *The Creativity Code*. London: Harper Collins

[2] *Interdisciplinary Science Reviews*, **45**(1), 1–70, (2020)

himself, for their constructively critical and scholarly reviews. This revised edition responds only partially to their wisdom (it does little, for example, to address the vast field of artificial intelligence, despite the vital questions that AI generates in the field of creativity, to which Andrew Hugill rightly draws attention in his review, apart from a brief extension to the discussion of Hannah Arendt in the final chapter, in the light of du Sautoy's book). Collectively, the reviews and the response paper cover the topics of: (i) historical contexts of 'creativity' and 'imagination,' (ii) multi-cultural framings of the questions, (iii) the differences (rather than similarities) of the creative process in science and art, (iv) their different social and institutional structures, (iv) computing and digital creativity, and (v) the theological strands of the discussion.

There is one topic raised in the *ISR* issue that has generated an entirely new chapter for this revised edition. Several reviewers pointed out that, for a book with 'poetry' in its title, the first edition rather under-delivered in that department. My original introduction even complained that Wordsworth's prophecy—that the 'day would come' when science would be as much subject to the poet's art as any other—had never been fulfilled. Then John Barnie wrote his *ISR* review on a personal account of deep poetic immersion in paleo-anthropology, Mary Peelen and Katrina Porteous kindly sent me copies of their own excellent science-inspired poetry, Malcolm Guite showed me the connection between Coleridge and science, I discovered Maria Popova's annual, *Universe in Verse* festivals, Bethany Green introduced me to Margaret Cavendish's atom poetry, and last year, the first journal dedicated to science-poetry, *Consilience*, was founded by Dr. Sam Illingworth. To hammer the message home, I received a kind invitation from Dr. Brian Volk to join him as a discussant in the 2020 Glen Workshop theme *Science, Poetry and the Imagination*. From all this generosity, the new Chapter 5, *Poetry and Theoretical Science*, offered itself as a bridge from the juxtaposition of prose fiction and experiment, to the comparison of mathematics and music in Chapter 6. I am additionally grateful to English literature scholars Bethany Green and Rosie McLeish (Cambridge University), and to theologian and environmentalist, Kaley Casenhiser (Yale), for poetry readings and discussions. Mary Peelen and John Barnie have kindly agreed to the reproduction of their science-inspired and inspirational poems, and the Carcanet Press Ltd, Manchester, have, with similar generosity, permitted me to include some by astronomer-poet, the late Dr. Rebecca Elson. I am grateful to all of these, as well as to Berkeley Art Historian, Italianist, and long-time interdisciplinary collaborator, Prof. Henrike Lange, who very kindly read the entire new draft, and to David Keplinger, Kaley Casenhiser, Mary Peelen, Brian Volk, Prof. Tom Stoneham, and Michael

Schmidt for reading the new chapter, as well as to the many kind readers of the first edition who have pointed out errors, typological and otherwise. Any that remain here are, as always, entirely my own responsibility. Sonke Adlung and his colleagues at Oxford University Press have been, as ever, uniformly encouraging and helpful.

York,
Festival of St Anselm, 2021

Contents

1 Introduction

Creativity and Constraint

Art is limitation; the essence of every picture is the frame.

G. K. CHESTERTON

Creativity, Inspiration, Passion, Form, Imagination, Composition, Representation—this powerful list of words leads a reader's mind inevitably into the world of the arts. Perhaps it conjures up the shaping of a block of stone into the form of supple limbs and torso, or layering darkly tinted oil-paints onto canvas to tease the eye into imagining a moonlit forest at night. Others may think of a composer scoring a symphony's climax—she summons the horns to descend as from a distant mountain peak to meet a harmonically ascending string bass-line in a satisfying resolution. A poet at his desk wrestles with meter and rhyme as he filters the streams of words, metaphors, and allusions that clamour for place on the page. The double miracle of art is not only that it allows humans to draw meaning from the world, but also that it reaches out to its listeners, viewers, and readers so that they may re-create for themselves something new and personal in response. Both by words and by images we are changed, troubled, made more aware, as art enriches us in small ways or great. To engage in art by creation, or reception and re-creation, is to exercise one of the capacities that make us human. Indeed, the academic study of art's products and process falls under the class of disciplines we call the 'humanities'.

The Poetry and Music of Science. Tom McLeish, Oxford University Press.
© Tom McLeish (2022). DOI: 10.1093/oso/9780192845375.003.0001

Experiment, Design, Formulation, Method, Theory, Observation, Hypothesis, Computation, Trial, Error—another list of words might lead to a different world of activity. These are more associated with disciplines we term 'the sciences'. Their energy seems to be of a different sort—we are not, perhaps, as emotionally moved by these terms; they do not suggest as much wild, unpredictable outcome. Are we encouraged to think, perhaps, of a laboratory setting—a careful mixing of liquids and a measuring of their temperature? Is the mental picture one of an observer carefully preparing a microscope, or calculating by computer the orbit of a distant planet? If the artistic associations are as likely to disturb as to excite, are the scientific associations more reassuring (the French cubist Georges Braque thought that, '*L'art est fait pour troubler, la science rassure*'[1])? Or do they disturb in a different way? Very likely this is a world that is unfamiliar and strange, less accommodating than the arts and, dare we admit it, less 'human' in some way (we do not class the sciences as 'humanities' after all).

But there are other voices that choose the same language to talk about art and science, and even in the same breath. Philosopher of science Karl Popper once wrote: 'A great work of music, like a great scientific theory, is a cosmos imposed upon chaos—in its tensions and harmonies inexhaustible even for its creator.'[2] This richly layered and dense commentary on music and science will need some background work to uncover Popper's meaning—its allusions immediately fail to intersect with the quite distinct word-lists that spring from usual talk of art and science. But it raises suspicions. Is a dualistic division into arts and science really faithful to our history, our capacities, and our needs? Does it spring from a deep understanding of what these twin human projects attempt to do—is it faithful, dare we ask, to their *purpose*? And if not, are we right to ask of our children,

[1] 'Art is made to disturb, science to reassure'.

[2] Karl Popper (1976 [2002]), *Unended Quest: An Intellectual Autobiography*. London and New York: Routledge.

'Are they on the science side or the arts side?' or to reinforce the well-worn narrative of C. P. Snow that are there 'Two Cultures'[3] at work in our late-modern world, non-overlapping, mutually incomprehensible, and doomed to conflict? If we are wrong to categorize culture, let alone people themselves, in this way then to make exclusive educational decisions based on such a dualistic assumption will be to trigger a process of atrophy in one or other aspect of those children's development, and in adult life to have closed off one or other world of expression, contemplation, creativity, enrichment—of complementary ways of being human.

Doubts intensify about a neat cultural divide if we take the all too unusual step of listening to an artist, or to a scientist, talk candidly about their creative journeys from early ideas to a finished work. For now, the language-clouds of the arts and the sciences start to collide and overlap. I have an intense memory of my first lengthy conversation with an artist (also a professor of fine art at my then home University of Leeds), comparing our respective experiences of bringing to light new work in art and in science.[4] Ken Hay spoke of his first experimental attempts to realize an original conception, of the confrontation of initial ideas with the felt constraints of material (paint and photographic print), of the necessary reformulation of the original concept, of the repeat of these frustrated essays not once but many times. I found that I could tell the story of almost any programme of scientific research I had experienced, in almost precisely the same terms. The discovery was mutual: if I had been surprised by the element of experimentation and trial in his artistic project, then he had not expected my story of science to speak so vitally of the role of imagination. Not only that, it became clear to us not only that the intellectual and technical histories of our projects mapped closely onto each other, but that our emotional

[3] C. P. Snow (1959 [1998]), *The Two Cultures*. Cambridge: Cambridge University Press.

[4] Professor Ken Hay's project is discussed in detail in Chapter 3.

trajectories of excitement, hope, disappointment, rekindling of hope, and resolution also found common expression. The more honest the story-telling in art or science, the more entangled and related became the experiences of emotion and cognition. Thinking and feeling are closer under the surface than in our public stories.

Why it is so much less common to discuss the long process of realization in art than to talk about the final article, composition, theory, or painting, is hard to say. The famous exceptions (such as the evolution of Picasso's *Guernica*,[5] the candid reflections of novelists Henry James[6] and Elena Ferrante[7]) underline the question. Maybe it has to do with the tradition of artisan and artist guarding carefully the 'secrets' of their trades, thereby to increase by mystique, as well as by wonder, the appeal of the finished article. Art has commercial value too—its finely honed techniques and formulae are secrets worth keeping, even though they are the vehicles, not the sources, of inspiration. Or perhaps there is less intention to weave a whimsical web of mystery around the production of art than a natural reluctance to admit too much of the false starts, errors, spilt ink, confused ideas, and dead ends that are the daily experience of any creative activity.

If art is shy about the sweat and tears of working out the form of an original idea, then science is almost silent about its epiphanies and moments of inspiration. Popper himself, celebrated for the most detailed modern outworking of a scientific method in his *Logic of Scientific Discovery*, wrote at length on how hypotheses may be refuted, but remained quiet on how they might be imagined in the first place. While acknowledging the vital necessity of such imaginative conception, Popper declared that, as it was essentially non-methodological, he had nothing to say

[5] Rudolf Arnheim (1963) *The Genesis of a Painting: Picasso's Guernica* Berkeley: The University of California Press.

[6] Henry James (1934 [2011]), *The Art of the Novel*. Chicago: University of Chicago Press.

[7] Elena Ferrante trans. Anne Goldstein (2016) *Frantumagli*. New York: Europa editions.

about it. There is some degree of logic and process in the testing and evaluating of a scientific idea, but there are no such recipes for conceiving them. Nobel Laureate, Sir Peter Medawar, lays some of the blame for our blindness to the role of imagination in science at the feet of John Stuart Mill's *System of Logic*, where he writes as if he believed 'that a scientist would have already before him a neatly ordered pile of information ready-made—and to these he might quite often be able to apply his rules.'[8] If science gathers to itself a narrative more weighted towards method, and art is more vocal about creative origins, then these retellings of partial truths will conspire to drive an illusory distance between them.

Some considerable historical work will be necessary to trace the origins of buried commonalities of art and science. Even the words we use to discuss them bear hidden references. A good example to start with is 'theory'. Seventeenth-century Puritan writer Thomas Browne was able to say in his 1643 *Religio Medici* that[9]

> . . . *nor can I thinke I have the true theory of death, when I contemplate a skull*

The ancient Greek Orphic priests seem to have used the word *theoria* to describe 'passionate, sympathetic contemplation'.[10] The explicitly visual, religious, and contemplative implications at the root of devising a theory will arise within a close examination of visual imagination in Chapter 3, and an unexpected relationship between scientific theory and poetry in Chapter 5.

The contrasting traits of silence within the community of science on its imaginative energies, and of art on its workaday reckoning with material reality, is not restricted to our own times. William Blake, the eighteenth-century poet, artist, and engraver, famously inveighed against what he perceived was the destructive

[8] Peter Medawar (1984), 'An essay on scians,' in *The Limits of Science*. Oxford: OUP.

[9] Kevin Killeen, ed. (2014), *Thomas Brown, Selected Writings*. Oxford: Oxford University Press.

[10] Francis Cornford (1991), *From Religion to Philosophy: A Study in the Origins of Western Speculation*. Princeton: Princeton University Press.

dehumanizing of 'natural philosophy', the term used for the quantified and experimental understanding of nature we would term 'science' today. He wrote of his own task:

> *in the grandeur of Inspiration to cast off Rational Demonstration . . . to cast off Bacon, Locke and Newton; I will not Reason and Compare—my business is to Create.*[11]

For Blake, inspiration has no place in Newton's work, and reason none in his own. There is some buried personal dissonance here, given what we know of his own painstaking technical developments in copper engraving. Yet he was not without cause for complaint against those early modern philosophers: John Locke, in his *Essays Concerning Human Understanding*,[12] had identified 'the imagination' as the source of false and fantastical ideas, as opposed to experience, the reliable guide to the true. Yet there are other voices within the nineteenth century, that witness to a very different vision. One is Ada Lovelace, poet and mathematical collaborator of Charles Babbage, who in an essay from 1841 wrote in powerfully metaphorical terms about the power of imagination in the sciences, and of the sense of exploration in pursuing them:

> *Those who have learned to walk on the threshold of the unknown worlds, by means of what are commonly termed par excellence the exact sciences, may then with the fair white wings of Imagination hope to soar further into the unexplored amidst which we live.*[13]

Other contemporary poets, including Wordsworth and Coleridge, as will become clear as we explore that century's reflection on the imagination, held a similarly positive view on its place in the sciences; Lovelace is significant in that her own work spanned both art and science.

[11] William Blake, *Milton* (1804), book 2, pl. 41; *Jerusalem*, ch 1, pl. 10.

[12] John Locke (2015), *The Clarendon Edition of the Works of John Locke*, Oxford: Oxford University Press, *An Essay Concerning Human Understanding Book II*.

[13] Ada Lovelace (1841), quoted in Sam Illingworth (2019), *A Sonnet to Science*. Manchester: Manchester University Press.

Yet it is Blake's and Locke's compartmentalized assignments of inspiration and rationality that I find at work today among British high-school students. When participating in 'general studies' discussions of science in society, or the importance of interdisciplinary thinking, I like to ask advanced students who have not chosen to study science subjects (when from their intellectual engagement with the material it is clear that they could master anything they wished) why they made that choice. Among the brightest of them, I never receive the complaint that the sciences seem too difficult, but rather that they appear to lack avenues for creativity and the exercise of imagination. The conversation sometimes also reflects the expectation of a more playful engagement with the humanities, contrasting with impressions of seriousness and narrowness in the sciences. I find this personally painful, and doubly saddening that these young people have been offered no insight into the immense fields for imagination offered by science, and that scientists have failed in communicating its call on creativity. As pioneer of science–art project, curator and commentator, Sian Ede writes:[14]

> *Compared with the cool rationalism of science with its material belief in wholeness, the theories employed by thinkers in the arts and humanities seem part of a playful circular game in which the truth is never to be privileged in one direction or another and is always out of reach.*

These echoes of Blake in the words of today's brightest young people are painful to hear. They speak to the urgency of a project that goes beyond the confrontational assumptions of the 'Two Cultures' to deeper levels of human motivation, desire, experience—one that recognizes the dual qualities of rationality and inspiration, of seriousness and playfulness, of imagination and constraint, but challenges their automatic alignment with the axes of humanities and sciences, exploring instead how they play out in both.

[14] Sian Ede (2005), *Art and Science*. London I. B. Taurus.

Admittedly, it has never been easy to speak with clarity about moments of imaginative conception. When inspiration eventually comes, a faithful articulation of the experience is fraught with difficulty. There is a wordlessness about those moments of vision that initiates the phase of craft within the creative process, that plants germs of energy, glimpses distant impressions of what might be accomplished, and perceives the direction of the road ahead. We know how to desire the moments that William Whewell called 'felicitous strokes of inventive talent',[15] but not how to summon them, and hardly to describe them. Here is Shakespeare struggling to do both (and succeeding) in his 100th Sonnet:

Where art thou Muse that thou forget'st so long,
To speak of that which gives thee all thy might?
Spend'st thou thy fury on some worthless song,
Darkening thy power to lend base subjects light?
Return forgetful Muse, and straight redeem,
In gentle numbers time so idly spent;
Sing to the ear that doth thy lays esteem
And gives thy pen both skill and argument.
Rise, resty Muse, my love's sweet face survey,
If Time have any wrinkle graven there;
If any, be a satire to decay,
And make Time's spoils despised every where.
Give my love fame faster than Time wastes life,
So thou prevent'st his scythe and crooked knife.

The poet longs for the return of his 'Muse', one of the ancient Greek mythological band, of personified inspiration recorded as early as Hesiod (*c.* 600 BCE) and that Plato invokes to explain acts of creation.[16] Shakespeare pleads for his Muse to sing him songs and

[15] William Whewell (1837), *History of the Inductive Sciences.* London: John W. Parker.

[16] In most ancient sources there are nine: Calliope (epic poetry), Clio (history), Euterpe (flutes and lyric poetry), Thalia (comedy and pastoral poetry), Melpomene (tragedy), Terpsichore (dance), Erato (love poetry), Polyhymnia (sacred poetry), Urania (astronomy).

guide his pen on the page before Time takes away all opportunity for further art. Yet, paradoxically, he makes this 'time between inspirations'—this ostensibly dry season of complaint—into its own sublime sonnet. The Muse's own song is muffled, and the sight of her hidden, by the humorous complaint of her absence. Unnoticed, the poet's imagined sole source of inspiration, the face of his beloved, is replaced by his rising outrage that the Muse refuses to come at his beck and call. In the compressed lines of the sonnet, the first stage of creativity is conflated with the second; the visitation of inspiration itself is brought together with the 'skill and argument' of the pen. The long labour of writing must then do battle with Time itself. Ironically, it is the wasting erosion of time that becomes the topic of the final work of art. The poet knows that even if inspiration comes, time is not his friend during the long process of gestation before a poem lies completed upon the page. No scientist can read this sonnet without stirring frequent memories of those dry days and weeks when ideas fail to come and fruitfulness all seems elsewhere.

After listening to Shakespeare on inspiration and labour in art, perhaps we ought then without delay turn to Einstein on creativity in science (why descend from the summit of Olympus before you have consulted all of its dwellers?). Here one of the foundation-layers of twentieth-century physics (we will meet another, Emmy Noether, in Chapter 5) writes on the two components of creativity:[17]

The mere formulation of a problem is far more essential than its solution,
which may be merely a matter of mathematical or experimental skills.
To raise new questions, new possibilities,
to regard old problems from a new angle, requires creative imagination
and marks real advance in science.

. . .

I am enough of an artist to draw freely upon my imagination.
Imagination is more important than knowledge. Knowledge is limited.
Imagination encircles the world.

[17] Albert Einstein and Leopold Infeld (1938), *The Evolution of Physics*. London: Cambridge University Press.

Both Shakespeare and Einstein, as they open the door into their workshops for us, albeit in very different forms, tell of two phases in the creative process. The first, 'the visit of the Muse', the 'creative imagination', or 'the mere formulation', is the inspiration, the genesis of an idea. The second is a longer, more directed process of developing the germ into its 'song' or 'solution'. As the comparative study of creativity in science and art unfolds, this narrative structure will fill in and acquire more fine structure, which will crystallize in the final chapter. But at this stage, it is intriguing that Einstein chooses to explain his knowledge of the wellspring of imagination—he 'draws' from it—by describing himself as an artist. He elevates 'imagination' as he demotes 'knowledge'. He wants to make clear that the greater task in science is the 'mere formulation' of the problem in the first place, rather than the application of methods to its solution (by 'mere' he of course means 'fundamental', 'elementary', or 'constitutive' rather than 'trivial'). The great scientist knows that we find our way to encircling the Earth, not principally by experiment, theory, deduction, falsification, or any of those important features of scientific method, but by imagination.

Thomas Kuhn famously coined the notion of 'paradigm shifts' to denote discontinuous changes in the scientific framework for understanding nature. They entail revolutions in entire sets of presuppositions and current mutually supporting scientific ideas. They typically witness the entry of new ideas not deducible from prior reasoning.[18] Classic examples are the Copernican revolution in cosmology and the shift from classical to quantum physics. Beyond identifying the growing dissatisfaction with the existing framework, Kuhn made no suggestions concerning the provenance of the new set of ideas—they are the protoplasm of his revolutions, but seed no methodology.

The formulation of the fruitful question, posed in the right way, constitutes the great imaginative act in science. It requires a

[18] Thomas Kuhn (1966), *The Structure of Scientific Revolutions*. Chicago: Chicago University Press.

developed sense of the current era of scientific thought, of timing. Historian of science and chemist, Lawrence Principe,[19] has pointed out the appropriateness of questioning the structure of the solar system at the turn of the seventeenth century, when Tycho Brahe's meticulous observations of planetary motion and Copernicus's inspired partial solution to the new paradigm had opened up a field of potential progress. Johannes Kepler's deductions, together with Thomas Harriot's and Galileo's new telescopic observations of the heavens, made asking about the dynamical consequences of gravity between the sun and the planets fruitful in a sense that it had not in any previous century. If the scientific imagination is fed by the creative and timely question, it also needs the nourishment of the discontinuous, of leaps in thinking that receive their impulse from some other source than the worthy process of logical deduction. A generation on from the establishment of the orbits of the moon and planets within a heliocentric structure of the solar system, Newton's great imaginative conception, drawing together the work of Kepler, Hooke, and others, was to contemplate a world in which the fall of an apple sprung from the same universal field of force as the monthly procession of the moon.

Einstein and Principe point us to the critical role of the well-formulated and timely question in science. Yet the silence of any formulation of 'scientific method' on this great creative act tends to mask its pivotal role, as well as to dull the perception of scientific creativity. Sharpening and highlighting the role of creative imagination to a degree faithful to the way science actually works would deepen the level at which the fundamental motivation for science is recognized within human culture. An account of science that accords appropriate weight to divergent reimagination of nature, as much as to convergent deductive method, also finds itself telling a much longer historical story than the exclusively post-Enlightenment narratives that dominate at present. A literature search for the question-form when addressed to the natural

[19] Lawrence Principe (2013), *The Scientific Revolution: A Very Short Introduction*. Oxford: Oxford University Press.

world, for example, lengthens the historical line over which we can map early stirrings of the desire to understand and contemplate nature. I have elsewhere written at length on the beautiful and profound ancient Hebrew wisdom poem, the 'Lord's Answer' of the Old Testament Book of Job.[20] A work of over a hundred verses, it assumes the unusual poetic form of the repeated question—but in the light of this discussion appropriately so, since its subject matter is the human understanding of nature. The origin of light, the formation of the coastlines and mountains, the provenance of hail and lightening, the ability of birds to navigate the Earth in their migrations—all appear in a grand cosmic sweep of its author's enquiry. One stanza from Chapter 38 in particular would have impressed a Newton or an Einstein:

Can you bind the cluster of the Pleiades, or loose Orion's belt?
Can you bring out Mazzaroth in its season, or guide Aldebaran with its train?
Do you determine the laws of the heaven?
Can you establish its rule upon earth?

The poet looks at the motions of the stars and constellations across the sky, even noticing that some cluster together while others, though visually similar in brightness and hue, are separated. The lovely Pleiades is one of the very few 'open star clusters' resolvable to the human eye, with up to six or possibly seven members visible to keen-eyed northern hemisphere observers during autumn and winter nights. An outstretched hand-breadth to the south-west of them lies the linear triplet of bright blue-white stars in 'Orion's belt', far further from each other than the members of the cluster. The presence of a strange class of 'law', to be obeyed not by humans but by the stars themselves, that might contain the statutes binding some stars closely together while others are far-flung, that oversee their regular and irregular motions, is an impressive creation of the imagination even now. The conjecture, buried in such ancient philosophy,

[20] Tom McLeish (2014), *Faith and Wisdom in Science*. Oxford: Oxford University Press.

that heavenly and earthly laws might be connected, is even more striking.

The presence of the creatively formulated question in as ancient a source as the Book of Job[21] (undatable other than to place it within the first half of the first millennium BCE) within the Semitic tradition, carries another salutary message to us late-moderns. Alongside the complex history of ancient Hellenistic science from 500 BCE, it surely erodes any idea that science is in any way exclusively modern, beginning rootless at the Enlightenment, and blowing away the cobwebs of centuries of darkness, magic, superstition, and alchemy. Sadly, much popular narrative of science history has it so, but claiming science as an exclusive property of the modern world removes the deep and slow cultural development of an imaginative and creative engagement with nature that develops, at least chronologically, alongside the story of art in its own multitude of forms.[22]

The timely question is not the sole province of science. It is surely not coincidental that the literary genre of the novel arises with Daniel Defoe, alongside early modern science, and that the genre of science fiction underwent an explosion in the last century. As Pat Waugh has pointed out,[23] the novel is the *experimental* medium of artistic creation par excellence. In the safe space of the novel, inhabited worlds can be summoned into existence, and their dangers and dark places explored. Questions of the relationship of human beings to time and space, to

[21] For a magisterial survey of the *Book of Job*, see the three-volume work by David Clines (Thomas Nelson publishers). We will encounter an explicit example of scientific inspiration drawn from reading it in the story of the rainbow, told in Chapter 8.

[22] For a more complete and integrated account of the history of science, see the now classic work by David Lindberg (2010), *The Beginnings of Western Science: The European Scientific Tradition in Philosophical, Religious, and Institutional Context, Prehistory to AD 1450, Second Edition*, Chicago: Chicago University Press.

[23] P. Waugh (2015), 'Beauty writes Literary History,' in Corrine Saunders and Jane Macnaughten, eds., *The Recovery of Beauty*. London: Palgrave MacMillan.

each other and to the Earth, can be teased out in both internal and external worlds of their characters. The novelist does not experience unconstrained freedom, however, but discovers the multiple moral constraints of the experimental form. Crucially, novelistic writing forces a more intense outwards gaze. For Iris Murdoch, novelistic writing enables an attention to 'the inexhaustible detail of the world, the endlessness of the task of understanding. . ., the connection of knowledge with love and of spiritual insight with the apprehension of the unique'.[24] Murdoch supplies us with another glimpse of commonality in the narrative that artists, creative writers, and scientists adopt when they are trying to articulate the deepest motivations for what they do. 'The endless task of understanding' and a focus on the 'inexhaustible detail of the world' are the shared delights and common labours of the physicist and biologist also. These novelistic hints are enough to tempt the much closer look at the relationship between experimental science and literature in Chapter 4.

Paying close attention to the stories of imagination and workmanship in the creation of art and science will be the first of this book's tasks in the project of reappraising science through the lens of the humanities. The second task must be a similar study of the way their creations are received by their respective audiences. Neither art nor science can exist in a solipsistic vacuum of their authors. Both must be listened to, observed, received, responded to. If the current public narratives of creativity are artificially divided into the imagination of art and the logic of science, then the framing of their reception is just as polarized. 'Science is not with us an object of contemplation,' complained social thinker, Jacques Barzun.[25] The impression is that art appeals to the response of emotion and affect, while science connects only to cerebral reason. Such a neat, Kantian, division appeals to a compartmentalized and fragmented structural view of culture, but it reinforces the picture of an artificial division of science

[24] Iris Murdoch (1998), *Existentialists and Mystics*. New York: Penguin.

[25] J. Barzun (1964), *Science that Glorious Entertainment*. London: Harper and Row.

and art into two realms. Perhaps this is where they do indeed divide—even if both draw at depth on a mysterious creative human energy in their production, it is conceivable that reading, listening, and hearing science and art nonetheless pursue distinct mental pathways.

On the other hand, if a distorted impression of creativity arises in part from selective silences on the part of their practitioners, then perhaps the same is true of their reception. Comment on the effect and the enjoyment of art is commonplace. It speaks of a healthy continuum from artist and performer to receiver and listener. We may not be able to paint or to sing like the great exponents of art and oratorio, but we are not silenced as a result from speaking, or even from critically appraising, paintings or performances. There is understood to be a 'ladder' of participation and reception in the arts. In music, for example, the lower rungs are occupied by those of us who enjoy concerts, who pick up instruments in the company of forgiving amateur friends. We would never presume to perform in public, but nevertheless can confidently express an opinion on which recording of a symphony we prefer. The upper rungs are occupied by the performers on those very recordings. Here is actor Simon Russell Beale talking about his response to the music of the romantic composer, Schubert:[26]

> *Schubert can make time stand still. In the last, miraculous months of his life, he expanded his vision of what music could do. His most experimental work is the slow movement of his B flat Piano Sonata. It is as if he has distilled the process of musicmaking. He takes a harmonic progression, explores it, changes a single note, explores it again; he breaks down a simple melody until only the bones are left and the music is suspended. The result is a play of pure sound, without external reference, that gives us a glimpse of eternity.*

This is a piece of exceptionally high-quality comment, to be sure, but it is not unusual for someone who is not a professional musician to write at such serious critical depth about a piece of music. Beale's testimony is also an example of the narrative we met in

[26] Simon Russell Beale (2012), *Ferocious, Tender, Sublime*, *The Guardian*, 19th March.

the case of artistic creators, stories that map rather remarkably onto the rare but honest stories of science's own creative process. Schubert is, as heard by Beale, experimental, exploratory, even reductionist and abstractionist. Yet he is also sublime. Beale describes an example of a deep property of music—its ability to reconcile us to the passage and structure of time by somehow suspending us from it. At the level of the abstract, music and mathematical science have long been compared, albeit inconclusively. Developing a comparison of their creative processes will be the concern of Chapter 6, but at this point the divergence in their apparent degree of communicability is stark.

It is harder to find comparable examples of reception and affect in scientific creation. But this is not because of a lack of inherent appeal to human desire and need. The 'ladder of access' that we identified in a creative art such as music is not (as observed by Barzun in different terms) present in our current culture in science. This was not always the case—Shelley, Coleridge, and Wordsworth all thought that science could and would inspire poetry (though Shelley foresaw that the inspirational beauty of science would be a hidden one). So, for articulated contemporary reception of science, we must usually listen to the scientists themselves. Here cosmologist Subrahmanyan Chandrasekhar describes in remarkable terms an example of the moments of transport for which science longs:[27]

> *In my entire scientific life, extending over forty-five years, the most shattering experience has been the realization that an exact solution of Einstein's equations of general relativity, discovered by New Zealand mathematician Roy Kerr, provides the absolutely exact representation of untold numbers of massive black holes that populate the universe. This 'shuddering before the beautiful', this incredible fact that a discovery motivated by a search after the beautiful in mathematics should find its exact replica in Nature, persuades me to say that beauty is that to which the human mind responds at its deepest and most profound.*

Chandrasekhar's 'shuddering before the beautiful' carries unmistakable resonance with Beale's 'glimpse of eternity'. It also,

[27] Subrahmanyan Chandrasekhar (1987), *Truth and Beauty: Aesthetics and Motivations in Science*. Chicago: University of Chicago Press.

perhaps unwittingly, resonates with accounts of aesthetics that have been driven beyond the simply beautiful to the 'sublime'. The cosmologist is speaking of the extraordinarily simple yet utterly strange idea of a 'black hole'. For many years, pure conjecture, observational evidence from stellar evolution, and highly luminous galactic cores has pointed increasingly to the inevitable existence of these bizarre and terrible objects. Black holes are places in the cosmos where the local presence of matter is so great that gravity generates its runaway collapse towards a point where density becomes formally infinite, surrounded by a finite region of space in which the tug of gravity is so great that no light can escape. Possessing a terrifying and austere beauty, these objects are as near to instantiated mathematics as one could imagine. They can have no other properties than mass, spin, and electric charge. All other attributes that their original matter once possessed are lost in its irreversible infall. The normal role of mathematics within theoretical physics is to provide approximate descriptions of natural objects, but in this case the attribution of a black hole's triplet of properties is complete.

The experience Chandrasekhar describes is a rarefied and extreme form of a precious wonder. Einstein put it thus: 'the most inexplicable thing about the universe is that it is explicable' and Eugene Wigner pointed towards it in the title of his celebrated essay *The Unreasonable Effectiveness of Mathematics in the Natural Sciences.*[28] The moment of connection of a constructed pattern of thought, mathematical, pictorial, or logical, with the deep structure of the natural world evokes an unparalleled experience of wonder. As Wigner pointed out, it also appeals through the inherent elegance of mathematics to the aesthetic sense. More than that, it seems to satisfy a need for creative connectedness, the act of understanding, of re-creating an internalized world patterned on the external. Such reaching out into the world in abstract thought is perhaps a flowering of a human response to the ancient

[28] Eugene Wigner (1960), *Communications in Pure and Applied Mathematics*, **13**, No. I. 1-14, New York: John Wiley & Sons, Inc.

questions of the Book of Job.[29] Such long perspective suggests that there are deep and complex commonalities in both the motivation and imaginative act of creation, as well as in the role of aesthetics, in the sciences as much as in the arts.

All this is not to deny a tradition of research and writing on the topic of creativity itself that includes scientific examples—far from it. A now-celebrated 1926 essay of Graham Wallas, *The Art of Thought*[30] introduced a four-stage abstract structure for 'the creative process' that we will make critical use of in the following chapters (finding it serviceable but wanting). Wallas himself draws deliberately on a genre of *The Art of . . .* works, initiated in modern times (itself an echo of Horace's ancient *Ars Poetica*) by writer Henry James in *The Art of the Novel*, and taken up by physiologist William Beveridge in the early twentieth century in his *Art of Scientific Investigation*. We will be examining both of these works at close quarters in Chapter 4. Major edited collections of essays and research from psychologists[31] on creativity have recently spawned a subfield of cognitive neuroscience. The genre has even produced a 'Cambridge Handbook' on creativity.[32] The laboratory-scale studies of the creative process and its context are informative, although by their nature they cannot encompass the extended and unconstrained creative projects that characterize substantial works of art and science. At a more selective level, there are surveys of creative individuals, including those representing fields of science and art—Howard Gardner's *Creating Minds*[33] is an excellent example. Anthony Storr has begun to tease out conjectural psychological patterns in the minds of very creative thinkers.[34] Other contributions to the creativity literature are listed for convenience in the bibliography of the

[29] Tom McLeish, *Faith and Wisdom in Science.*

[30] Graham Wallas (1926 [2016]), *The Art of Thought*. Tunbridge Wells: Solis Press.

[31] *e.g.* Robert J. Steinberg, ed. (1988), *The Nature of Creativity*. Cambridge: Cambridge University Press.

[32] James C. Kaufman and Robert J. Steinberg, eds. (2010), *The Cambridge Handbook of Creativity*. Cambridge: Cambridge University Press.

[33] Howard Gardner (1993), *Creating Minds*. New York: Basic Books.

[34] Anthony Storr (1993), *The Dynamics of Creation*. London: Penguin Books.

present work. But there is still a clear issue of misunderstanding in the perception of science within popular culture, as an activity that calls on minimal imagination.

There is little discussion of the way that imagination plays out in the experience of the thousands of people engaged in the scientific and artistic work that adds colour to our communities and national lives. There is also almost total silence within the educational formation of scientists on the topic of imagination, of the creative formulation of questions and hypotheses, or of the experience of scientific ideation. There may not be a method for this most vital of all scientific processes, but there are accounts, practices, and a communal experience that ought to be more widely and openly shared both within and without the scientific community. Furthermore, as the explorations in this book of visual, textual, and abstract creativity in science and arts will show, the entanglements of experience between their artistic and scientific examples are so close that a renewed interdisciplinary dialogue promises a fruitful dividend for both cultural communities.

I have suggested elsewhere[35] that, because of the 'missing rungs' in scientific ladder of reception, it is lamentably less common for non-practitioners of science to experience the intensity of aesthetic response to a new understanding of nature, than for the scientists whose professional training has taken them to the ladder's higher footholds that still exist. But it is not impossible, and could be as common as the learning of a new tune or appreciating an unfamiliar painting for the first time. In a moving personal example, a friend told me of the moment when, gazing up at the moon one evening, he suddenly understood how its phases worked. A life of familiarization with the monthly cycle of crescent, half, full, and gibbous moon was not equivalent to 'seeing' how these shapes served as the signature of an illuminated orb. On that moonlit night shortly after sunset he allowed the two-dimensional screen of the sky to become, in his mind, a vast

[35] Tom McLeish, *Faith and Wisdom in Science.*

three-dimensional structure. The moon became a solid sphere, illuminated by a much more distant sun from different angles on different days, as seen from the centre of its orbit on the Earth. The celestial geometry and its circling dynamics found a home in his imagination—releasing an experience of pure joy. He described feeling present to the world in a deeper sense than before, and knowing that this stronger relationship was, once found, not going to be lost.

The genre of 'trade book' science writing has recently produced examples of exquisite care and intelligent communication that seem to be painstakingly replacing those removed rungs in the public ladder of science-engagement, giving their readers some of the experience of personal revelation, such as my friend's personal reconstruction of lunar astronomy. Nobel Prize-winning physicist, Frank Wilczek's lovingly articulated story of symmetry in art, philosophy, and physics authentically communicates a physicist's delight in the discovered layers of nature's patterned structure.[36] Former editor of the scientific journal, *Nature*, and now science writer, Philip Ball, has demonstrated how molecules make a story as human as it is material.[37] Television producer and writer, Simon Singh's account of the three-century-long search for a proof of Pierre de Fermat's famous 'last theorem' is a thing of beauty and emotion, and an opening to the common gaze of a world of pure mathematics that engages without patronizing.[38]

Experiences of such reception in science or in art, achieve at their most profound such an intensity of emotion and of felt transformation, that they must draw our exploration to a third level of parallel comparison—that of the human function of creative engagement with nature and, if we dare talk of it, of purpose. A nest of questions confronts us here: why do art, and early science, arise in prehistory? What do they achieve socially and psychologically today? Where do art and science appear, both

[36] Frank Wilczek (2015), *A Beautiful Question*. London: Allen Lane.

[37] Philip Ball (2001), *Stories of the Invisible: A Guided Tour of Molecules*. Oxford: OUP.

[38] Simon Singh (1997), *Fermat's Last Theorem*. London: Forth Estate Ltd.

explicitly and hidden, in the complex of cultural narratives? How do they receive, and provide, value and virtue? The humanities discipline of theology comes to our aid here, for no other reason than that it is comfortable with the category and narrative of purpose. Recent writers have attempted to articulate a 'theology of' music (Begbie[39]), of art (Wolterstorff[40]), of science (the present writer[41]), and found that this trailhead leads to a fruitful landscape within which such questions of purpose can be attempted. Critic and literary scholar, George Steiner's view of art in this regard elicits a striking reception when read from the perspective of science. In his moving critique of the humanities in late-modernism, *Real Presences*, he writes:[42]

> *Only art can go some way towards making accessible, towards waking into some measure of communicability, the sheer inhuman otherness of matter.*

speaking at the same time of deep need, and of powerful satisfaction. Writing in despair over a 'broken contract' of art with nature that he sees as vital to reverse, Steiner describes a human condition 'out of joint' with the world in which humans are immersed. There is a gulf of otherness, of strangeness, with which, for reasons we do not understand, we remain uncomfortable. The divorce is painful. The paradox is heightened when we reflect that the 'sheer inhuman otherness of matter' is the very stuff of which our bodies are composed. Art may indeed go some way towards a reconciliation of the human with the material. The creation by paint on canvas of a visual illusion, for example of standing before a riverbank picnic illuminated by mottled sunlight, is achieved only by a deep understanding of the received visual cues by which we reconstruct our surroundings. Or, if another contrast imposed by the 'inhuman otherness of matter' on

[39] Jeremy Begbie (2014), *Theology, Music and Time*. Cambridge: Cambridge University Press.

[40] Nicholas Wolterstorff (1987), *Art in Action Toward a Christian Aesthetic*. Grand Rapids: Eerdmans.

[41] Tom McLeish, *Faith and Wisdom in Science*.

[42] George Steiner (1989), *Real Presences*. London: Faber and Faber.

the human is the appalling possibility of eternity in the face of our own temporality, then the 'glimpse of the eternal' afforded by a Schubert sonata is also doing some reconciliatory work. But so too, surely, is the science of the night sky, be it the humble reimagining of the illuminated moon or the severe simplicity of a mathematical black hole that finds connection with a myriad of unseen gravitational wells hidden within the whirling and immense galaxy overhead. Their utter inhuman otherness is to some degree woken into Steiner's 'some measure of communicability' by the conceived form of mathematical understanding. And if this is the role of art, then what other category remains for science? The final chapter will reprise a discussion of purpose in human creativity.

Exploration of a possible parallel purpose at the deepest level for art and science will steer our trajectory into headlong collision with those who have perceived an irreconcilable antithesis between the two. To navigate these stormy waters will need some historical perspective, for an oppositional framing seems to reawaken, at least in the modern period, with each generation. Forty years previous to the late twentieth-century combatants of the 'Science Wars', public intellectuals engaged in angry words over the 'Two Cultures'. But half a century before C. P. Snow and F. R. Leavis locked horns, a gentler but equally incisive debate, anticipating some of the later rancour between the arts and the sciences, was engaged by Matthew Arnold and Thomas Henry Huxley.[43] Before them, romanticism drove home with force the charge that science does precisely the opposite of (at least narrative and poetic) art in the meeting of human creative need. Blake's dismissal of reason as the antithesis of creation was by no means a solitary one. In his long poem narrating the story of the mythical serpent *Lamia*, John Keats complains of science—for him 'cold philosophy':

[43] For a discussion of this debate see E. S. Schaeffer (1994), How many cultures had Lady MacBeth? in L. Gustafsson et al., eds., *Science and the Powers*. Hasselby Castle: Swedish Ministry of Science and Education, pp. 136–92.

Do not all charms fly
At the mere touch of cold philosophy?
There was an awful rainbow once in heaven:
We know her woof, her texture; she is given
In the dull catalogue of common things.
Philosophy will clip an angel's wings,
Conquer all mysteries by rule and line,
Empty the haunted air, and gnomed mine
Unweave a rainbow.

In similar commentary vein, Edgar Allan Poe called science the preying 'vulture of the heart' whose wings are 'dull realities', in his *Sonnet to Science.*[44] The question of why Blake and Keats, Poe, and many others among their nineteenth-century contemporaries perceived science to be the means of desiccation, of demystifying, of replacing wonder by measure, and then took up the tools of their trade at their highest energies to inveigh against it, becomes central to understand the origin of the 'Two Cultures' divergence.

Historical locus is important. Retrospective projection of arguments from our own times, such as simplistic assurances that the romantic poets had nothing to worry about concerning the draining of wonder from the world, will not get to the root of their disquiet. In any case, the same concerns arise today in different forms. There are shifts in the role of science particular to the romantic period that may be significant. A change in nomenclature—for example, 'natural philosophy' becomes 'science'—carries with it many etymological undercurrents of meaning. A Greek declaration of 'love of wisdom of natural things' (*philo—Sophia*) is slowly replaced by a Latinate claim to knowledge (*scio*) (Wordsworth's critique '*we murder to dissect*' uses the term 'science' where Keats and Poe retained 'philosophy'). Science departments and syllabuses began to appear

[44] Poe, Edgar Allan. *Poetry and Tales.* New York: Library of America, 1984. p. 39

in the universities. William Whewell coined, around 1836, the term 'scientist', which gathered currency first in America and then in Britain throughout the century. Faraday and Maxwell both rejected the label, insisting on the older 'natural philosopher', yet the adoption of 'scientist' was complete by the end of the century. Momentously, the discoveries and theories of geology (Lyell's gradualist and ancient formation of geological strata) and of biology (Darwin's evolution by natural selection) were utterly transforming understanding of relationships between the human race and other species on Earth. The period of romanticism swept in a fragmentation of disciplines and a further distancing of 'the inhuman otherness of matter' unprecedented in thought.

Writings within the stormy cultural change of a different cultural period to our own can easily be misread, however. In order to gauge the story of science's apparent offensiveness to the poetic in the nineteenth century, we will need to look harder at the context of the criticism. To taste just one example, Wordsworth's contrast of the scientist with the poet, explored in the preface to the second edition of his and Coleridge's *Lyrical Ballads*, although much more positive than Keats, centred on the solitude of science, its lack of communicability with anyone other than the lonely investigator. The passage is worth quoting at length at this point as later chapters will have cause to return to it.[45]

> *[Science] is a personal and individual acquisition, slow to come to us, and by no habitual and direct sympathy connecting us with our fellow-beings. The Man of Science seeks truth as a remote and unknown benefactor; he cherishes and loves it in his solitude; the Poet, singing a song in which all human beings join with him, rejoices in the presence of truth as our visible friend and hourly companion. Poetry is the breath and finer spirit of all knowledge; it is the impassioned expression which is in the countenance of all Science. Emphatically may it be said of the Poet, as Shakespeare hath said of man, 'that he looks before and after.' He is the rock of defence for human nature; an upholder and preserver, carrying everywhere with him relationship and love. Poetry is the first and last of all knowledge—it is as immortal as the heart of man. If the labours of Men of science should ever create*

[45] Wordsworth, William (1800). *Lyrical Ballads with Other Poems* (2nd ed.). London: Printed for T. N. Longman and O. Rees.

> *any material revolution, direct or indirect, in our condition, and in the impressions which we habitually receive, the Poet will sleep then no more than at present; he will be ready to follow the steps of the Man of science, not only in those general indirect effects, but he will be at his side, carrying sensation into the midst of the objects of the science itself. The remotest discoveries of the Chemist, the Botanist, or Mineralogist, will be as proper objects of the Poet's art as any upon which it can be employed, if the time should ever come when these things shall be familiar to us, and the relations under which they are contemplated by the followers of these respective sciences shall be manifestly and palpably material to us as enjoying and suffering beings. If the time should ever come when what is now called science, thus familiarized to men, shall be ready to put on, as it were, a form of flesh and blood, the Poet will lend his divine spirit to aid the transfiguration, and will welcome the Being thus produced, as a dear and genuine inmate of the household of man.—It is not, then, to be supposed that any one, who holds that sublime notion of Poetry which I have attempted to convey, will break in upon the sanctity and truth of his pictures by transitory and accidental ornaments, and endeavour to excite admiration of himself by arts, the necessity of which must manifestly depend upon the assumed meanness of his subject.*

It is neither the practice nor the insights of science that Wordsworth sets apart as irrelevant to human sensation and sensibility, but their remoteness from common currency. In far richer language, he has identified the same missing lower echelons of science's 'ladder of access' that still prevent all but the most athletic practitioner from climbing it. Poetry had not been widely inspired by science, runs the *Preface*, but only because it has no access to it at the level of contemplation that would make contact with its 'divine spirit'. Then as now, only someone steeped in the learning of a scientific discipline might 'shudder before the beautiful', as we might all long to do as we read the cosmologist's account of the mathematical connection of black holes to the night sky above us. It turns out that the story of mutual inspiration of poetry and science is a complex one, as the encounter with ancient Hebrew nature poetry in Job, has already suggested. It is also richly informative of the creative process, and is the focus of Chapter 5.

A more optimistic nineteenth-century voice that can still be heard, albeit more quietly than the romantic poets, is that of John

Ruskin. His Oxford lectures *The Eagle's Nest*[46] attempt a unified cultural view of art and science. Ruskin's frame is constructed from the cerebral and practical aspects of wisdom—*Sophia* and *phronesis*—that Aristotle conceived as complementary. He teases out the practical within the process of creating art from the mystical provenance of inspiration. He is writing as late as it is possible to do so without suffering from the loss of the vocabulary of Wisdom from resonances with science. 'Can anything be more simple, more evidently or indisputably natural and right, than such connection of the two powers?' Ruskin asks his student audience in the third lecture. We need to confront the reasons why, in the century in between then and now, his voice and Wordsworth's hopes have been lost in the clamour of Keats' and Poe's complaints.

As it becomes clear that we have suffered from an arbitrary disciplinary set of divisions, and that these have blurred our comprehension of creativity itself, then other re-oriented ways of arranging our subject suggest themselves. The pictorial art that Steiner had partially in mind, the poetry and prose of Wordsworth, and the wordless wonder of a Schubert sonata, urge a categorization of creativity that recognizes stronger affiliations than structures within the vast domains of art and science themselves. Visual imagination is common to artist and scientist alike; the metaphor for understanding as 'seeing' is so ubiquitous and powerful that we fail now to notice its metaphoric status. The textual work of writing—both poetry and prose—calls on a different, sequential, and syntactical imagination, but again one that is common to literature from the domain of science that reinvents the natural. Finally, there emerges, in both art and science, wordless and picture-less domains of invention whose very existence we still strain to comprehend: the fields of music and of mathematics. The four central chapters (Chapters 3 to 6) that follow will

[46] John Ruskin (1905), *The Eagle's Nest*. London: George Allen.

each deal with one of these (the 'textual mode' is divided between novelistic prose writing in Chapter 4 and poetry in Chapter 5), experimenting with holding the artistic and scientific examples in close juxtaposition. Rather than construct parallel accounts of the creative narrative in art and science, experience suggests that they tell their stories together within these three realms of visual, textual, and abstract.

The cultural history of the nineteenth century also suggests reappraising the relationship of art and science through another lens: that of theology, among the other humanities. There are, for example, close parallels between the strained relation of science and religion, and that between science and art. The divisive romantic period saw, for the first time, strident claims of conflict along both axes. Consider, for example, the critical language of the romantic poets alongside the later appearance of conflictual narratives such as Andrew Dickson White's *A History of the Warfare of Science with Theology in Christendom* (1896). Now recognized as polemic rather than history, of which it makes at best a highly selective reading, its constructed message still circulates today through chains of references, without the vulnerability of its own flimsy support.[47] The drivers of the oppositional narratives for both art and religion to science, as received a century later, also bear similarities: both are served by presenting science as a cultural newcomer, and as a competitor for cultural territory occupied previously by art, the humanities, or religion. The story of conflict between science and religion is equally served by a superficial understanding of science itself and the scientific motivation, as much as by a distorted view of religion. It is enhanced by banishing talk of creative imagination from science. Furthermore, the entire argument survives only by banishing all teleology, all talk of purpose.

[47] For a comprehensive cultural account of the Draper-White thesis, see James Ungureanu (2019), *Science, Religion, and the Protestant Tradition: Retracing the Origins of Conflict*. Pittsburgh: University of Pittsburgh Press.

I have attempted to subvert the three prior conditions for presenting the case for conflict in the book *Faith and Wisdom in Science*, where I traced the endeavour we now call 'science' back through the renaissance, medieval, and late classical worlds into the wisdom writings of our Old Testament texts, in parallel to its philosophical roots in ancient Greece. That journey led to a recasting of the oppositional 'geometry' of theology *and* science to the mutually encompassing and twinned relationships described by a 'theology *of* science' and a 'science *of* theology'. In resonance with Ruskin's proposal that Wisdom frames art and science together, the narratives of ancient Wisdom, such as Job's nature poetry, appear as the very tributaries of science. The parallels of mis-construal, of mutual suspicion, and the projection of science through a filter that removes its roots in longing creativity within the pain of distance from the world—all these strongly suggest that we play with equally fresh geometries of relationship between the humanities, arts, and the sciences. Such a more consonant perspective focuses on the teleology, cultural and theological, of both. A comparison of poetic imagination with that of scientific theory, in Chapter 5, will require more theological thinking, and a discussion of common purposes and ends will draw threads together in the final, eighth chapter.

A journey into the purpose of science, and of art, must learn from the misunderstandings and the mutual pain of fragmented disciplines. It must, finally, move from *talk* about relationship into a practice of it. If we do find familial fellowship between science and art in a deeper reappraisal, then we will surely notice a structural imprint of their shared cultural DNA as we proceed. Returning to our first perspective—the comparative practice of creative imagination—suggests the lines of a possible framework. No art results from unconstrained exercise of imagination. The poet's vision and communicated emotion take shape within the constraining form of sonnet or quatrain. The composer lets thematic material expand, combine, and develop within sonata form or rondo. The painter conjures with light, colour, and representation, but only successfully when she observes the material properties of oil on canvas, or of watercolour on board. It is the tension between imagination and constraint, of idea within form, which focuses creative energy into artistic creation itself. The greater the imaginative impulse, the tighter the form is needed to channel and shape it.

Seen in this light, science no longer looks quite so strange, for if its task is to reconceive the universe, to create a mental map of its structure, the interrelationships of force and field, of the evolution of structure and complexity, then what task could possibly call on higher powers of imagination? If its goal is to understand the patterns of matter from the earliest moments of time to its closing aeons, from the smallest fluctuation of space-time to the immensities of the cosmos, and to reconcile all this inhuman otherness to the finitude of our minds, what could demand a greater act of human creation? But what greater form, what more focusing constraint, could be supplied than the way we observe the universe to be? If writing a sonnet is the collision of creativity within constraint of expressing within a tight form, and with new potency, the human experience of the world, then science also becomes the conception of imagination within constraint. We re-create the universe by imagination within the constraint of its own form.

In this sense, science is not simply *like* poetry, it is construed as a form of poetry. Perhaps this is the sense in which Shelley claimed in his *Defence of Poetry*,[48] that 'poetry comprehends all science'. He knew, as did Einstein letting his mind wander through worlds of imagination, or as Shakespeare awaiting his Muse, that no one can 'say I *will* write poetry . . . the greatest poet even cannot say it.'

Cousinly creativity with constraint—that is a starting hypothesis for a journey through art and science. It will be one with a listening ear. We need to spend time in the workshops of artists

[48] Percy Bysshe Shelley (1840 [1839]), *Essays, Letters from Abroad, Translations and Fragments*. In two volumes. Edited by Mary Shelley. London: Edward Moxon.

and of scientists, and look without prejudice at the way their work is, or could be, received emotionally as well as cerebrally. We will need to stand back from our own time and look at longer narratives, and at other ways of differentiating disciplines. Reflection from the high medieval centuries will join as a continuous conversation partner to contemporary voices. The journey will require some close, even technical, readings of great creative examples of art, music, mathematics, and science. The choice of which imaginative voices in all these avenues we closely listen to will be a personal one, but will be guided by the requirement that they should have reflected on the process of creativity itself. In such company, our journey will explore the hope that science might re-weave a rainbow in a way that Keats might have recognized as poetic, true, and constitutive of the human.

Creative Inspiration in Science

In the case of all discoveries, the results of previous labours that have been handed down from others have been advanced bit by bit by those who have taken them on, whereas the original discoveries generally make an advance that is small at first though much more useful than the development which later springs out of them. For it may be that in everything, as the saying is, 'the first start is the main part,' and for this reason also it is the most difficult; for in proportion as it is most potent in its influence, so it is smallest in its compass and most difficult to see; whereas this is once discovered, it is easier to add and develop the remainder in connexion with it.[1]

ARISTOTLE

Where do new scientific ideas come from? Aristotle was thinking on a much wider canvas than the endeavours we call 'science' now—he later goes on to illustrate his point through examples in rhetorical technique. In contrast to the work of Plato's Muses, Aristotle defines the generative force of ideas, *poiesis*, as an act of the innate imaginative mind, but did not solve the problem of 'the first start.' Our curiosity into the process of creation has not slackened since. Yet, as we saw in the introduction, we understand almost nothing of it. To test a great idea in science is easier than to conceive of it in the first place. It is not simply easier—it is even possible to codify the practice of examination, modification, or refutation, once a new idea is grasped. Look up any popular definition of 'scientific method'—it is exclusively to this second stage that it refers. Such recipe-book accounts of science describe

[1] Aristotle, *De sophisticis elenchis* **34**.183b17–184b9, trans. W. A. Pickard-Cumbridge Oxford: Oxford University Press (1992).

The Poetry and Music of Science. Tom McLeish, Oxford University Press.
 DOI: 10.1093/oso/9780192845375.003.0002

something like this: ideas are sharpened into 'hypotheses' that have sufficient power to make predictions for the outcome of experiments. These are then performed and, depending on the results, the hypotheses are discarded or allowed to live to fight another day.

In education, if not in more nuanced twentieth-century discussion of the philosophy of science, this pattern of *Conjectures and Refutations*, described in methodological detail by Karl Popper, is portrayed as the way that science proceeds. Yet even as a faithful account of the process of 'normal science'—as Thomas Kuhn termed its daily work of incremental progress—it has severe drawbacks. Subsequent philosophers of science have pointed out that theories are not quite as easy to refute by a refusal at the first fence of experimental agreement. They have a way of circling around with minor modifications and trying the jump again.

But the greatest lacuna of any theory of scientific method is in its silence on where ideas come from in the first place. The small but vital 'first start' of Aristotle is the creative beginning of a new road that, once its potential geography is perceived, can be laid down mile by mile by the labourers. But how is the idea of the new start conceived in the first place? From where does the notion of possibility for a new road come? No one has written a method for the generation of new ideas. The 'scientific method' only tells half the story that we need to complete the arc of new understanding. In its classic articulation by Popper,[2] the 'logic of scientific discovery' explains in detail how we evaluate theories, construct critical experiments to test hypotheses, and identify what adjustments or additions might be permissible without a complete overthrow of the theory. This narrative has never framed recommendations for the creation of hypotheses in the first place, however. Like the early internal combustion engine, it knows how to keep turning over but contains no means to self-start.

[2] Karl Popper (2002 [1934]), *The Logic of Scientific Discovery*. London: Routledge.

As we have already seen, the strange silence of science on its intellectual germinations, its sensitivity around talk of their conception, has damaging consequences. Once the results of science are portrayed as dehumanizing in their effect, or as the enemy of the poetic impulse, the reluctance for creativity-talk does the other half of the work in portraying it as a recipe-based activity for lovers of logic and sceptics of wild imagination. In spite of the personal knowledge that the creative first step is as essential in natural philosophy as in art, and recognized as such from Aristotle to Einstein, there is still a false story circulating that imagination is not a strong requirement for the scientist. The distorted and dry narrative is as prevalent in public media as in education. What harm this does in all its ramifications is impossible to assess; it certainly turns away some young people from enjoying a potentially fruitful and rewarding path in science. It contributes to the dualism of the 'Two Cultures' paradigm that still hangs so leadenly in our cultural skies.

The disappointment of the bright young people I encounter, who have not chosen to study science because they could not see, 'any channel for imagination' or any call on their 'creativity' were they to tread that path, condemns the generation that precedes them in two ways. We stand accused first for our public silence on the way that science draws so deeply from those essential wells of human energy. We are equally culpable for what we have said to those we educate—for our delineation of a science curriculum that manages to persuade most of those who follow it that science is about the accumulation and regurgitation of fact, when before us extends a universe of intricate structure and mystery. We have failed to communicate the adventure of reimagination and learning, of nature's invitation to peer into the world's multiple levels of structure, to extend our sense of vision through imagination, to think about matter and form in entirely new ways.

To persuade a scientist to recall and describe the moments when she or he realized that their imagination has been excited by a beautiful new idea is like unlocking a treasure chest.

I enjoy these conversations immensely—the description of discovery is often reluctant at first, but the remembered moments of revelation are worth waiting for, even if some of these memories will have matured with age. A sparkling array of formerly hidden truth is open to a shared gaze, there is an immediate impression that as delights are plucked one by one, more will be found underneath—and overall a strong sense of personal discovery. More than that, there is an emotional as well as intellectual appreciation, an aesthetic reward, and a sense of gift and privilege. These experiences are quite different from the 'logic of scientific discovery'—although they can be experienced even when immersed within the process of methodological work. The coming into shape of a radically new idea can occur in the routine of experiment, of calculation, of careful processing of data, in solving the myriad daily challenges of progressing a research project—they often seem to arrive unsought, as from unconscious layers of the mind. They seem as much gifts as discoveries, and although surely dependent on the extended labours of routine work, their own dynamic is special. This chapter will explore in more detail the creative experiences of scientists, some very well-known publicly, others only within their specific fields, but all attempting to articulate the scientific imagination. The continual question is of the origin of new ideas, especially those that radically transform our understanding of an aspect of nature.

A private moment of discovery

As we have noticed, there is a thinner record of the experience of creative discovery in science than one might hope for, but there are a few candid and courageous counter-examples. One we have already met—indeed we were alerted in the introduction to explore imagination and affect in science by Chandrasekhar's emotional realization that black holes must proliferate in the universe in a mathematically exact way. Richard Feynman, Nobel Prize-winner for the quantum theory of light, describes a different but equally precious memory of creative conception, a culmination of a long period of thinking about the puzzling data from experiments on subatomic particles in the early 1950s. The recent developments in cyclotrons—machines that collide particles together at great energies—were providing enormous amounts of new data that indirectly indicated what the unknown forces between them might be. The problem was that no current mathematical law of force made sense of the measurements. As is normal in science, some of the experiments turned out in retrospect to be more trustworthy than others, but at the time it was unclear which should be granted greater weight.

Long before, Isaac Newton had been faced with data on the orbits of the moon, planets, and comets and realized that these might encode a general pattern of the gravitational forces between them. Kepler's deductions of the elliptical shape of Mars's orbit, and of the mathematical relation linking planetary distances to the sun with their orbital periods, published in his revolutionary *Astronomia Nova* of 1609,[3] were decisive. Kepler's results, the invention of new mathematical methods (the calculus), and the imaginative application of it to the mutual orbits of two bodies, led Newton to his celebrated inverse-square law of gravity. The further two bodies were from each other, the weaker would be the force of attraction between them, and the way in which the force dropped with distance would determine the geometrical form of the orbits. The 'inverse-square' label simply describes this law of weakening attraction with distance: removing two bodies to twice their original distance reduces their attractive force by a quarter (½ squared), to three times the distance by a ninth (1/3 squared). Jupiter, five times further from the sun than the Earth, feels just one-twenty-fifth of the sun's gravity as does the Earth. A beautiful moment saw the astronomer Edmond Halley calling on Newton to ask him if he knew of a law that would lead to elliptical orbits—for by 1684, observations had indicated that comets, as well as planets, followed just this geometry of paths

[3] Johannes Kepler, *New Astronomy*, trans. William H. Donahue, Cambridge: Cambridge University Press (1992).

around the sun. Newton had already calculated that elliptical orbits follow directly if gravity weakens with the inverse-square law of the separation of a planet from the sun.[4] The paths of planets, visible to plain sight in the sky, were from that moment on signs of a universal field of force, entertained in the imagination, and not 'seen' in any other way.

Nearly three full centuries later, Feynman had been searching similarly for the law of interaction between newly discovered particles called 'pions'. The essential data had only recently become available because the particles themselves required high energy collisions between the more familiar protons to be created, and existed stably for only a very short time before decaying to more stable particles. Both production and detection required technology developed in the 1950s. One of the pathways of disintegration called 'beta decay' was proving especially puzzling. Like the task of reimagining gravity that had faced Newton long before, in this case too, a visual conception of a field of force subsisting in the space between the particles needed the accompaniment of a mathematical description—the analogy of the inverse-square law of gravity. Although the framework for this physics would have to be quantum mechanical rather than the classical paradigm in which Newton's gravity worked, the notion of a field of attractive force survives the transformation. Feynman writes of the night that he started seriously working with a theory that would, to his astonishment, predict a difference between left- and right-handed versions of the field of force:

> *That night I calculated all kinds of things with this theory. The first thing I calculated was the rate of disintegration of the mu and the neutron. They should be connected together, if this theory was right, by a certain relationship, and it was right to 9 percent . . . I went on and checked some other things, which fit, and new*

[4] Although it seems that Newton had misplaced his earlier work, and had to repeat the calculation before he could reassure Halley of the link between elliptical orbits and the inverse square law of gravitational attraction, in the end a happy circumstance led to the writing of his magisterial *Principia*. The tale is well told by, for example, Gale E. Christianson (2005), *Isaac Newton*. Oxford: OUP.

> *things fit, new things fit, and I was very excited. It was the first time, and the only time, in my career that I knew a law of nature that nobody else knew. (Of course it wasn't true, but finding out later that at least Murray Gell-Mann—and also Sudarshan and Marchak—had worked out the same theory didn't spoil my fun.) The other things I had done before were to take somebody else's theory and improve the method of calculating, or take an equation, such as the Schrödinger Equation, to explain a phenomenon, such as helium. We know the equation, and we know the phenomenon, but how does it work?*
>
> *I thought about Dirac, who had his equation for a while—a new equation which told how an electron behaved—and I had this new equation for beta decay, which wasn't as vital as the Dirac Equation, but it was good. It's the only time I have discovered a new law.*[5]

I have read this extraordinary piece of writing over and over again. Its explosive yet layered prose holds so much together. It communicates at once the thrill, the elevated pulse-rate, the fevered and sleepless labour of plucking one jewel after another from the chest that Feynman had just opened. He had imagined and made concrete an elegant law of force deep within the atom, that itself provides a lens through which a mass of blurred and confused observations become suddenly clear. Yet at the heart of his first forays of exploration into this wonderful object there is a sense of contemplation beneath the excitement: 'I knew a law of nature'. *I knew a law of nature*—the bald statement carries the force of a focused gaze across the divide between mind and the material world, and a sense of personal privilege that one more bridge has been constructed across it, and that this one was *mine* to build. Feynman is careful to distinguish the special nature of this work from the 'normal' science (in the case of a theoretical physicist) of 'calculating and explaining'. This is a moment of gift, and uniquely precious.

There is a solitary specialness to the instant, but his reflections are also drawn to the community of colleagues on which he

[5] Richard P. Feynman (1992), *Surely You're Joking Mr. Feynman?* London: Random House, p. 250.

depends. Dirac's famous mathematical representation of the electron comes to his mind, for it, too, constitutes a moment when the most common particle of all (the electron is responsible for all of chemistry, the properties of materials, and the electrical conductivity of metals) was imagined with a much richer geometry than ever before. Dirac's creation was also a moment at which clarity replaced confusion and consistency contradiction—for the first time the quantum mechanics of atomic structure that had burst upon the physics of the early 1920s appeared consistent with Einstein's enduring framework of special relativity from 1905. With the same stroke, the electron's property of 'spin', which endows atoms with their essential extended structures and their magnetic properties, appeared naturally and unforced. Miraculously, Dirac's idea of this higher geometry for the electron also predicted 'antimatter': the existence of a 'mirror' partner for every known particle, but with an opposite set of properties. Within a few years new experiments had begun to find these antiparticles, first conceived through mathematically structured imagination. We can see why Feynman called Dirac's work 'vital', and why that is the class of experience into which he places his humbler discovery of what became known as 'V-A coupling' in beta decay.

There is an interesting account, which Feynman also gives in his honest and amusing biography, of the struggles that led up to this moment of illumination. He had suffered an experience very common to young researchers—everyone else seems to know so much more, to be so many steps ahead, to write and speak at a level of sophistication that they could never aspire to. I recall the coffee time blackboard conversations when I was a new research student in physics. The professors and post-doctoral researchers seemed so *quick*, so mathematically recondite, nimbly navigating conceptual oceans in minutes that still remained obscure to me hours later. It is all too easy to feel intimidated by this common experience, almost afraid to venture out onto uncharted seas of thought at all. Changing the metaphor, to attempt a new brushstroke, however slight, on our scientific canvas of nature requires enormous confidence, yet all the experiences of a young scientist among his or her more knowledgeable peers tends to sap that confidence. Nevertheless, just listening in to the to-and-fro of ideas, whether or not they are consciously grasped at the time, can plant seeds of new insight that later germinate. Such experiences, especially when followed later by the apparent upwelling of a convincing idea that appears as a gift from elsewhere, reflect the role of unconscious thought—an aspect of scientific creativity we have already glimpsed and will need to explore further. But as with all shy creatures, we tend to detect its movements only in the periphery of our vision. When we turn to stare at it full on, the experience of unconscious ideation is apparently nowhere to be seen.

More examples will be needed to gather together what we can of such fleeting experiences of the unconscious in creativity, for whatever we think about the entire Freudian inheritance for the analysis of subconscious mind today, Freud's remaining legacy is an acknowledgement that our conscious mind constitutes only a fraction of the thinking whole. If it has been shown that much of our emotional life is buried in memories, stances, or conditioned responses of which we are not consciously aware, as well as the demonstration of legion automated systems within the unconscious brain that we need to function, then how much of our creative capacity might also function below the limit of our self-perception? As Peter Baofu writes about the unconscious mind:[6]

> *The subtle distinction between the conscious and the unconscious—and their indicated relations—reveals something interesting in the context of an ontological principle in existential dialectics, namely the openness-hiddenness principle, in that both openness and hiddenness have a major role to play in understanding reality in the world.*

[6] Peter Baofu (2008), *The Future of Post-Human Unconsciousness*. Newcastle: Cambridge Scholars Publishing.

There seems to be something in the metaphor of 'bringing hidden things to light' that speaks not only of the act of illuminating what is still hidden in nature itself, but also of bringing into conscious apprehension new understanding that is latent within our minds, of hauling out into daylight the treasures lying unseen in the depths of our mental caves. If we are to grasp the process of scientific creativity, then we need to understand how to access these depths and how to nourish them in ways that best furnish their environment for the germination of ideas.

There are inspiring accounts of creativity that take as their exemplars the greatest in their fields—Howard Gardner's *Creative Minds*,[7] for example, takes us to the high altitudes of Einstein and Freud in the sciences, and of Stravinsky and Picasso in the arts. This, and its companions, are wonderful reading (many are listed in this book's bibliography), but leave one wondering: how much can one learn of the common experience of creativity from the most extreme examples, from the greatest exponents of new paradigms? Might there be a qualitative difference in mental strategy at such elevated levels that fails to map onto the way that science—or art for that matter—works for the thousands who strive in laboratories or studios, but without the profile and fame of Gardner's heroes? We will need to pay attention to much smaller stories than the heroic tales of relativity or cubism, as well as to such great creative narratives. The advantage of exploring more everyday science will be personal access—conversations in hushed tones with scientists who feel uneasy about acknowledging the overwhelming emotional experience of seeing a mental shaft of light fall onto a puzzling aspect of nature for the first time—this is precisely the intimate context in which we might get behind the polished hagiographies of the greats.

A conversational engagement with the creators of everyday science will also help to overcome the difficulty of articulating, or at the very least reporting, the role of the unconscious, intuitive,

[7] Howard Gardner (1993), *Creative Minds*. New York: Basic Books.

and illogical, the 'first half' of the full 'scientific method' that we are chasing. I will start the conversation on the experience of creativity in the humble field of small science by sharing a personal story from my own scientific journey. As in Feynman's case, this will involve quite a 'deep dive' into a story of research, but as we found there, like putting together a jigsaw puzzle, the detailed contours at the boundaries of each piece are as important as the picture they constitute.

Entangling thoughts

I experienced a puzzle sinking (into the subconscious), and surfacing (into the conscious) as a new thought, quite early on in my first ever research programme. I was trying to make progress with an unexplained finding in the flow of complex liquids composed of long-chain molecules. Both the personal experience, and the story of the background science, bear on the question of scientific imagination, so it will be helpful to delve into its context—the 'soft matter' science of polymer dynamics—to a degree of depth. The story adds to what we have already heard about the emergence of simple and powerful conceptions of complex material behaviour. This time it draws on several thinkers in communication, and on the interplay between their conscious and unconscious thought. It also furnishes an example of the intense interplay between visual and mathematical imagination. First, we need to explore some of the background of 'polymer physics', itself a late twentieth-century story of highly imaginative science.[8]

I had become fascinated with the very puzzling fluid behaviour of molten plastics, noticed by researchers in both academic and industrial laboratories. Plastics are materials whose properties are commonplace to us when in solid form—they constitute much

[8] For a lay introduction to the new field of 'soft matter' in which this topic, and other examples in 'liquid crystals' and biological physics lie, see: Tom McLeish (2020) *Soft Matter—A Very Short Introduction*. Oxford: Oxford University Press.

of our constructed material world as implements, furniture, casings, wrapping, textiles, insulation, tools, and coatings. But as molten fluids on their way to becoming the fibres, films, and other solid forms of their final state, they exhibit extraordinary properties of elasticity and fluidity simultaneously. Melted cheese behaves similarly—thinking of cutting out a hot pizza slice will give a notion of the behaviour. A string of polymer melt (or cheese for that matter) may be extended (messily) to any desired length as a highly viscous fluid, but at the same time if it is let go, will spring back to recover some of its extended length. These 'viscoelastic' fluids seem not only to explore material properties somewhere between the elastic solid and viscous fluid, but also to possess an internal 'memory' for shape. Wait long enough and they will flow into any form with completely fluid nature, but excite them rapidly and they will recoil and vibrate elastically.

The study of unconventional fluids such as viscoelastic polymer melts belongs within the wonderfully interdisciplinary and classically termed field of *rheology*. Heraclitus, the Ephesian pre-Socratic thinker of the fifth century BCE, thought deeply about change and flow—it was he who pointed out (at least as recorded much later by Simplicius[9]) that one cannot step twice into the same river (because the water would be different the second time) and for whom *panta rei*—everything flows. Encompassing equally the science of pastes (such as unset mortar and cement), granular flows such as cascading sand, the delicate formulation of inks, paints, and other coatings, and biological fluids endowed with sticky, slimy, and stringy properties by their constituents, rheology brings together chemists, physicists, engineers, mathematicians, and biologists. Originally concerned with phenomena at the observable length-scales of the flow phenomena themselves, it has been driven downwards into smaller length-scales towards the microscopic structure of its strange fluids in search of explanations for their unusual flows and ways to engineer them at even the molecular level.

[9] Francis E. Peters (1967), *Greek Philosophical Terms: A Historical Lexicon*. New York: NYU Press, p. 178. Commentary on Aristotle's Physics, 1313.11.

The elastic flows of molten plastics, with their unexpected ability to remember previous states for a finite time while in flow, are an ideal case in point. The motion of some internal structure must set this 'memory time'—this much is clear—but the central puzzle is the extraordinary duration of this elastic memory; many seconds is not an unusual scale. This is an aeon in terms of normal molecular motion—a water or oil molecule will tumble head-over-heels at the same temperatures in a million-millionth of a second. If this single molecular somersault were accomplished at the rate of a single human pace, then a one-second memory time in a flowing polymer melt would correspond to a march lasting 30,000 years.

The special molecular structure of plastics provides the clue to their glacially slow internal dynamics, for their molecules are quite different from the simple combinations of a few atoms with which school science is familiar (the three-atom molecules H_2O of water or CO_2 of carbon dioxide). The molecules of plastics are giants—enormously long chains of simple few-atom repeated units—these are the concatenated molecules termed 'polymers'. Powerful microscopes have now visualized the entangled mess of molecular noodles that a molten polymer resembles when magnified over a million times, but when German chemist Hermann Staudinger first proposed, in 1922, the existence of what he called 'macromolecules', he was to face years of criticism and ridicule from fellow scientists. Less than a century ago, not only were long-chain molecules unimagined and unimaginable, but even when proposed by one who saw further than others, they seemed ridiculous. One reason was the sheer size of the monster molecules compared to even the most complex organic molecules known at that time, but even more objectionable was the notion that their molecular weight would be variable and rather ill-defined: a typical polymer material might mix shorter with longer macromolecular strings. Chemistry was an exact science and would brook no blurring. Sometimes the obstacles to our scientific imagination are of our own making, our blinkers donned deliberately. Staudinger's great intellectual move was to remove those obstacles to insight, rather than to dream of anything wildly discontinuous from structures known before.

The macromolecular hypothesis promised to become a powerful explainer. Like a plate of noodles or spaghetti, the polymer molecules within the tangled melt find it all the harder to move the longer they are. Polymers cannot simply flow past each other like the much simpler and smaller molecules of simple liquids like water or oil. Their long string-like forms would have to pass through each other to do so, a dynamic move only possible if they could magically be cut and reformed (wonderfully, there exist proteins in bacteria that perform precisely this process on the natural macromolecule of bacterial DNA when it needs to disentangle itself into the two 'daughters' during cell-division). At first sight, such a massively complex system of entanglement and tortuosity would seem to be forever beyond comprehension or quantitative calculation. Could a human mind be expected to discern pattern or structure in such a mess?

An imaginative way to think about the problem simply was conceived by the French physicist (and later Nobel Laureate) Pierre-Gilles de Gennes, in 1972. He was thinking about the related system of a 'polymer network', in which the long-chain molecules possess occasional chemical attachments to each other, so that nothing flows at all, but forms a weak solid. This is, for example, the way to think of the molecular structure of rubber, or the weak solidity of a jelly. De Gennes then supposed one molecule *not* to be attached to the network of all the others, but instead free to wander through its random three-dimensional latticework. This special string-molecule would have only one way to move through the network, by sliding along its own contour—the tortuous path created by the neighbouring strings and their cross-links at every turn. To return to the noodle (or spaghetti) analogy, this is the very motion that we generate when we eat them politely, winding up a few at a time with a fork. Each string of pasta is pulled out of the mess by one end, the rest of the string following its leader through the noodle-labyrinth along its original path, until it is wound up on the fork.

The difference between the macroscopic plate of noodles and the molecular-scale mess of polymer molecules is that, at the molecular scale, everything has a natural motion of its own—no external winding-forks are required. This is an example of the incessant random seething and wriggling known as 'Brownian motion', named after the Scottish botanist Robert Brown. He was the first to investigate the phenomenon systematically, in the 1820s, through the effect it has on microscopic 'colloidal' particles. Much larger than molecules, so big enough to be seen though an optical microscope, colloids are yet small enough to pick up the random buffeting from the molecular world underneath. Einstein finally showed in 1905 that Brownian motion is the microstructural manifestation of heat—the higher the temperature of a material, the faster is the random motion of all its constituent atoms and molecules, and the more violently they will affect larger particles suspended in their fluid or gas. One consequence of Brownian motion is 'diffusion'—no particle or molecule will ever stay fixed in a fluid, as the molecular buffeting will randomly move it this way and that from any initial position. So, after opening a perfume bottle so that the volatile molecules responsible for the scent start to enter the air above the bottle, it is not long before the exquisitely sensitive molecular receptors in noses several metres away detect that some have wandered that far already. This point in the story of macromolecules in motion—of Einstein, Brownian motion, and colloids—marks a pivotal moment in the history of science itself, for until his explanation of colloidal diffusion through contact with the molecular world, it was still possible to deny the existence of atoms and molecules as anything more than convenient fictions. At the turn of the twentieth century even some leading scientists preferred a continuous theory of matter over a discrete one. After 1910, there was none who held that view.

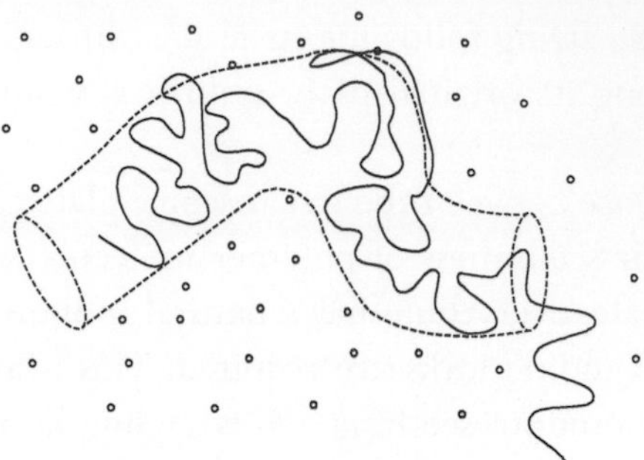

Figure 2.1 The meshwork of neighbouring polymer chains (represented as dots in the diagram) surrounding a given chain (curved line) creates an effective tube-like constraint around it (dashed line) so that its random Brownian motion is confined to a slithering back and forth along the tube – called 'reptation'.

De Gennes' conception of the tangled polymers' Brownian motion took a special form. The motion of a free polymer chain in a network along its own contour reminded him of a snake slithering in reptilian fashion along a tunnel or tube, inspiring the coining of a new word for this special dynamic mode, which he called 'reptation'. The idea is illustrated in Figure 2.1. Though no instruments were available at that time that could visualize the motion and verify the reptation hypothesis, it did have measurable consequences for how fast a polymer chain could diffuse through a host network. The pressing question was how, if at all, this imaginative simplification or something like it might survive. when all the polymer chains were free to move simultaneously, rather than just one.

Just a little later in the same decade, a physicist, Sam Edwards, and a chemical engineer, Masao Doi, working together at the Cavendish Laboratory in Cambridge, realized that the same conceptual key—to think about one molecule at a time, constrained in its motion into a tube by its many neighbours—would also unlock this problem even in the melt case, where no cross-links tie the stringy structure together permanently. The point is that each chain can only move along its own contour even when all other molecules are also in motion, and only when each molecule has crawled out of its straightjacket of entangling constraints—fortunately they may all do this simultaneously—can the composite material flow a small amount. The polymer chains are all trapped in effective 'tubes' constructed from the hedge of their near-neighbours; these survive just long enough, as those neighbours themselves slither away, to insist that each chain moves by the random reptation to-and-froing along their self-made tubes.

It's an extraordinary picture—but more than that, it powers the calculation of numbers for the prediction of experimental outcomes. Now we have the sight before our minds' eyes of how all these millions of molecular strings must move (and, recall, no 'scientific method' taught us how to arrive at that). It is quick work to derive what the 'memory' time of such a fluid must be—just the duration of an escape attempt of an average polymer chain from its original tube. Everything conspires to create a tube-escape time that increases as the *cube* of the length of the polymer chains—double the length, and the emergent memory time increases by eight times. When it first came to light, this relationship did not require any new experiments to check, for it lay very close to the data that had long been known to polymer scientists in universities and in the plastics industry alike.

I remember a conversation with an experienced chemical engineer, who had worked in the industry for decades, on his reaction to the appearance of this 'Doi-Edwards tube model' for entangled polymers. Over the years, from the 1950s to the 1970s, he explained, all sorts of more or less mathematical models had been constructed in an attempt to explain the viscoelastic properties of polymer melts and how they varied with the length of the molecular chains themselves. None was especially convincing, or clearly linked to the molecular structure, or was powerful enough to make new predictions. If one property was perchance correctly accounted for, another would miss the mark entirely. Crucially, however, none of these models had identified the *topological*[10] interactions of the polymers as the most important aspect of the physics. Now, however, that everything seemed to fall into place at once—the steep growth of the memory time with increasing polymer length, and even the way that the viscosity was found to 'thin' as the flow rate increased. Most impressive of all the consequences of entangled 'tube-theory' was that polymers of different chemistries would flow in the same way, an example of a phenomenon known as 'universality', common when material behaviour is dominated by physics at length-scales much larger than those of atoms.

All these new ideas sprung in pictorial and mathematical form from a vividly imaginable picture of tangled, coiled strings filling space and wriggling past each other under the relentless drive of Brownian motion, yet never one string crossing another. Although there were quantitative details of the predictions that looked a little untidy, the power of a single idea, that perceived simplicity within complexity and generated unprecedented insight, was overwhelming. My engineer colleague talked of a 'fog lifting', of new light bathing a previously dark field—in other words, using metaphors that spoke of a sudden perception of the relations between objects that had previously been encountered only individually. He came from a privileged and hard-won position—only years of study could have given anyone such a panoramic knowledge of the rich phenomenology of polymer melts, and at sufficient internalized grasp that, once he had understood the new conceptual picture, knew instinctively that this was the right way to think about them.[11]

[10] A topological property of an object is one that cannot be changed through stretching, squeezing, and twisting alone, but only by cutting or gluing. So a ring doughnut and a mug with a handle are topologically identical, but different from a ball. Polymers are topological because they can only pass through each other by cutting and re-attaching one of them.

[11] The engineer was Bill Graessley, whose experience is articulated at length in a chapter we co-authored in a book celebrating Sam Edward's imaginative contributions to physics: W. W. Graessley and T. C. B. McLeish (2004), 'The

The strangeness of star-shaped molecules

My own first experience of the strange dance of scientific creativity belongs to the next episode in this story of entangled polymer fluids. It is a very ordinary piece of science, yet it reflects the delicate interplay of visual imagination, mathematical abstraction, and the role of the subconscious, and so has some evidence to contribute towards the question of scientific creativity. Something else was known about the flow of polymer melts that had remained a complete mystery until the advent of the tube concept—the bizarre effect of 'branching' in the structure of the polymers. Since the earliest days of polymer chemistry in the 1930s, ways of generalizing the structure of giant chain molecules by adding branch points had become possible, realizing very different fluid properties. One early laboratory experiment, focused on a special case of apparently extreme simplicity, constructed special polymers in which each molecule contained just one branch point from which several 'arms' radiated. Soon to adopt the name of 'star polymer', for obvious reasons, these starfish-like structures set into stark relief just how extraordinarily powerful the effect of chain branching could be. Compared with linear chains of the same size, they showed huge increases in the viscosity of the star-polymer melts, sometimes by thousands of times or more. Yet, once the star structure existed, adding more arms changed nothing at all.

These extraordinary and counterintuitive findings remained unexplained for many years in spite of their industrial importance. The material known as 'low-density polyethylene', from which most plastic sheeting is constructed, is just one example of a polymer whose molecules were known to be highly branched, but whose emergent behaviour could be treated only empirically. The core-variable of topology proved finally to be the key here, as it was for clearing the fog around ordinary linear

Doi-Edwards theory,' in D. Sherrington. P. Goldbart, and N. Goldenfield, eds., *Stealing the Gold*. Oxford: OUP.

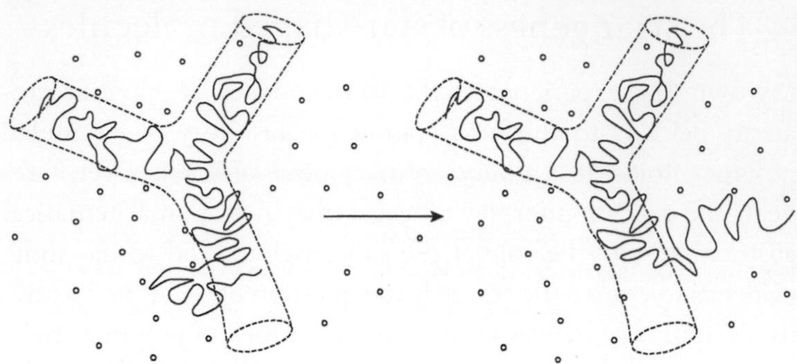

Figure 2.2 The retraction and breathing motion along the tube constraints permitted to a star polymer in an entangled melt.

polymers—it is, after all, the essential difference between linear and branched polymers and the key to entangled 'tube-physics' in the first place. The visual thought experiment of entangling strings develops in the scientist's mind's eye—still concentrating on one polymer, but now in star-form—the picture is sketched in Figure 2.2.

Comparing Figures 2.1 and 2.2 suggests that the reptation mode of the snakelike linear polymers is impossible for the stars. The zoological analogy now is not so much a snake in a drainpipe as an octopus in a fishing net: rather than pass through the mesh wholesale, the animal must withdraw its arms from their entanglements and shuffle a short distance through the net before the retracted arm re-threads a new set of meshes, and another arm is withdrawn. To be sure, an entangled star polymer is not alive, making such conscious moves as an animal, but because of the continuous, random activation of all its moveable links by Brownian motion all of these liberating processes must eventually happen by chance. 'Eventually' is the dominant term—the process of randomly folding up an entire arm so that it retracts to the branch point is calculable, but very much slower even than the random 'reptation' of a linear chain to-and-fro along its tube.

At first the result was thrilling—and a triumph of the theoretical picture of topologically uncrossable tubes trapping each arm of a star polymer in the same way that they trap the length of a linear one. And remember that experiments had shown just this effect—that the time over which the melt of star polymers could flow was typically vastly longer that the linear case, and grew very strongly with the length of the arms. Even more powerful was the evident result that adding length to each arm was much more effective in slowing the whole escape process down than adding more arms, for their attempted retractions and re-engagements happen independently of each other, and in parallel. No other approach even remotely accounted for these qualitatively striking and mysterious results from experiments on the carefully synthesized test materials. The breakthrough had been made by a radical reimagining of the key constraints on the molecules, essentially from an old picture of continuous drag to a new one of topological barriers. The leap had to be an imaginative one to some degree because of scientific culture—the mathematics of motion is still dominated by the set of tools invented by Newton and Leibnitz in the seventeenth century for that purpose—the calculus. To work properly, calculus requires its objects to behave smoothly, continuously, like planets in orbit through the gravitational attraction of their star, or the flow of water over a stony stream-bed. Topology is, on the contrary, highly discontinuous—in the same way that a mug either has a handle or doesn't, one chain is either above, or below, another.

Yet there was a problem with the star melts. On close inspection the slowing down actually predicted by the tube idea was far too great. It is not as rare in science as one might think to hold in your hand an idea, a model of a part of the universe, that appeals deeply by its elegance and persuasive simplicity, as well as by its apparent power to explain a puzzling behaviour, yet whose actual predictions are factors of a million or more distant from measurements. There are current ideas in high energy physics that, while mathematically and physically lovely, and of such power and elegance that physicists are unguarded about their intense desire that they be true—but still seem currently to be in disagreement with observation by the unspeakably large number of 10^{120} (itself vastly more than the number of atoms in the known universe).[12] The shortcoming of the topological tube ideas for branched entangled polymers was a minor peccadillo, compared to such gargantuan transgressions of disagreement, but this was my first opportunity to face the scientist's dilemma: does one follow the logic of the numbers and ditch completely the idea that produced such poor results, or follow the aesthetic of its compelling power and seek out the source of the discrepancy?

I thought a great deal about this problem. Perhaps the way out lay in the huge sensitivity of the predictions to details of the constraints. One can conceptualize a hierarchy of escapes from the tube in this way: most of the time, the random breathing motions of the chain just result in the end of the chain jittering around the end of the tube it occupies. Just occasionally, say one random jitter in ten, carries the free end one whole tube diameter's worth up inside the tube. When this happens, it is nearly always followed by an immediate return of the chain-end to its normal position. It is as if Sisyphus's punishment had become exponentially more frustrating: his rock can roll back down to the bottom of the mountain at any step of his climb. But again, just one attempt in another ten, this does not happen, and a further improbable fluctuation carries the chain end one more entanglement spacing up into the tube, Sisyphus's rock one more pace up the mountain. This argument can be repeated—each step along the way is ten times rarer than the last one. So, the time we would expect to wait for the chain to retract three entanglement lengths into its tube would be a thousand times that of a single jump; for six entanglement lengths this is a million times as long.

[12] I refer to the elegance, but bewildering multiplicity of 'string theory.' For a lay review see Brian Greene's (2003), *The Elegant Universe: Superstrings, Hidden Dimensions, and the Quest for the Ultimate Theory*. New York: W. W. Norton and Co.

This repeated multiplication of timescales is why they are said to increase 'exponentially' with the length of the star-polymer arm.

The basic structure seemed to be on the right lines, but lacking a significant piece of the picture. The imaginative leap that led to that missing physics emerged from a series of events that resonates with many others' experiences of discovery. At that time, I was making regular trips to a fibre manufacturing company in Coventry, England. Anyone working at the soft matter theory group in Cambridge during the 1980s was introduced to the practice of strong industrial collaboration by its inspirational leader (and inventor of those entanglement tubes), Sir Sam Edwards. He had been a student in Princeton of Julian Schwinger, Nobel Laureate for his co-discovery with Feynman and Tomonaga of the quantum theory of light. Later on, Edwards was one of the pioneers of the theoretical physics of polymers and other complex fluid materials by noticing that the mathematical methods developed by people like Schwinger to treat the physics of electrons, light, and magnetism could be adapted and applied in such radically different fields of science. His exploration of transposition across apparent conceptual distance brought Edwards another discovery: the immense value to fundamental science of working with industry. Such collaboration brings a scientist face to face with the real behaviour of complex materials, and with fresh questions and ways of thinking. As we noted by way of example, the distinctly different flow behaviour between melts of branched and linear polymers had first been noticed by industrial scientists, not in a university laboratory. This is by no means the only example of a complex phenomenon arising in the relatively messy world of commercial chemical engineering that has led to a deep and simple piece of physics.

A very tiring day of discussions with a colleague at our collaborating company had addressed some applied problems that the slowly emerging new picture of polymer melts might help to solve. On the way home we started to discuss the 'simple' problem of the star-polymer melt experiments. I confess that I was too tired to follow my companion's rapid-fire thinking—and even too tired to ask him to repeat himself or to slow down. I lost the thread of his argument very quickly and, unable even to nod convincingly, I think he gave up on me before long. We arrived back in Cambridge quite late and I think I must have gone straight to bed and to sleep.

The next morning felt different. The answer to the star-polymer problem was somehow obvious. The first theory was clearly incomplete. The random breathing motion of the star arms in their tubes had to be the fundamental reason for their much slower flow than linear chains. However, it was not yet 'self-consistent'. The tube constraints around each arm could not be fixed, as in a network, for they arose from identical neighbouring copies of the retracting arm, after all. They too were undergoing the same sequence of reconfigurations and re-tracings through the mesh of entanglements. Each time a single chain retracted past a neighbour, it not only freed itself from that constraint—it also liberated the other chain from it. By virtue of thinking about only one chain at a time, we were counting its own escape but not the simultaneous setting free of its neighbours. We had even forgotten that this reciprocity had been considered before in the case of reptation of linear chains, but had turned out to have only a small effect. However, the very different and exponentially slow dynamics of the stars would amplify it hugely. The star-arm Sisyphus was not on his own, but surrounded my myriads of others on neighbouring hills, and the further up some of them managed to roll their rocks, the easier the task became for all the others.

The strange experience of that morning was multiple, for not only was the physics clear, but the way to treat it mathematically was also somehow transparent as well. It required just a step-by-step modification of the recursive multiplying of timescales we have already outlined, counting how frequently we would expect deeper and deeper retractions. But now at each step we also had to count how many entanglements had been lost by other chains rethreading themselves. There were no extra assumptions to make and no unknown numbers. The slow dissolving away of the topological constraints between chains effectively speeded itself up. Using this approach, the result could be calculated within a single page. The central idea worked exactly like the successive multiplying by ten we followed through for a fixed network of entanglements, but now the numbers successively reduced. By the time the outer half of the arms had lost their entanglements, I was multiplying by five rather than ten. Eventually, the chain of accelerated retractions reached the deepest one of all. It was surprising, and also very exciting—this longest time came out of the calculation as if the effective length of the tubes around all arms were divided by exactly three—I already knew that this was precisely the effect needed to agree almost perfectly with the experiments. By good fortune I was spared the anxious wait for experimental confirmation, since the measurements, and their puzzling outcome, had been published some time before.

Underground rivers of the mind

It is rare to find scientists talking about the process, or experience, of creation in print, but after extensive conversations I have found that my little story of star polymers is by no means unusual. After weeks of making no progress at all, one morning everything becomes clear. The fog cleared simultaneously at visual, concrete, and formal levels of thinking. Physics often works at these different levels: a picture (in this case a distinctly visual impression of the mass of inter-threaded star-polymer arms mutually disentangling themselves) is connected with a mathematical description that brings structural form and measure to the visual substance. More usually it takes considerable time to pass from the visual or intuitive idea to the mathematical form. The reason that, in this case, both appeared at once I am sure was due to the conversations of the previous day, in spite of their apparent incomprehensibility to me, as much as to the weeks of apparently fruitless thinking before that. I have no conscious recollection of the content of the industrial discussion any more than I have of my colleague's animated ideas on the way home that I failed to grasp in my exhaustion. However, I am quite sure that without that night-train monologue, I would not have received the sudden realization the next morning of what was happening in the entangled fluids. There had been more unconscious progress than conscious.

The role of the unconscious in the scientific imagination is rarely acknowledged, but a persistent curiosity around scientists about where ideas come from unearths the common experience of an unarticulated and unconscious transport from impasse to clarity. Michael Berry, a mathematical physicist who has spent a career of remarkable productivity and insight at Bristol University, even has a jocular name for the conscious surfacing of unconscious reflections, when a fog of murky incomprehension is dispersed by the entry of a clarifying insight—he calls those epiphanies 'claritons'. He describes one such moment:[13]

> *I was on a train in January 1985 in Germany, and it was so cold that there was ice inside the windows. I had some insight about the connection between quantum mechanics and chaos and was working this out, suddenly it popped into my head that this is all relevant to the Riemann Zeta function, and I remember the exact moment it happened. Of course, I was prepared, because I had read an article about it 8 years before, but I hadn't thought about it since, I don't know how it happened. It wasn't visual.*

Berry provides another account of current connection in the construction of a new and mathematically beautiful idea, but he is clearly aware of a palpable contribution of deeper-lying and longer-running narratives. In my case, the 'seeds' that others had planted germinated overnight, perhaps because of a prepared soil through much earlier puzzling. In Berry's account, there are other contributory structures that must have been somehow in his mind long beforehand. Some of the ingredients within an act of scientific creativity have been circulating deep beneath conscious thought not for days or weeks but years, and within scientific communities more widely for decades. Following this

[13] M. V. Berry, recorded discussion for the Imagination Institute, Kings College, Cambridge (2016).

idea through in the running example of the star-polymer problem, it is one of a class in physics that initially appear intractable because of circular logic. We cannot work out what one polymer is doing, because that depends on what its neighbours are doing, and their dynamics depends on their own neighbours. Without knowing something about one component, how do we break into such a vicious logical circle? The creative key that unlocks them is an example of a long story that begins with the apparently very different science of magnetism, but which keeps recounting itself in the mind of every physicist today.

Magnetic materials present a similar challenge as the mutually entangled polymers: I apply a magnetic field to a material whose atoms act as tiny magnets themselves; how strongly do they line up with the applied field? This would be a straightforward problem if there were just one atomic magnet, as then the only magnetic field it experiences would be the externally applied one. But there are thousands, even millions of atoms that surround any one of them. Each contributes a tiny field of its own, and that field depends on how it is aligned. So, to know in which direction any one atom is pointing we really need to know about all the others first—another frustrated logical circle that seems to lock out all comprehension. Such fearsome 'many-body' problems have been around in physics for a long time. Their recursive structure appears in human endeavour as well. We experience an all too human form of them when we play games of strategy (for fun, or in real life). How we play our next move, or our next card, for greatest advantage, depends on how our opponents will play. But they are faced with a similar quandary of first-guessing. How do we progress?

A beautiful and simple answer to problems like this was first suggested by the French physicist Pierre Weiss in 1907.[14] His

[14] Pierre Weiss (1907), 'L'hypothèse du champ moléculaire et la propriété ferromagnétique'. *Journal of Theoretical and Applied Physics*, **6**, 661–90. The idea of a 'mean' or 'molecular' field was independently suggested in the same context by Pierre Curie.

was the self-referential problem of magnetism—how could he understand and predict the behaviour of one atomic magnet within a vast collection held within a crystalline solid lattice, when each depends on the behaviour of its neighbours? A clue to thinking theoretically about the problem was provided by a familiar, but nonetheless astonishing, property of some *ferromagnets* (the most common example, bestowing the 'ferrous' name of the class itself, is iron). There are two ways in which all the atomic magnets can be made to line up: the first is to apply an external magnetic field (have you ever magnetized a needle by stroking it repeatedly with a strong permanent magnet?). But another is to heat it up and cool it slowly from a very high temperature. Below a temperature characteristic of each material, a spontaneous alignment of the atomic magnets sets in—and the entire solid becomes a permanent magnet of its own, *without the application of any external aligning field.* At low temperatures, each atomic magnet is aligned in just the same way as it would be under a uniform external field at high temperatures.

Weiss reasoned that if the effects were the same, then perhaps, as far as each atom was concerned, the causes might be too. Could he conceive of a 'molecular field'—the sum of all the magnetic fields from all the other atoms brought together at the site of any one of them? An atomic magnet would not be able to distinguish the source of this field from any other—including an external one. If it were large enough, then the local atomic magnet would align with it—*and this decision criterion is the same for all the atomic magnets in the solid.* The invention of this object, the molecular-average or 'mean' field, as if it were something new and independent, rather than mere shorthand for the appalling sum of all other atomic magnetic fields saving one alone, is a brilliant conceptual move, as well as a deft sleight of hand—it breaks the vicious circle of requiring knowledge of the alignment of all atomic magnets before one can predict the behaviour of any one of them. Furthermore, the conception of the molecular field immediately leads to the insight of spontaneous collective behaviour, for there can now clearly exist, in principle, *two* solutions to the coupled magnetic physics. If all the directions of atomic magnets in the solid are entirely random, then there are as many pointing in any one direction as in the opposite direction and the average field felt anywhere is zero. If this is true then there is no tendency to align, and this non-aligned, zero magnetization state is self-consistent. But now think about the opposite case in which all the atomic magnets align in the same direction. Now there will be a finite 'mean', or average, field created by their joint contributions at the site of any one of them. We need to ask if this field is large enough to counter the randomizing effect of heat—the thermal fluctuations that will tend to knock the atomic magnets out of alignment. If it is (and note, this must be true below some low temperature as the thermal motions die away) then this solution—the uniformly aligned one—is *also* a solution of the coupled many-body physics. To ascertain which solution is the one observed in nature is now a question of stability. If a small alignment, with its equally small field contribution, is introduced into the random, zero-field case, does this perturbation 'run away' to the fully aligned case or re-randomize? At a stroke, the imaginative creation of an emergent object, the average molecular field summed from all other atoms or 'mean field', has cut the knot of self-referential states, and given birth to the insight of a collective phenomenon—ferromagnetism.

The 'mean field' concept, although no more than a mental construction that cuts a circular problem in half, has taken on an extremely powerful role in physics in the century since Weiss's creative move. Mean fields now participate in the integrity of our discipline's conceptual fabric. From magnetism in its many guises, the idea arises as well in fluid mechanics, the quantum theory of solids, the optical switching phenomenon of liquid-crystal displays, and the transitions of melting and boiling between the 'phases' of solids liquids and gases, to list just a few examples. Correcting the shortcomings of the mean-field approximation, such as the omission of local fluctuations of the field around its mean value, has itself generated an entire industry within the mathematical physics of 'phase transitions'. The notion of a mean field, the effective local sum of interactions of any one microscopic physical object with its myriad neighbours, runs like a deep underground river in the thought of any physicist. The original 'tube' model of Edwards and de Gennes is really an instantiation of the mean-field idea applied to the topology of interacting strings. So, when in the case of entangled star polymers, the mutual disentangling effects of each chain between all its neighbours becomes dominant, the idea of some sort of uniformly changing 'molecular field of entanglement' was ready to surface for anyone who was motivated to think about it in the context of twentieth-century physics.

The creation of the radically new can often disguise a hidden continuity with earlier ideas, which have themselves been transposed from one domain to another. The case of entangled fluids called not only on innovation and transposition, but also combination of structures whose histories developed on different timescales. The tools of quantum field theory from the 1950s and the chemical engineering and materials physics of the 1970s, drew on the radical idea of the polymer of the 1920s, and ultimately from the theories of magnetism and of Brownian motion from the twentieth century's first decade. Such 'constructive interference' of scientific stories is happening today, but requires increasingly deliberate acts of communication, as we will see.

The creativity of the new—and doing biology in a physics lab

If it is at all possible to talk about 'more imaginative' scientists, or those responsible for a greater degree of creative output, then it must be worth noting any practices or habits that characterize them. One recurrent and surprising pattern of behaviour that this study will encounter at many turns is a curiosity that drives creative scientific thinkers *out* of their area of expertise, sometimes to explore rather distant territories. This tendency to trespass, from physics to biology, or from chemistry to electronics, from quantum mechanics to computing, initially presents as paradoxical, because of the deep and single-minded expertise required to make any progress in science at all. Howard Gardner, on whose *Creative Minds*[15] we have already drawn, reflects throughout his study of creative artists, scientists, and politicians on the operation of a 'ten-year rule'[16]—the length of focused study in a field before which it is generally not possible to produce a work of significance and novelty.

> *I have been struck throughout this study by the operation of the ten-year rule. Should one begin at age four, like Picasso, one can be a master by the teenage years; composers like Stravinsky and dancers like Graham, who did not begin their creative endeavors until later adolescence, did not hit their stride until their late twenties.*

If so much depth within a field of expertise is essential, we might anticipate that time taken elsewhere only subtracts from the central path to achievement. Yet the very prevalence of the botanic metaphor of 'cross-fertilization' ought to point us to the importance of two currents in creativity: the destination but also the detour, the depth of expertise but also an admixture of experience from a breadth of intellectual voyaging, a close community of like-minded and similarly schooled peers but also regular encounters with people from different fields of questions and methods.

Jan Vermant is a Belgian scientist now working in Zürich, an expert in the rheology of complex fluids, such as the molten polymers we have already encountered, and to which he brings an experimental physicist's eye for careful measurement. He ensures regular encounters with other fields of science by the simple, and generous, expedient of welcoming any from outside his department to use the specialized equipment he has acquired and built within his laboratory. Every day, he also takes time to walk

[15] Howard Gardner, *Creative Minds*

[16] Also referred to as the 'ten thousand hours rule'

around the lab a few times. One day he found a group of colleagues from his university's school of bioengineering having trouble with an instrument designed to measure 'surface tension.' This is a property of a liquid surface responsible, for example, for the spherical shape of drops and bubbles. He went over to offer some assistance, and found that the visitors were interested in the way that bacteria grow and multiply on surfaces—a rich pattern of behaviour known as 'swarming.' Their description was so fascinating that Vermant determined to observe it directly and asked if he could look at these systems under the microscope on the adjacent bench.

The pattern formation (Figure 2.3a) that he saw in the bioengineers' bacterial swarms immediately reminded him of instabilities in ordinary flow-fronts between two non-mixing fluids that he had learned about in presentations over the years by researchers from Caltech (an example is given in Figure 2.3b). Inspired by the similarity between the bacterial swarm and the non-living fluid instabilities more familiar to him, he began to read up on the explanations proposed so far for the patterns and the phenomena, quickly concluding that the ideas had never been examined critically. He contacted the group of biologists who had been simply borrowing his apparatus with the proposal that they set about testing suggestions for the underlying rules by which the bacterial colonies grew and changed form.

The connection between very different disciplinary communities was essential to the creation of a new and fruitful research programme. The biology of swarming, with an accumulated experience of its delicate phenomenology, dependence on understanding how to manipulate different species of bacteria, conditions of growth, and how to measure the structures of their biofilms—all this was a rich possession of the bioengineering community. But to connect the observations to a fundamental set of laws governing the local replication rate of bacteria and their production of a surfactant (a soap-like molecule which in great numbers can drive surface flows)—those required the experimental and mathematical tools of physics. The combined team began doing experiments looking at the spreading velocities

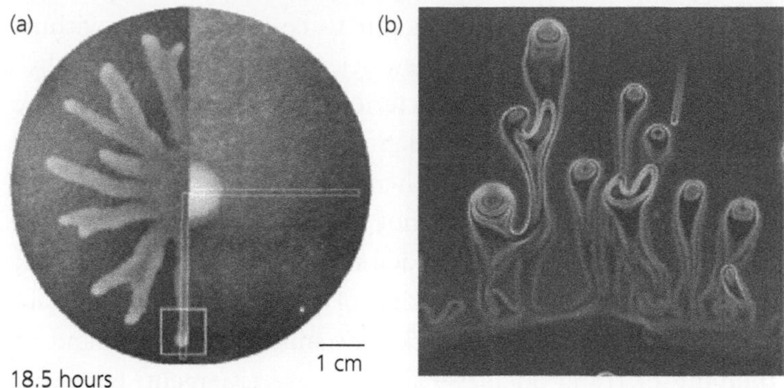

Figure 2.3 (a) A growing and branching bacterial swarm from Fauvart *et al.* (with permission from Royal Society of Chemistry). (b) Example of dynamical system governed by coupled set of fourth-order nonlinear equations: Interferometric image of an instability triggered by surfactant transport along the surface of a microscale aqueous film (with permission from Prof. S. Troian, Caltech).

of the bacterial films with a hydrodynamic analysis. They worked out how to measure the height profiles of the films, realizing their significance. Later on, the experimental programme of physics was modified by uniquely biological techniques, because living organisms—even relatively simple ones when behaving according to relatively simple physical rules—can be modified in a controlled way by making genetic changes. Such deliberate mutations can be designed to test the consequences of those physically expressed rules. The new bacteria, and then their surfactant product, could be labelled with small chemical groups that fluoresced in characteristic colours under the illumination of the microscope. The combined team could compare images rich in the specific information of where the labels were, and at what density, with quantitative predictions of models for the flows, borrowed from fluid mechanics but modified in minimal ways to reflect biological functions, such as the *in situ* manufacture of surfactant by the bacteria.[17]

[17] M. Fauvart, *et al.* (2012), 'Surface tension gradient control of bacterial swarming in colonies of Pseudomonas aeruginosa', *Soft Matter*, **8**, 70–6.

Vermant's experience turns out to be one example within a wider landscape of vibrant new scientific ideas generated by a global encounter between bacteriology and the physics of rheology. The background to such recent history of creativity is worth recounting. The rich phenomena of fluid flow are familiar to us, though they may not always elicit the wonder that they might. The turbulent eddies and whorls of rising smoke or cascading water, the breaking of waves, the bizarrely stable vortex rings of the party-trickster's smoke, the dance and co-merging of rising bubbles—all these are emergent, large-scale consequences of rather simple small-scale laws of fluid flow. These laws bear in turn simple mathematical representation, as differential equations. The history of rheology began at its outset to work on these two levels—the macroscopic patterns of flow and the local mathematically described laws of the fluid. During the twentieth century, the classes of fluids themselves became more complex, adding to the small molecules of simple fluids (such as oil or water) more intricate components, such as the 'macromolecules' of polymers, the colloidal particles too small to perceive with the eye yet giant on the molecular scale, or long rod-like molecules that created a local field of orientation—the 'liquid crystals' of electronically switchable displays. Now it became necessary to think about fluids at three levels simultaneously: to the local and the macroscopic levels was added a third 'mesoscopic' or 'middle' stratum at which the intermediate structures were operative. Affected by the twisting, shearing, and elongation of the flow (from above) and by the constant thermal agitation and lubrication of the solvent molecules that surrounded them (from below), the polymer, colloid, or liquid-crystal entities affected both in turn, extending their influence both upwards and downwards in the cascade of scales, and creating new types of emergent fluid behaviour. We have already worked through a classic example in the case of entangling polymers—the spatial network of entanglements exemplifies the 'mesoscopic' layer of structure, and generates the strange elastic forms of flow, the new emergent property.

Even this rich world of complex fluids shared a simplifying feature in common: while the new mesoscopic ingredients may introduce intricate forms of interaction and flow, such as local orientation and memory, yet they were all passive ingredients. The only generators of motion were the random thermal Brownian buffeting from the molecular world below, and the directed flow of the macroscopic fluid from above. From the point of view of rheological physics, a bacterium sits just at this intermediate mesoscopic length-scale—one can imagine a fluid composed of myriads of suspended bacterial particles behaving rather like the colloidal suspensions that, under the optical microscope, appear structured with similar sized particles. But there is a revolutionary difference, for some bacteria can *swim*. Unlike the passive colloidal particles, they are able actively to cause motion in themselves and in the surrounding fluid. They metabolize biochemical 'fuel', powering tiny molecular motors in their outer membrane, in turn attached to extended paddle-like structures called 'flagella', the bacterial equivalent of propellers.

It is dawning on the rheological community, even as this chapter is being written, that every aspect and each example of passive fluid flow has at least one possible partner in an alternative world of 'active' fluids. The spreading and branching films of Vermant's project constitute an active version of the flow of a passive fluid across a flat surface under a force, like the course of raindrops down a windowpane. Even the turbulent flow of wind and water with its ever-bifurcating structure of eddies also has a counterpart in bacterial films where the swimming strength of the organisms is higher.

Colour Plate A displays in a striking visual manner the structure of experimental and calculated turbulent flows from a dense medium of swimming bacteria. The intricate and interwoven twists and turns of the fluid are driven internally by the active 'particles', but give rise to structure at a much greater length-scale than the size of the bacteria themselves. The task of visualizing the phenomenon appropriately is a challenge in itself but, in the figure, the collaboration of scientists from Cambridge,

Oxford, Dusseldorf, Paris, and Princeton has found, through a combination of contours and colour, a direct way to communicate an immediate grasp of its emergent complexity.[18]

Here, science moves, as it always must, beyond mere observation to another form of visualization, an explanatory layer of unseen structure. The lower panels of the figure represent thinking at the other two 'layers' of complex fluid dynamics. Panel c of the figure presents the results of a computation of a mesoscopic model for the active fluid. Individual 'bacteria' are represented in this simulation by simple rod-like structures that possess just the two properties of mutual repulsion, and the exertion of a constant swimming force along their own length. The rest is simply calculation of the consequences. No more detailed account than this is taken of the complexities within a bacterium. It is somewhat astonishing that a model of the intermediate elemental structures, on such parsimonious lines, is able to reproduce the complex features of the emergent flow structure. It is impossible to deduce the salient features of the underlying physics from the fluid flow alone—creative imagination and a theoretical scalpel are required: the first to create a sufficient image of reality at the underlying and unseen scale; the second to whittle away at its rough and over-ornate edges until what is left is the streamlined and necessary model. To 'understand' the turbulent fluid is to have identified the scale and structure of its origins. To look too closely is to be confused with unnecessary small detail; too coarsely and there is simply an echo of unexplained patterns.

A remarkable element of this work is that, for its authors, the questions had not been answered by a demonstrated connection between the levels of macroscopic experiment and mesoscopic calculation alone. The mathematical tradition of description at the level of local laws of fluid motion, irrespective of the underlying structure, also demanded an answer. The fourth panel in the Colour Plate A displays another calculation, this time simply

[18] H. H. Wensink, et al. (2012), 'Mesoscale turbulence in living fluids', *Proceedings of the National Academy of Sciences*, **109**, 14308–13.

of an equation for the flow of a fluid. Its mathematical form uses the same elements as for a normal watery fluid, but they are supplemented by new terms that reflect the presence of 'bacteria', but expressed as fundamental laws of a fluid element. An analogous process of dreaming-up, then paring down, operates within this area of mathematical physics as well. Indian theoretical physicist Sriram Ramaswamy, one of the originators of this 'continuum' approach to active matter, has written a helpful review of its elegant framework.[19] Here the visual and cognitive play with the symbols of mathematical physics, and the signifiers that come from years of working with their role in local descriptions of fields of velocity, magnetism, or concentration of particles, constitute an imaginative toolkit of building material that we will encounter again.

If the case of entangled polymers suggested that creativity can arise from the combination of ideas on different historical tracks, then the tale of active matter tells a story of thinking simultaneously at different spatial scales. It also exemplifies the power of reconnecting science's fragmented sub-disciplines.

A conversation about creativity in science

The moment of awareness that a new scientific idea has surfaced into consciousness, such as the emergence of clarity about the cooperative slow-disentanglement problem, or the realization that a bacterial film might be thought about as a problem in physics rather than biology, are examples of Bristol theoretical physicist Mike Berry's 'claritons' (he would wish me to report that these 'particles of thought' can also suffer annihilation—the arguments that destroy these lovely ideas are the dread 'anti-claritons'). The railway-carriage example quoted earlier of a deep moment of mathematical realization came from a wider and structured conversation about the role of imagination

[19] S. Ramaswamy (2010), 'The mechanics and statistics of active matter', *Annual Review of Condensed Matter Physics*, **1**, 323–45.

in science, conceived and arranged by the Philadelphia-based *Imagination Institute* and run in Cambridge, UK, in the summer of 2016. As well as Berry, the meeting included astrophysicist Ashley Zauderer; Princeton engineer, Naomi Leonard; the UK's Astronomer Royal and former President of the Royal Society, Martin Rees; the inventor of 'meta-materials' at Imperial College, London, John Pendry; soft matter theoretical physicist and Lucasian Professor of Natural Philosophy at Cambridge, Michael Cates; Harvard particle physicist, Melissa Franklin; Cambridge and Bristol mathematicians, Herbert Huppert and Jon Keating; and science-writer, Philip Ball. The conversation, urged on by psychologist and Institute Director, Scott Kaufmann, spun out the threads of experience from these contrasting thinkers towards questions that have already arisen in our journey into the ill-illuminated topic of the creation of scientific ideas:

What are the modes of scientific thinking, especially in physics?
How do the processes of visual and formal thinking combine?
What external circumstances lead to new ideas?
Is the nexus of scientific creativity in the individual, in a pair in conversation, or in a wider group?
What role is played by aesthetics in science?
What personal and institutional behaviours prove conducive to creativity?
What is the nature of personal experience in scientific creativity?

We have already drawn upon this remarkable meeting of minds, and will do so again. Eavesdropping on its table-talk will feed our search for wisdom on the origins of ideas too—these are scientists talking candidly about the way they experience their most creative work, a brief counter-example to the normal shroud of silence we drape over the shady first stage of scientific creation.

If this entire topic is a personal one, and difficult for scientists to talk about, then most personal of all are those uncanny moments of transition—Berry's claritons—between the unconscious and conscious—yet all gathered at the conversation had experienced them. Pendry spoke of a mathematical, but strongly visual, dream featuring the geometry of a twisted path in the plane of complex numbers.[20] The following day's careful calculation showed this structure to be the key step in solving his current problem. Huppert talked of a problem in fluid mechanics—the form of a 'separation line' in flow within a channel that he had been wrestling with for two weeks. Deciding to experience the phenomenon in the laboratory once more, the solution emerged into his mind just as the new experiment was about to start. Cates referred to a recurrent experience of colleagues calling on him with theoretical puzzles to which he had not the faintest notion of how to reply helpfully, or the slightest suggestion of a solution. Yet on several occasions, although he had consciously forgotten the original question, a solution was ready, apparently effortlessly, before his mind sometime later. The strong consensus was that the experience of interplay between conscious and unconscious thought was the commonplace—rather than the exceptional—experience of scientific creativity.

A related duality began to emerge in the conversation—the exchanges between conscious and unconscious seemed connected to another complementarity between the personal and the communal. Finding one's creative milieu and avoiding areas in which one is mentally uncomfortable appeared as a general experience. Mike Cates talked about 'choosing a field to fit your brain'—when a thinker has that right, then 'it's the way you can think comfortably and fast, and within your own comfort zone, with reflexes that are attuned to what you are trying to achieve.' One of the most creative thinkers in the field of soft matter, the world of random Brownian motion, and complex molecular structures such as the entangled polymers, Cates has a special artfulness for the

[20] Readers will be familiar with the horizontal axis of a 'real number line' of negative numbers to the left of zero, and positive to the right. Mathematics and mathematical physics have found very fertile ground in extending this to a 'number plane' by supposing a vertical axis of 'imaginary numbers'—multiples of the square root of -1. A 'complex number' on the plane has both real and imaginary components in the form of horizonal and vertical distances from zero.

mathematically elegant and incisive choice of model. His team of collaborators, for example, connected the complex molecular system of a 'forest' of polymer chains grafted to a surface (such as finds application in some highly lubricated hard-disk read-heads) to an elegant formulation of the theory of classical motion. He admitted that he found physics attractive because, 'a small amount of knowledge gets you a long way'. Knowledge on its own, however, is clearly not enough to re-create these mental models of nature. Berry talked of the mental analogy of 'green fingers'—not, he insisted, the same thing as mental prowess, but more an innate feeling for the right choice of problem.

The group was quick to point out that these 'comfortable zones' of personal thinking are not simply personally defined but also created within the community, both near and far, within which one thinks and works. Such local communities of practice in turn require contact from more distant zones to produce novelty. Comfort needs some discomfort to be creative—Berry insisted that that attraction of science is not simply the challenge of the puzzle: 'no, no,—it's the connection; I believe it's the *connection*. I remember Sir Charles Frank reminded us in his retirement speech that physics is not only about the nature of things but about the connection between different natures of things, and that is very abstract and deep.'

Naomi Leonard also insisted that it was the excitement of making connections that, for her, drove science onwards. As well she might: Leonard has, among other achievements, brought classical ideas of mechanics into contact with animal flocking behaviour and fluid mechanics to engineer autonomous underwater gliders. She has extended this approach more recently in collaboration with a choreographer to explore how dancers employ the fast mental algorithms native to flocking or herding animals to synchronize their movements, and used that work to explore how machine–human interaction might be optimized. It is profoundly unusual to encounter a researcher in active communication with so many ideas, but not coincidental that what emerges breaks the moulds from which all of them come.

The experience of producing the qualitatively new from a combination of the old proved a common experience, if not quite for everyone at the breadth of Leonard's interdisciplinary span. Mathematician Jon Keating echoed the older story of quantum field theory and polymer physics, relating the experience of importing a familiar idea from condensed matter physics into a current puzzle within quantum computing, a bold programme that seeks to apply the strange notions of parallel, superposed states from quantum mechanics to the task of computation. A 'quantum state' of a particle can combine two or more of its classical states, which in our normal world of experience would be mutually exclusive. An electron that, classically, may be situated at either one site or a second may assume a quantum state that can be described mathematically as a superposition of both. The great promise of quantum computers is that some problems of intractable complexity and length when coded within ordinary hardware may be solved exponentially faster by the use of specially prepared quantum states of atoms, interacting in prescribed ways. The idea is still in its infancy and faces several very considerable technical challenges before any of its wonder-computational dreams can be realized. One of them is the need to prevent the special quantum states from dissipating into their environment before they have completed their computational task, an effect known as 'decoherence'.

Surprisingly, the path to a possible solution came via thinking about the physics of waves—at first apparently at a long conceptual distance from the information science of logical operations, quantum or classical. Think for a moment, Keating suggested, about waves spreading across the expanse of a wide and calm lake, perhaps from the bow wave of a boat. Here is another case in which a form of energy (the motion of the waves) is dissipating widely, rather than being confined locally. Suppose that it were possible to switch the wave behaviour from the usual widely propagating form to another in which the wave energy is kept within localized confines. For a long time now, it has been understood that waves may travel as far as their continuum (the surface of the lake in the case of the water waves, or air in the case of sound waves) can take them, providing that continuum is smooth, but that this is no longer the case when disorder or roughness is present. Recall ripples on a smooth but shallow lake travelling off into the distance, then imagine what happens when the bottom is strewn with large boulders whose peaks come close to the surface. As the water shallows to almost nothing at these places, waves passing those points are partially reflected and scattered sideways and backwards. As a result, the ripples from a large splash no longer travel right across the lake, but now tend to stay within an area confined by the subsurface obstacles, bouncing between them rather than escaping to a far lakeshore. The same effect arises for quantum-mechanical waves of matter, such as electrons, in disordered (rather than smoothly crystalline) solids.

If the waves are of light, then we have the phenomenon of opacity, or 'milkiness'—very little light passes right through the glass of milk but continually bounces between the tiny droplets of fat and protein suspended in the watery supporting fluid. Keating noticed that a similar notion of disorder was mystifying researchers thinking about its consequences for quantum computing. The key conceptual step—the analogy of the localization from condensed matter wave theory—was missing in the quantum computing case. What seemed 'obvious' to a researcher in the first field became 'tremendously creative and insightful' in the second.

At every turn of the conversation about scientific creativity, tracing the origin of new ideas to their conception seemed always to find a meeting point of difference, or an active tension between ostensibly irreconcilable pathways. The creative interplay between conscious and unconscious thought, the connection between previously separate domains, the deliberate exposure to the discomfort of unfamiliar fields of knowledge, even the physical change of environment or mental state and its ability to unlock the pathways of idea-generation into conscious thought—all these returned to a pattern of duality set at the outset by the tension between imagination and form, by idea and observation.

A final duality emerged as the conversation moved on to the way that scientific thinking is embodied—to a generalization of language. Here we came across another creative tension—the pictorial and the symbolic. Some confessed that they thought predominantly in pictures, others in symbols, be they linguistic or mathematical. Martin Rees recalled a technical conversation between two astrophysicists that captures perfectly how these very different modes of thought are brought to bear on precisely the same problem. Leiden astronomer Hendrik van de Hulst, a very pictorial thinker, was explaining an astronomical idea to the great mathematical cosmologist Chandrasekhar, drawing the structures he was thinking about on the board in front of them. 'Chandra' remained baffled for most of the explanation, until his expression cleared and he announced, 'Oh I see—you're really saying that this interaction matrix is antisymmetric.' The one thought in pictures, the other in mathematical abstraction. Translation between them required an intuitive leap of its own.

If scientific creation is a re-creation of patterns and structures observed in the world, within a mental and communicable map of them, then representation of those structures is an essential link in the creative process. But has the essentially visual mode of imagination now given way to mathematical representation? Is van der Hulst an example of an old methodology in decline, and Chandrasekhar the harbinger of the new? The very commonplace invocation of the visual as a metaphor for understanding itself—'Oh I see!'—must in any case provoke further reflection on the visual wing of the palace of the imagination. Representation, symbolic or pictorial, also holds the promise of linking the subconscious and conscious sectors of creation. But science does not hold a monopoly on pictorial representation of the world. The visual arts live and thrive by visual representation and its communication.

Bringing the survey of artistic and scientific confessions of the first and this chapter together underlies the common experience of different modes of creative thought. Rather than 'scientific' or 'artistic' creativity, people's language speaks of complementary classes of visual, narrative, and abstract thinking that nourish both scientist and artist. These three domains of imagination will govern the structure of the following chapters. So first, to explore the territory of visual creativity, we must wander out of the laboratory for a while and into the visually stimulating surroundings of artist's studio, architecture, and the starry sky.

3

Seeing the Unseen

Visual Imagination and the Unconscious

So much of fire as would not burn, but gave a gentle light, they formed into a substance akin to the light of every-day life; and the pure fire which is within us and related thereto they made to flow through the eyes in a stream smooth and dense, compressing the whole eye, and especially the centre part, so that it kept out everything of a coarser nature, and allowed to pass only this pure element. When the light of day surrounds the stream of vision, then like falls upon like, and they coalesce, and one body is formed by natural affinity in the line of vision, wherever the light that falls from within meets with an external object. And the whole stream of vision, being similarly affected in virtue of similarity, diffuses the motions of what it touches or what touches it over the whole body, until they reach the soul, causing that perception which we call sight.

Plato, Timaeus

Seeing is not a simple matter of 'taking pictures with the eye.' In his great work of natural philosophy, the *Timaeus*, Plato describes a theory of vision that was to dominate both Occidental and Arabic thought for a millennium and a half, yet to us today it seems strange, counterintuitive, even illogical. Known as the 'extramission' theory, it assumes that 'seeing' is an active, not passive, experience. Rather than light from the objects around us streaming *into* the eye, this ancient idea posits a substance (the 'stream of vision' in the quotation above) streaming *out* from the eye, making contact with objects in the external world. The earliest source of this notion of extramissive vision we know of is found within the

The Poetry and Music of Science. Tom McLeish, Oxford University Press.
 DOI: 10.1093/oso/9780192845375.003.0003

fragmentary textual legacy of the fifth-century BC Sicilian thinker Empidocles.[1] Adopted by the great physician of antiquity, Galen, it dominated the alternative 'intromission' theory for centuries, in spite of objections from Aristotle. We might well wonder from a twenty-first-century perspective, when the inwards streaming of light into the eye seems so evident, why it took so many centuries to establish this 'obvious' hypothesis, and why in ancient times the favoured model was built on a diametrically opposite scheme.

A substantial refutation of extramission and a nuanced argument for intromission seem to have been mounted first by the scholar known in the west as Avicenna.[2] Reading the great tenth century Persian philosopher is a helpful antidote to the blindness we suffer due to the dubious benefits of hindsight. Avicenna refutes two other theories before introducing his own: Platonic extramission and an Aristotelian form of intromission. Aristotle held that the intervening air was set in motion by the coloured light of an object, and this motion caused an image of the object to appear in the eye, whereas Avicenna wanted to maintain an instantaneous transport of the impression of an object to the eye. Alert to optical anatomy, he also disagreed with Galen in the identification of the central perceptual organ within the eye, holding that the optical nerve was essential—more so than Galen's preferred vitreous humour.

Avicenna's work was extremely influential in the European medieval enlightenment of the later twelfth and thirteenth centuries, via the twelfth-century 'Translation Movement,' which included Arab commentaries on Aristotle, as well as the original scientific works of 'The Philosopher.' Among other imaginative reconceptualizations of the natural world, this period of

[1] We know this, not directly from any surviving text, but through Aristotle, On the Senses and their Objects, 437b23–438a5.

[2] Ibn Sina, c. 980—June 1037. The key passage is found in his *de Anima* III5 and 8, but has been beset by problems of translation, at least since its treatment by Albert the Great, see D. N. Hasse (2000), *Avicenna's De Anima in the Latin West*. London: The Warburg Institute.

rapid intellectual development saw optics becoming a central theme in a new approach to science. Ignited by the rediscovery of Aristotle's natural philosophy, through Arab translations and commentaries, this renaissance of the high middle ages included the application of mathematical and geometric demonstrations to the understanding of nature, as well as early experimentation. The foremost pioneers of mathematical thinking in the first and second halves of the thirteenth century were the English polymaths Robert Grosseteste and Roger Bacon, respectively. Grosseteste, in the 1220s, the first natural philosopher to rationalize a three-dimensional theory of colour,[3] also suggested that refraction was responsible for the phenomenon of the rainbow. He seems to have held a theory of vision that subtly combines the extramission and intromission views. The relevant passage from his *c.* 1225 work on the rainbow *De iride*, gives us a glimpse into the complexities of vision that we simply pass over today.

> *Therefore we start by declaring that perspective is a science that is founded on visual figures, and this subalternates to itself the science that is founded on figures contained by radiant lines and surfaces, whether these radiants are projected from the sun, or from stars, or from some other radiant body. And it should not be thought that this going forth of visible rays is an imagined supposition merely, without basis in reality, which is the opinion of those who consider the part only and not the whole. But it is to be known that sight is a luminous and radiant substance like to the nature of the sun, the radiation of which, conjoined with the radiation of the external luminous body, entirely completes vision.*
>
> *Thus natural philosophers dealing with that which belongs to the natural and passive part of sight say that sight happens by intromission. But mathematicians and physicists who study what is beyond nature, dealing with the part of sight that is beyond nature and active, say that sight happens through extramission. This part of sight that happens through extramission Aristotle treats clearly in the last book on animals, saying 'the inner eye sees remotely; for its movement is not broken up or consumed, but the power of sight goes out from it and passes straight to the things seen.'*[4]

[3] Greti Dinkova-Brun, et al. (2013), *Dimensions of Colour: Robert Grosseteste's De Colore; Edition, Translation and Interdisciplinary Analysis*. Durham: Durham Medieval and Renaissance Texts.

[4] R. Grosseteste, *De iride*, translation courtesy of Sigbjørn Sønnesyn (personal communication).

This rich passage gives us a glimpse into the complex and divergent thinking about vision that dominated from ancient times until the dawn of the modern era. Behind the dual system that Grosseteste proposes is the idea, shared by most classical, Arabic, and European thinkers alike, that perception via the senses of the animated human soul possesses an active aspect. No sense, they reasoned, could be entirely passive. Augustine, for example, sustains a lengthy discussion of how the ear and mind must perceive actively as well as passively in the process of hearing.[5] In the sensing of anything, the soul (the word in Latin is *anima*, in Greek *psyche*—and to connect with our own late-modern categories might in most cases be translated *mind* rather than *soul*) must make decisions, must 'reach out' into the world and connect with what it sees, hears, or touches there. To be 'animate' in this sense is what differentiates what happens when light falls into an eye, from the process of light falling onto a stone.

Grosseteste criticizes the theory of intromission as over-simplistic, and attributes it to a disciplinary narrowness. Only philosophers with a restricted interest in passive phenomena might reasonably construe such a theory of nature, he claims. The Platonic sense in which he invokes both mathematics and physics to rectify such narrowness demonstrates that, even in the thirteenth century, there were the seeds of rival worldviews, but also that the 'mathematical' covered much more then than it does today. Music, as we shall see, was a mathematical discipline from antiquity to the early modern era, and joined astronomy, geometry, and arithmetic in the 'quadrivium' of higher disciplines.

Plato introduces this pattern of the curriculum in his *Republic*, and it became the later pattern of higher learning in the medieval schools. Above all, mathematical ideas *connect*—they relate points with lines (geometry), the motions of the heavens with the cycle of the seasons on Earth (astronomy), the harmonious

[5] See Chapter 6 on the long story of mathematics and music and Augustine's *de Musica*.

relations of musical sounds through the whole-number ratios of their vibrations. Number itself provides the key to perception for Augustine, when patterns of numbers in the external world match those learned or innate in our minds. So, a mathematical thinker at the crossroads of ancient, Arabic and medieval thought, such as Grosseteste, will wish to supplement an intromissive theory of vision with a more active relational role for the mind.

Whatever its truth value, this is sophisticated stuff, and not only for its intrinsic levels of critical thought and perceptive anticipation of purely receptive/passive theories of vision. For although the physics of vision may be entirely intromissive to our current understanding, the complete science of visual perception, including its essential cognitive component, possesses a resolute and extramissive aspect as Grosseteste maintains. Even the possibility of visual art, as now understood, is permitted only by an active component of seeing—humans interpret the visual world by an exercise of pattern matching between impression and expectation. Art critic and writer Ernst Gombrich makes the point in detail in his *Art and Illusion*.[6] The deception that a two-dimensional representation of a figure or a landscape can elicit an immersive experience of the original object depends on a formal decision to perceive, a developed partnership between artist and viewer. Ultimately, the possibility of visual illusion and *trompe l'oeil* rests on the active processes of vision, inventing the perception of structures not present in the external world. All acts of visual perception—of 'seeing'—rest upon acts of visual imagination.

We are tricked into the illusion of motion in the objects of a diagram when there is none, for example by exploiting the apparent motion that we ourselves create by continual rapid eye movements. The effect is normally cancelled out in the cognitive

[6] E. Gombrich (2002), *Art and Illusion A Study in the Psychology of Pictorial Representation*. London: Phaidon Press; also see David Summers (2003), *Real Spaces: World Art History and the Rise of Western Modernism*. London: Phaidon Press.

chain of vision, so that we perceive a steady image, but carefully constructed patterns can confuse the corrective pathway. Even as fundamental an idea as 'motionlessness' is one that we project onto our visual world, rather than passively receive from it. So, an idea that draws on roots far earlier than Renaissance art—extramission—mounts a critique of one of its core tenets, that of 'central perspective.' Visual perception, even of static images, is a dynamic and projective process as much as a receptive one. Strong demonstrations of the 'extramissive' psychology of vision include illusions of motion in carefully constructed images. To take a fascinating and illuminating example, consider Colour Plate B. This is the 'rotating snakes' illusion first recorded by Ashida and Kitaoka.[7]

The illusion of motion, a gentle apparent rotation of the circles in the peripheral field of vision, is very striking for most observers of this and similarly constructed pictures. Yet of course no motion is present. The illusion exploits our visual system's special capacity to detect motion (thought to have evolved originally as an aid to survival from predators). It is possible to confuse this sensitive system by the similarity of pattern at slightly different scales, when the eye's natural movements alight on them. The local presence of stripes is important because it seems to be the mechanism of detected edges or borders between areas of different tones that excites the relevant perceptual trains. Irrespective of the fascinating details by which the illusion works, the point here is that we are ready to ascribe motion to a visual scene even when given complete yet static information. It is an example of the constructed nature of much of what we 'see,' and evidence in turn, that our visual imaginations are more powerful that we might believe. It also points out the possible distancing and dialogue that we can maintain in our minds—we are critically aware that the motion is an illusion even while experiencing it. Is it

[7] A. Kitaoka and H. Ashida (2003), 'Phenomenal Characteristics of the peripheral drift illusion'. *VISION: The Journal of the Vision Society of Japan*, **15**, 261–2.

possible that the power of visual construction necessary even for normal perception might be recruited in the service of creative acts in science and art?

The visual metaphor within the scientific imagination

There are indeed lessons for scientific creativity from theories of visual perception. Contemplation of the process of vision will not lead us far from the imaginative connections between art and science, if for no other reason than the visual has always held a prior place as the metaphor for thought itself. 'Oh, I see it now!' is the commonplace response to the first grasp of understanding. Here is Carl Friedrich Gauss on understanding in mathematics:[8]

> *Anybody who is acquainted with the essence of geometry knows that [the logical principles of identity and contradiction] are able to accomplish nothing by themselves, and that they put forth sterile blossoms unless the fertile living intuition of the object itself prevails everywhere.*

Here 'intuition' is translating the German *Anschauung*—a visual gaze, or a capturing of a scene displayed before the onlooker. The importance of visual thinking emerged in every example we considered in the last chapter, as well as revealing an interesting tension with thinkers who were resolutely non-visual. A counter-tradition can be traced to Plato for whom, particularly in regard to mathematical truth, visual thinking falls short in rigour and is suspect through the very illusory power that makes art possible.

Modern examples of Plato's critique are not hard to find, from critical theory in humanities,[9] to mathematics, for example, '[the diagram] has no proper place in the proof as such, for the proof is a syntactic object consisting only of sentences arranged in a

[8] Quoted in W. Ewald, ed. (1996), *From Kant to Hilbert. A Source Book in the Foundations of Mathematics*, Oxford: Clarendon Press Vol. 1: 300

[9] Late-modern philosopher Emmanuel Levinas develops this critique, as we will see in Chapter 8.

finite and inspectable array.'[10] A precious memory from my own experience of pure mathematics courses at university was the moment the lecturer admitted, 'I think a teeny little diagram might help at this point. . . but I'll rub it out quickly in case someone comes in.' This mathematical prioritization in value of the textual over the visual or diagrammatic has not always held, as we saw in the last chapter in two approaches to astrophysics. On Gauss's side we ought also to invite Isaac Newton, who formulated the proofs of his dynamics in his great work, *Principia*, using the semantic and algebraic logic of his newly invented calculus, yet presented them for formal scrutiny in the form of geometric reasoning, replete with diagrams.

The early twentieth-century mathematician Jacques Hadamard was fascinated with the mental worlds of mathematicians, investigating by means of a questionnaire sent to any he knew well—including Albert Einstein, who made a richly reflective response:

> *The words or the language, as they are written or spoken, do not seem to play any role in my mechanism of thought. The psychical entities which seem to serve as elements in thought are certain signs and more or less clear images which can be 'voluntarily' reproduced and combined. There is, of course, a certain connection between those elements and relevant logical concepts. It is also clear that the desire to arrive finally at logically connected concepts is the emotional basis of this rather vague play with the above-mentioned elements. But taken from a psychological viewpoint, this combinatory play seems to be the essential feature in productive thought—before there is any connection with logical construction in words or other kinds of signs which can be communicated to others.*
>
> *The above-mentioned elements are, in my case, of visual and some of muscular type. Conventional words or other signs have to be sought for laboriously only in a secondary stage, when the mentioned associative play is sufficiently established and can be reproduced at will.*[11]

[10] N. Tennant (1986), 'The withering away of formal semantics?' *Mind and Language*, **1**, 382–8.

[11] From Jacques S. Hadamard (1945), *A Mathematician's Mind, Testimonial for an Essay on the Psychology of Invention in the Mathematical Field*. Princeton: Princeton University Press.

The visual metaphor is overwhelming here: 'physical,' 'clear images,' 'muscular,' 'associative play' are the signs that point to the priority of a visual and concrete imagination over language, prior even to the use of mathematical symbols. Perhaps the most widely communicated example of visual thinking in science is Einstein's early *Gedankenexperiment*[12] of the attempt to catch up with a ray of light. All of his reflective language applies to the wonderful description of the oscillating but apparently stationary fields of electricity and magnetism that an observer would perceive if travelling alongside light were possible. The vibrant strength of this picture impressed on the young Einstein that this action, although permitted by the Newtonian theory of motion, was in flat contradiction to Maxwell's recently formulated theory of electromagnetism. All oscillating electromagnetic fields move, for all observers, at the speed of light. The visually imagined contradiction led directly to the special theory of relativity in 1905.

Gauss's notion of *Anschauung*, and the role of the visual in formulating physics of Einstein or Newton, responds to another question raised by scientific creativity. For in as much as visual perception is not consciously computed, but is an unconscious perception of patterns, it is also a path into the duality between conscious and non-conscious thought in imagination that appears wherever we look for the origin of scientific ideas.

Mathematical theory-painting

An instructive example of visual representation emerged during the *Imagination Institute* conversations.[13] At one point, the theoretical physicists developed a surprising conversation about the visual and symbolic use of mathematical notation. The discussion did not pivot, as might have been expected, around processes of logical calculation or of deriving quantitative predictions from

[12] 'Thought experiment.'

[13] See Chapter 2 for an account of the meeting.

theories. Within the more intimate confines of the conversation, the dialogue turned instead to the use of mathematical symbols in creating theories in the first place. Participants spoke of the way that mathematics became a creative plaything in the initial formulation of models—the formulation of the first equation on a paper. In the latter half of the twentieth century some physicists have even started to use the term 'theory' in this very restrictive way—to refer specifically to the mathematical expression that defines the dynamic behaviour of a model. Depending on the context, this quantity might be the total energy (when its expression is known as the 'Hamiltonian,' after the great nineteenth century Irish mathematician George Hamilton), or the 'action' (when the mathematical object is the 'Lagrangian,' named after the eighteenth century Italian and French mathematician Joseph-Louis Lagrange). From these powerful functions of all the variables in a model of the system, the emergent physics is implied and, in principle, calculable.

But how does one write down a 'theory' by its formal Hamiltonian or Lagrangian? What should be included and what left out? Some readers will be surprised that there seem to be choices to make here, given the established laws of atomic interactions. Yet such decisions always arise when a model is built from 'coarse-grained' variables—such as the strategy of taking single grains of sand as the smallest unit in a theory for forces in a sand pile, or simple strings in the case of polymers, rather than to start (needlessly) with all their atoms. Such an approach is actually even adopted when atoms are themselves theorized, for they too are composite. Molecules, atoms, nucleons, quarks—these are the milestones that mark the inward journey of the last century in physics.[14] Even the 'standard model' for all known elementary particles and their interactions requires writing down a Lagrangian *ex nihilo* in a mathematical form that includes many

[14] For a beautiful account of this inward and downward journey of scale see Frank Wilczek (2015), *A Beautiful Question*. London: Allen Lane.

arbitrary numbers. The same mathematical ingredients occur in 'fundamental' theories for microscopic membranes, or the rod-like molecules of liquid crystal materials used in advanced displays. There are terms for kinetic energies, others for attractive and repulsive forces, others for the incessant Brownian 'kicks' from the thermal background. Theoretical physicists will have developed through experience a 'feel' for the different terms. One of the Imagination Institute participants reported from personal experience, 'What I personally find—and I do need equations to think—then I need a piece of paper in front of me and I'm pushing symbols around on the page and things fall into place.'

To listen to, and especially to watch, the construction of models from the palette of mathematical ingredients draws comparison with the making of preparatory sketches for paintings. Both draw on an established visual heuristic, using a developed set of symbols that assist cognition in a very efficient representation. Both enjoy a plasticity of construction and enable a visual-cogitative process of creation. Furthermore, the symbol set, its operations and meaning, are shared with a community—the creative process can be entered into by more than one participant. From the Imagination Institute conversation once more:

> *You do it at the blackboard as well. This is something which is unique to math and physics, that, you know, a conversation between two physicists without a shared place to write things . . . actually you think well let's go find somewhere where we can have a blackboard or that whiteboard.*

The effectiveness of an established notational system for theoretical physics was illustrated by an amusing incident recalled by mathematician Michael Berry. At one point in a course that he attended on quantum mechanics, the lecturer decided to change all the conventional notational symbols for quantities in a mathematical derivation (substituting, for example, the symbol normally used for Planck's constant, h, with that usually representing momentum, p). Although it was in every other sense as meaningful as before the substitutions, the students found the new version bordering on the incomprehensible. Mathematical notation also points beyond visualizable to entirely abstract structures—another 'mode' of creativity that is the focus of Chapter 6.

There is another strong parallel between this type of visual symbolic manipulation in mathematical physics, and the construction of a (representational) painting. Throughout the process of visual and graphical manipulation the thought process is continually in tension between the representation and the thing represented. Here is the canvas, the paint, the arrangement of colour, but there is the scene which the painting is to conjure up and interpret in the minds of viewers. The mathematical symbols on a whiteboard similarly possess an internal relational structure as a 'theory,' which can be criticized in its own right. But they also refer to phenomena in the world, be they the dance of electrons at the surface of an illuminated metal, or the fluctuations in a cell membrane as a protein-complex 'self-assembles' within it. The physicist creating a theory continually refocuses between two visual planes. In closer view is the 'canvas' carrying the formulae, whose symbols and terms are added, removed, modified, strengthened, and reduced. Beyond lies the physical world depicted and modelled, in imagination and as experimental data. The distant plane supplies the constraints on the reinvention occurring at the close level of theory, which must possess a form of reciprocal 'transparency' onto the represented world. Neither art nor physics employs a creative process of one-way traffic from nature onto a symbolic plane, but a dynamic tension and communication between the representation and the represented.

The ancient aesthetic of active seeing

Comparing scientific creation to the broader tradition of visual perception, and especially to visual art, suggests that we need to explore parallels between the creation of scientific models of the world and artistic representations of it. Modern discussion of artistic aesthetics often begins with Kant's *Critique of Judgement*,[15] where the philosopher finds himself continually presented with the task of disentangling art and nature:

> *Genius is the talent (natural endowment) that gives the rule to art. Since talent is an innate productive ability of the artist and as such belongs itself to nature, we could also put it this way: Genius is the innate mental predisposition (ingenium) through which nature gives the rule to art.*

Here Kant uses 'genius' to mean simply 'creative ability,' rather than the contemporary usage as 'high-performing individual,' and his 'nature' includes all organisms, including—and especially here—the human. How close does the notion that 'nature gives the rule to art' lie to the scientific desire to grasp the rule of nature? Kant claimed, for the oppositional stances of art and science to nature, that science can never be truly original in the same way as art, but at the same time admits the strong constraints that art receives from nature. Both scientist and artist find themselves acutely engaged in active visual perception, and in the metaphorical use of it. One of Kant's great themes, perhaps the most important architecture of his thought-world, was the delicate interplay between the external world of phenomena and the internal, mental world of impressions, reason, judgement, and aesthetics. His resolution of the opposed scientific philosophies of empiricism and idealism, in conflict since Aristotle's departure from Plato, turned on just this reciprocal relation between internal thought and the external world. Art historian Martin Kemp looks at Kant's duality through the lens of evolutionary thought:[16]

> *It seems to me that the evolution of the human brain (and, at lesser levels of complexity, animals' brains) has equipped us with the means to set the exterior structures and inner constructs in ceaseless dialogue.*

[15] I. Kant (1996), *Critique of Judgement, in Critique of Pure Reason.* trans., Werner Pluhar. Indianapolis: Hackett (sect. 46).

[16] M. Kemp (1996), *Art Journal*, **55**, 29.

Kemp is talking not just about perception, but of its deeper extensions that both art and science know. This sense makes explicit the dialogue between projection of imagination onto the world, and reception of impressions from it, in turn interpreted through inner mental structures. It recalls Thomas Browne's 'true theory of death,' which though triggered by material objects, requires mental completion.

Such reciprocal perception as a metaphor for thinking and reasoning has a long history, one we inherit from earlier thinkers whether we know their names or not. Behind Robert Grosseteste's thirteenth-century discussion of vision itself, for example, lies a much deeper philosophy of knowledge that also extends into the metaphorical grasp of the natural world, five centuries before Kant. Here the medieval Oxford scholar is commenting on Aristotle's scientific method presented in the *Posterior Analytics*, in the then-recent Latin translation by James of Venice. James introduced the word 'sollertia'[17] to describe the power of mental sight in its ability to reimagine the natural world:

> *Sollertia, then, is a penetrative power by which the vision of the mind does not rest on the surface of the thing seen, but penetrates it [the thing seen] until it reaches a thing naturally linked to itself. In the same way as corporal vision, falling on a coloured object, does not rest there, but penetrates into the internal connectivity and integrity of the coloured object, from which connectivity its colour emerges, and again penetrates this connectivity until it reaches the elementary qualities from which the connectivity proceeds.*[18]

At this high watermark of resurgence in natural philosophy, arguably the point of departure for trains of thought that led

[17] Aristotle's Greek is *agchinoia*, which might also be rendered 'acumen.'

[18] P. Rossi, ed., Robertus Grosseteste, *Commentarius in Posteriorum Analyticorum Libros*, Unione Accademica Nazionale Corpus Philosophorum Medii Aevi, Testi e Studi, ii, p. 281, quoted by R. W. Southern (1986), *Robert Grosseteste: The Growth of an English Mind in Medieval Europe, 2nd edition*. Oxford: Clarendon Press, trans. Sigbjørn Sønnesyn (personal communication); Aristotle's Post. An. II.19 is also in the background here, where the emergence of general understanding from particulars of sense-perception is described: 'It is like a rout in battle stopped by first one man making a stand and then another, until the original formation has been restored.'

to the rise of early modern science three centuries later, the metaphor of vision for understanding follows in exact parallel the relational and reciprocal structure of optics. We see by 'reaching out' to structures in nature by the power of mind, building inner patterns that reflect the outer. They match external connectivity by an internal mental image with the same internal relationships. Grosseteste's choice of the example of colour is a happy one, for he was the first to rationalize, in just 400 words within a mathematical jewel of a treatise on colour—his *De colore*[19]—the three-dimensional space within which perceptual colour dwells. He was centuries short of the detailed work on retinal anatomy and biophysics that revealed the biological reason that combinations of red, green, and blue lights can span all perceptible colours, or even of Thomas Young's brilliant nineteenth-century deduction that there must be three distinct types of wavelength-dependent receptors in the human eye. Yet he correctly imagines that the mathematical structure of colour reflects a mental connectivity as much as the physical operation of light within coloured materials. The neurology of the eye is indeed an extension of the brain itself—in this sense we think as we see.

Our next stopping point in the story of the visual scientific imagination is in the fourth century, in the company of the Cappadocian theologian Gregory of Nyssa. We know that the writings of this remarkable and influential thinker, in particular his *De officio hominis* (*On the making of man*), were familiar to Grosseteste through his commentaries on the Psalms and commentary on the creation stories in Genesis—his *Hexameron*.[20] Gregory records a remarkable death-bed dialogue with his sister Macrina (whom both Gregory and his brother Basil the Great call 'The Teacher') in another remarkable text, *On the Soul and the Resurrection*. The greater part of the discussion is a debate on the reality of 'the soul' (again, in context the word—the Greek is *psyche*—might better

[19] See Greti Dinkova-Brun, et al., *Dimensions of Colour*.

[20] R. W. Southern, *Robert Grosseteste*.

be translated for contemporary readers as 'mind' for the purposes of that treatise). Macrina's final and decisive move against Gregory's position (assumed for the sake of the argument, that mind is merely an epiphenomenon of matter) is that it 'penetrates to something below the visual image.' She chooses the very example featured in the introduction of the phases of the moon: we do not assume that the appearance of a waxing and waning object describes a reality of a celestial disk growing and shrinking, but understand that the moon is a sphere passing through successively different angles of illumination by sunlight as seen from the Earth. It is mind that performs this task—'the mind that sees' below the surface of phenomena, or in Macrina's words:

> *You see what the eye does teach; and yet it would never of itself have afforded this insight, without something that looks through the eyes and uses the data of the senses as mere guides to penetrate from the apparent to the unseen. It is needless to add the methods of geometry that lead us step by step through visible delineations to truths that lie out of sight, and countless other instances which all prove that apprehension is the work of an intellectual essence deeply seated in our nature, acting through the operation of our bodily senses.*[21]

These words would not be out of place in the introduction to a contemporary text on the cognitive neuroscience of perception. They elegantly describe the experience that any scientist recognizes as the task of re-creating a mental image of the world we experience.

If Gregory of Nyssa testifies to the analogical power of vision from the perspective of eastern late antiquity, then Augustine of Hippo speaks for the west of the same period. In his *De Trinitate* (*On the Trinity*),[22] written over the first two decades of the fifth century, he outlines a three-level theory of knowledge as a metaphorical extension of vision. First, *corporal vision* is the pure sense that

[21] The text can be found online at http://www.newadvent.org/fathers/2915.htm

[22] Aurelius Augustine (1963), The Trinity, in *The Fathers of the Church: A New Translation*, *45*, trans. Stephan McKenna. Washington, DC: The Catholic University of America Press, p. 344.

humans share with animals. Second, *intellectual vision* (*spiritale*) is unique to humankind, and included the ability to recast and rearrange physical reality within the imagination. It is here that Augustine gives us the idea of the 'mind's eye.' The highest, and third, level of vision, the noetic, or *sapienta* (*wisdom*), is the ability to extrapolate to realities beyond those sensed. This threefold epistemology recurs repeatedly within Augustine's writing.[23]

The name *sapienta* given to Augustine's highest form of sight provides a clue to the ancient source of the analogy. Tracing the notion of a deep, abstract but penetrating form of seeing further back still, to the foundational texts with which Macrina, Gregory, Augustine, and Grosseteste would be familiar, points to the 'Wisdom' literature of the Old Testament as strongly as to works of Plato and Aristotle. The most closely parallel Biblical source is again to be found in the book of Job, whose nature-poetry beckoned towards ancient sources in the introduction. But Augustine's 'wisdom-seeing' points to the earlier 'Hymn to Wisdom' (of Job chapter 28), which employs the metaphor of a miner gazing up at the jewels and mineral seams of the Earth, unlike all other creatures seeing 'from below.' The Hymn uses the miner's gaze to illustrate the particular and penetrating ability of humans to 'see' into the structures of the world. We will revisit this important text in the final chapter, but note here that its natural wisdom points beyond itself to higher senses still.

The late classical and medieval thinkers were unanimous that sense perceptions can awake higher faculties into a grasp of underlying reality when mathematics and geometry are also summoned to the task of deeper seeing. Their insight suggests a question of whether such philosophical and theological thinking on light, colour, meaning, and metaphor permeated the practice of

[23] Augustine's 147th epistle, for example, takes one Paulina's question of how Old Testament passages can talk of 'seeing' God (e.g. Moses in *Exodus*) yet also affirm that 'no-one has seen God'. The question draws Augustine into the three metaphorical aspects of 'seeing'. Augustine *Letters*, Vol. 3, trans. Sister Wilfred Parsons SND. New York: The Fathers of the Church, pp. 131–64.

visual art itself in the intellectually expansive thirteenth and fourteenth centuries. The newly minted universities of Paris, Oxford, Padua, and Bologna constituted a tight network of scholarly communication that knew nothing of today's lack of communication between art and science.

One artist nearly contemporary to Grosseteste who unmistakably translates such 'deeper seeing' into his work is the Italian painter Giotto, outstanding in both artistic innovation and self-reflective practice. Colour Plate C displays details from the Arena Chapel in Padua, which he painted (and probably built) for the newly rich moneylender Enrico Scrovegni in the first decade of the fourteenth century. Giotto filled the entire interior with almost forty scenes from the lives of Mary and Jesus. One of the scenes, the *Adoration of the Magi*, contains an early and strikingly realistic depiction of a comet, possibly from the 1301 appearance of Halley's comet (Colour Plate C (A)). The chapel's embedded commentary on issues of naturalism and optics, on matter and illusion, has recently been brought to light by Berkeley scholar Henrike Lange,[24] who surveys the entire prisma of intellectual, liturgical, and historical settings in which Giotto was able to formulate a theology of history in painting and architecture.

Lange demonstrates that the entire chapel is modelled geometrically on the bay of an ancient Roman triumphal arch, from the overall geometry of form to the placement of the mock reliefs. Responding to the Arch of Titus in Rome with a Christian critique of ancient pagan idolatry, the architectural correspondence bears metaphysical meaning. As the Arch of Titus displays a relief with the apotheosis of the ancient emperor in the vault, so Giotto responds with images of the servant lives of Mary and Jesus at positions in the Arena Chapel corresponding exactly to the

[24] Henrike C. Lange (2015), *Relief Effects: Giotto's Triumph*. PhD Thesis, Yale University; Henrike C. Lange (2022), "Giotto's Triumph: The Arena Chapel and the Metaphysics of Ancient Roman Triumphal Arches." *I Tatti Studies*, **25** (1); Henrike C. Lange (2022), *Giotto's Arena Chapel and the Triumph of Humility*. Cambridge: Cambridge University Press.

placement of the emperors' military victories in his triumphal arch. The walls depict, in both cases, the subjects' living deeds, and the vaults their spiritual rewards. Lange points out that the message of 'The Triumph of Humility' becomes evident through a visual system that first installs and then crushes its own perfected illusionism.

The deployment of optical illusion in the chapel is ubiquitous and significant. Painting transforms the flat walls, by the use of light and shade, into a highly modulated ensemble of faux-marble reliefs, including frames, ledges, and fictive marble panels. Emphasizing the effectiveness of the illusion, the part of the mural shown in Colour Plate C(B)—a detail over the large *Last Judgement* on the west wall—borders on a real window frame with actual stone relief. But at the same time, Giotto asks visitors to the chapel to see through the veil of the physical world to an eternity beyond, as two angels over the apocalypse peel away the surface of the chapel's firmament. Significantly, just behind them can be glimpsed something mysteriously metallic—possibly closed doors or the embossed covers of closed books, behind which we can imagine some unattainable true vision of eternity. Giotto realizes, in art, the successive layers of 'awakenings' from materiality to immateriality similar to Grosseteste's and Macrina's vision. Their ideal of sublimation of the sensual into the intellectual and spiritual follow Augustine's levels of seeing, from the physical visual sense, to awakening the mind's eye, towards the discovery of wisdom.

We will meet with this rich underlying seam of philosophy from late antiquity and the middle ages again when we take up the intriguing interplay of cognition and emotion in imaginative thought in Chapter 7. Before that, we need to attend to some first-person accounts of visual imagination at work in art and science. Ancient and medieval theories of perception, inner sight, and imagination are fascinating in themselves, but require the critical assessment of contemporary experience to assess their value today.

The creativity and constraint of a visual project

I worked for a number of years in the physics department at the University of Leeds in the UK. Thanks to its well-attended senior common room, I had the regular opportunity to talk with staff from other departments, including professors in the department of fine art. Unusually for a university department, all were artists themselves as well as teachers, including the current head, Professor Ken Hay, who works in mixed media of paint and photography, digital media, and film. Here was the opportunity, touched on in the introductory chapter, to ask a close colleague about the process of artistic creation, that in fine art, more than in science, tends to be eclipsed by discussion of the inspiration and final product. What did the stages in between feel like? Ken chose a recent project to illustrate his experience: a series of painted canvas backgrounds to gruelling black and white photographs from the World War II Battle of Stalingrad. The photographs on their own are bleak, grainy uninterpreted historical records of wounded and dying soldiers suffering in intense cold. They are very uncomfortable to look at and leave an onlooker with emotional or intellectual revolt at the needless misery, irresolution from the absence of context, and a viewer's position of powerlessness. Ken wanted to provide at least part of an interpretation, to add 'background' in both visual and metaphorical sense, thereby giving an onlooker more layers of emotional material to work with.

He then embarked on the story of the project, explaining that his original concept for the series simply 'didn't work,' in spite of his experience with the genre. The theoretical conception of the project had motivated initial experiments in the studio, but only when Hay was able to inspect the concrete results was he able to judge where his original assumptions fell short. In material form, the first attempt did not achieve the goal of assisting and evoking a response from the onlooker. The combination of photographs and backgrounds was not, as he had hoped, a visual and metaphorical whole greater than their sum. The central problem was that this first formulation of painted backgrounds had tried to lead towards too narrow a set of imposed interpretations, rather than provide material for the viewer to collaborate by forming their own. So he set off designing backgrounds that were more abstract, yet still responded in visual ways to the photographs. The cycle of concept, material experiment, evaluation, rejection, reformulation, and new artistic experiment, was repeated several times before the final result satisfied, holding the gaze longer than either abstract patterning or black and white image would have done on their own. As in all visual art, the longer the attention is held, the more reflections are permitted to surface. Ken had captured the gaze dynamically by passing the attention continually from foreground to background. A work of transformation had been achieved.

The parallels with a scientific project will be as obvious by now as they were to me at the time. Ken and I decided to mount a collaborative exhibition of parallel projects in art (his Stalingrad project) and science (the entangled star-shaped polymer project of Chapter 2) at the British Association for the Advancement of Science meeting, held in Leeds that year. The result was very different from the genre of 'Sci-Art' exhibitions, for there was not the faintest notion that my visual scribbles of random molecular strings and tube-like constraints might present any aesthetic quality at all, nor that Ken's real works of art were inspired by any scientific object. Rather, we were attempting to suggest a comparative process of interplay between the conception of ideas and the constraints of the external world, within a parallel narrative of the way that artistic and scientific projects develop.

Two strong parallel elements particularly intrigued us. The first was the commonality of cyclic processes in the way we worked. Both artistic and scientific projects revolved through a sequence of abstract formulation of an idea, a subsequent trial in the form of material instantiation, circling in turn to a modification of the original idea. Beginning with a material world that refused to align with our preconceptions, working and reworking with the material corrected and redirected the imagined ideas, both in the art and in the science.

The second parallel was the role of the visual imagination itself in its relation to non-visual ideas. The tragic notions of wasted lives, the pain of conflict, the appalling injustice of empire-building's distant weight bearing down on the needle-point of individual human beings trapped in the freezing snows of Stalingrad, the unique horror of the holocaust and the difficulty of empathy over the separating spaces of distance and time—these are not visual ideas in themselves, but in Ken's project the visual motifs carried and developed them. This is true for the development of the project in the minds and work of both artist and audience.

In macromolecular science too, the conceiver of newly imagined forms in the microscopic world needs to excite the visual imagination of a research-community to motivate a mental connection with an unseen realm. The notion of 'stress' in an elastic fluid is a highly abstract one, as is its emergent connection between the macroscopic and microscopic scales of structure. Ultimately, the theoretical model of this connection was written in mathematical language, but the path to the model and, for most scientists, the comprehension of it, is navigated through the visual constructs of entangled molecular strings, networks of constraints, and the continuous seething dance of random thermal agitation.

The visual is an indispensable tool for scientific 'seeing' in the metaphorical sense of comprehension, and like painting, must be reworked in the slow process of reconciling our imperfect notions to the patterns of the world around us. But does the visual operate in the creation of scientific ideas in the first place? This was the question with which we left the Imagination Institute's Cambridge discussion at the close of the last chapter, so I will record here a memory of the visual birth of an idea that I found myself recalling as part of that conversation as we explored this very personal experience of scientific imagination.

A scientific experience of a visual idea

The background of this personal example of visual ideation was coloured by the increasing interest taken by soft matter physicists, from the mid-1990s, into biological phenomena. We had spent decades experimenting and theorizing long polymer-chain molecules, colloidal particles subject to strong Brownian random motion, and self-assembled membranes of various synthetic kinds. Yet the most cursory glance at a biological cell detects examples of all of these structures: giant molecular chains of DNA and other 'biopolymers,' colloid-scale particles of proteins in their 'globular' form, and self-assembled cell membranes. The basic ingredients of cell biology and soft matter physics look remarkably similar.

That recognition of fundamental form was enough to suggest that the methods from polymer, colloid, and membrane science might have something to say about biological structure and dynamics. However, physicists involved in this mini-movement during the 1990s were treading with caution. Biology may not have the strong tradition of quantitative mathematical modelling and theory that chemistry or physics has acquired, but life science applies itself to systems of enormous complexity and specificity. Examples of structure development in embryonic life—'morphogenesis'—supposed by some early twentieth-century physicists to be captured by simple differential equations, turned out instead to be controlled by ferociously complex networks of genes and the carefully orchestrated expression of the proteins they coded for. Recorded proceedings of joint conferences between biologists and mathematicians in the 1960s, likewise, make it clear that the communities were simply talking past each other by using different languages.

A more hopeful characteristic of this new collaborative movement was the recognition that working intensively together on an almost daily basis would be vital for real progress. So it was that I found myself at a graduate school in the French alpine village of Les Houches, perched high against the south side of the Chamonix valley, opposite the Mont Blanc massif itself. The 'École de Physique des Houches' was established after the Second World War by French astrophysicist Cecile DeWitt as part of a strategy to revive science in a war-torn and exhausted country; for over half a century since, it has been highly successful, running high-level courses across a broad range of theoretical physics. Aimed at a global audience of graduates, that summer the main course was pitched at researchers in soft matter physics or in molecular biology who wanted to work with colleagues from their sister discipline.

The whole school was a wonderfully stimulating experience. I learned for the first time about the intricacies of molecular biology, experiencing at times a strange and almost physical excitement, a sense of visceral wonder and thrill. A distant memory recalled that these internal frissons were similar to those I had felt in childhood when learning about astronomy, my first scientific love. Then it was the vast structures of the outer planets, the splendid diversity of colour in the pantheon of stars—now the delicate control of protein synthesis by the information-carrying molecule of DNA, that excited my wonder. The renewed experience was a lesson in itself—how is it possible pursue a career in science for nearly twenty years yet gradually to lose that sense of childlike wonder that had driven its choice in the first place?

Enraptured amazement notwithstanding, I found the highly complex and evolved molecular machinery we were learning about very difficult to comprehend. A particular class of event fascinated me (see Figure 3.1 for a schematic aid): a certain protein molecule A will only bind to DNA (polymer B) when a small signalling molecule C is also bound to A, but at a different site on its surface. The phenomenon of C communicating its presence

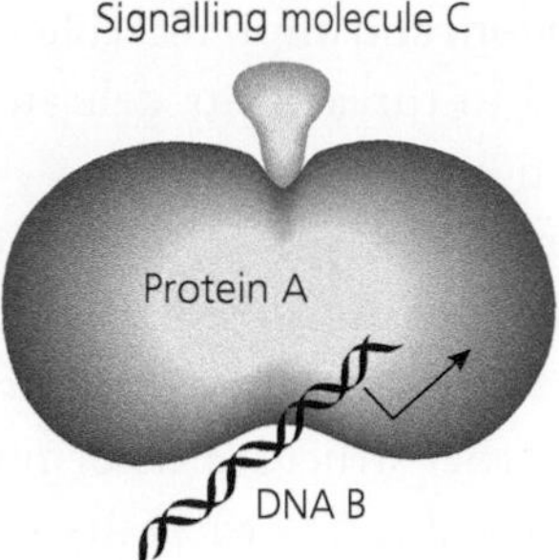

Figure 3.1 A schematic coarse-grained 'cartoon' of a large protein molecule (A) binding to a site of the helical polymer DNA (B) only when a third, small molecule (C) binds elsewhere on the protein.

across the gulf of the large protein molecule is termed 'allostery', from ancient Greek for 'other space.' Might the mystery of this effect be clarified by physical as well as biological insight?

More familiar to the physical sciences than the biological—although the insight was originally a gift to physics from a botanist—is, as we have seen, the thermal Brownian motion of the molecular world. Yet since the discovery of the double-helix structure of DNA through the deciphering of X-ray scattering patterns from its crystals, the notion of structure rather than dynamics has been dominant as an explanatory pathway within molecular biology. Crick and Watson themselves, at the end of the short paper in *Nature* with which they announced their famous discovery, pointed out in one weighty sentence that DNA's double-helical structure suggested immediately how the biopolymer might encode, and replicate, the genetic code from parent to offspring. Unwinding the two helices and then allowing them to 'template' new partners is indeed the pathway of genetic information between generations. So was born the molecular-biological mantra, 'structure determines function.' Yet at the nanoscale, the thermal energy dictated by the temperatures of living organisms is large compared to energies required to contort the structures of folded proteins, or of DNA itself. This means that the double-helix is continually bending, untwisting and writhing in random and unpredictable ways. Such structural fluctuations seemed to threaten its delicate binding, switching, and activating functions.

The transmission of the binding-state information across allosteric proteins had been the subject of continuous investigation since the 1960s. The canonical theory for how the signal crosses the protein follows the 'structure determines function' narrative and invokes an ability to switch its structural form from one shape to another. These 'protein-quakes' are initiated by the binding of signalling molecule C, then propagating across the protein, opening up the binding site for DNA segment B (Figure 3.1). There is evidence that such structural switches occur in many allosteric proteins. But as I was thinking about this model, the nagging issue of thermal fluctuations kept tugging at my thoughts—was there a softer, gentler way to send messages between the binding sites, one that didn't have to combat the Brownian motion, but worked with it?

A picture swam into my mind's eye as I listened to the graduate school lectures, a visual impression of the protein excited, like a microscopic ball of quivering jelly. The image presented another property: the effective 'stickiness' of a binding patch to its binder would not only depend on the number of attractive bonds it presented, but also on the local violence of the fluctuations. Too much random jiggling at a potential binding site would simply cause an approaching potential binder molecule to be thrown off.

The pieces of this very visual physics came together in my mind as a sort of slow-motion movie. The fluctuations at the binding site for B were at first enough to rebuff all attempts at binding there—the DNA was simply kicked off the protein whenever it came close. It might stick for a moment to the site especially designed for it, but the stormy sea of thermal motion was always too much for it, and before long a large ripple travelling through the protein would unceremoniously kick away the section of double helix—until, that is, the signalling molecule C approached the other side of the protein, wafted there by the same sea of random motion. Constructed of rather stickier stuff, this molecule attached in spite of the waves, but now I was able to discern a remarkable calming effect that spread out from the place on the protein where it had attached—for the local stiffening was able to reduce some of the long-wavelength waves as well as simply local ripples. The whole protein became less turbulent, including its other binding region facing the DNA. Now when the DNA attempted once more to bind, like a boat trying to moor up at a jetty in rough seas, it found the waves had calmed just enough to lasso the mooring post, and was successfully attached.

It was exhilarating as the little movie-of-the-mind answered all the questions that the clash of biological and physical concepts had stimulated—the signal to bind *had* been transmitted across the protein, but no structural change had been required, only a calming of fluctuations. Furthermore, and with a tangible aesthetic appeal, it had not been necessary to overcome the perpetual thermal noise, but instead the protein had used the pattern of dynamical structural waves as a sort of medium through which to send a message. I thought it a very beautiful idea and became quite excited at the prospect of this subtle information-carrying potential offered by thermal fluctuations.

Having the idea, 'seeing' it in operation, and 'feeling' that it must somehow be true in nature might have been the work of a day or so, but working out where to take the idea from there, how to confront it with the constraints of the real molecular world, and communicating it in the appropriate scientific community, that was a much longer labour. It was to take more than five years of developing, first a very rough mathematical form,[25] then a more detailed theory, before I found some biologist collaborators and some funding, which finally allowed us to test the idea with real proteins. The apparently effortless process by which the visual idea surfaced seemed a very different one from the hard

[25] R. J. Hawkins and T. C. B. McLeish (2004), 'Coarse-grained model of entropic allostery', *Physics Review Letters*, **93**, 098104.

work of shaping, crafting, experimenting, and communication that followed.

Early in the subsequent phase of turning the fluid concept into a more crystallized theory, we found that the core idea had surfaced before, most notably in a single remarkable paper by Alan Cooper and David Dryden almost twenty years earlier.[26] Further digging revealed a yet earlier version in an obscure journal article from a Japanese theoretical biologist twenty years before that, but the idea had not taken root following either of these publications. Such simultaneity of independent creation of a scientific idea is not uncommon. It points to a common circulation of foundational ingredients from which new ideas spring. In this case, the background essentials seem to have included: (i) the appreciation that thermal fluctuations must be ubiquitous in the function of molecular biology, (ii) previous experiences of mathematically relating examples of thermal fluctuation to the entropy of macromolecules, (iii) a vague familiarity with the way that computing systems can be built from logical 'gates,' and that these might in principle be built from soft systems rather than electronics. The foreground stimulus in my case was clearly provided by the beautiful Les Houches lectures.[27] Although my memory of the specific visual aids is now poor, the extremely visual 'molecular movie' that presented itself to me that afternoon was in all likelihood co-inspired by the rich variety of graphical representations of proteins in molecular biology.

Possibly the most significant environmental condition for the ultimate flourishing of this idea (at the third historical attempt) was the growing interdisciplinary conversation between physicists and biologists. These two very different communities, with different methodologies and distinctive ways of formulating even what a scientific question looks like, create a sort of 'electric' potential difference, a creative tension in dialogue. Sparks may jump

[26] A. Cooper and D. T. Dryden (1984), 'Allostery without conformational change. A plausible model'. *European Biophysics Journal*, **11(2)**,103–9.

[27] These were given by UCLA physicist, Robijn Bruinsma.

such high-tension gaps when the two poles are brought close enough. This historical context is not irrelevant to the visual nature of the ideas that then emerge, for when two communities first meet, their more developed terminology is mutually incomprehensible. In these circumstances communication tends to begin with pictures and shared metaphors. The visual imagination is both the conduit and cradle of new ideas.

Conversations on creativity with visual artists

The long history of thinking about the visual and its experience in art and science that we have surveyed thus far has pointed to a dialogue between mind and the material that the image negotiates. In classical and medieval times, the duality of visual imagination and perception found expression in the delicate complicity of intromission and extramission.[28] Artists such as Ken Hay talk of the meeting point between visual idea and visual experiment. Perhaps this two-way, cyclical travel in the experience of creation in art and science is the most surprising finding in an exploration of the creative process. To some, the discovery that art suffers from the initially ill-understood constraints that drive this reciprocal process of invention in much the same way that science does also confounds intuition. How do other artists talk about the creative interplay of the visual idea and the material?

Vanessa Chamberlin is a portrait and still-life artist working in London, using mostly oil and pencil.[29] She describes a central interest as 'the boundary between representation and abstraction.' Conceiving in the abstract is a necessary goal in science, but as we have seen it is often achieved through the visualization of a representative structure. Asked about whether she feels the need to negotiate constraints in the realization of an idea, she didn't

[28] Another, related and deep duality between the rational senses and emotions was also developed in medieval philosophy; see Chapter 7.

[29] Examples of Chamberlin's work can be found at https://www.vanessachamberlin.co.uk.

have to ponder: 'Art is not free—it's a battle between what the artist conceives and the canvas.' So strong was the notion that the artist's internal life and her materials were 'complicit' in the act of art-making, that we arrived at the statement that 'matter co-creates with the artist.' The artist suffers from a sort of blindness, avers Chamberlin, 'Our eyes are dirty but we think they are clear—we just see the world through a blindness.' But by determining to see through the imperfect senses, by allowing that sense to reignite the imagination, and crucially to engage with the materiality of art, the dullness may cede to clarity. The sense of a need for reconciliation of broken relationship, but an inner, rather than outer, one also emerged—'Art exposes the lack of reconciliation to ourselves; the painting determines the course,' was her way of describing her personal journey through art-making.

Chamberlin has, over the course of her creative development, considered the similarity between artist and prophet, and found an encouraging commonality in American theologian Walter Brueggemann's writing:[30]

> *The prophet engages in futuring fantasy. The prophet does not ask if the vision can be implemented, for questions of implementation are of no consequence until the vision can be imagined. The imagination must come before the implementation. Our culture is competent to implement almost anything and to imagine almost nothing. The same royal consciousness that make it possible to implement anything and everything is the one that shrinks imagination because imagination is a danger. Thus every totalitarian regime is frightened of the artist. It is the vocation of the prophet to keep alive the ministry of imagination, to keep on conjuring and proposing futures alternative to the single one the king wants to urge as the only thinkable one.*

The sheer difficulty of imagining what is not, or not yet, perceived resonates with what artistic and scientific imagination have in common. For Chamberlin, the way to first imagination is 'to give all and *search* for art—in every cup of tea in the kitchen!' I was

[30] Walter Brueggemann (2001 [1978]), *The Prophetic Imagination*. Minneapolis: Fortress Press.

reminded of the advice that Leonardo is reported to have given to younger painters, to take their starting point from even the chance outline of a damp stain on the wall 'because by indistinct things the mind is stimulated to new inventions.'[31]

The indistinct, the blurred, fuzzy, half-conceived outlines of an idea becoming steadily sharper, the false assumptions initially made of what the object is, of which parts lie before or behind others becoming clearer through steady gaze—this is the active and passive journey by which mind and material engage in the hard labour of co-creation.

If the London-based artist Vanessa Chamberlin's creative project is rooted in 'the boundary between representation and abstraction,' then she finds common cause, though a different starting point, with a contemporary artist from the other end of England. Graeme Willson was based in the Yorkshire Dales town of Ilkley, near Leeds, for over thirty years.[32] His work ranges from large commissions for public places, including murals, to ecclesiastical themes, contemporary still-life, abstract, and geometrical art, and has used watercolour, acrylic, and more recently stained glass. Works commissioned for York Minster and Leeds Corn Exchange are known to a wide public. His starting point is frequently the human figure, especially the face, but embodied within his art these facial forms can find themselves engaging with architectural (especially classically inspired) motifs, natural forms, abstract geometry, and a characteristic use of colour.

Willson describes the abstract forms of colour, line, and shape that he uses to frame his representational figures as having 'an autonomous life of their own'—language that springs from his experience of working within the tension of representation and the abstract. He also talks of 'order' and 'chaos': 'Sometimes I try

[31] Leonardo DaVinci (1952), *Notebooks*, Irma A. Richter, ed., Oxford: Oxford University Press, p. 174.

[32] Examples of Graeme Willson's work can be accessed at http://www.graemewillson.co.uk/introduction.html His input was critical to the first edition of this book, and his death in 2018 was a huge loss to British art.

and generate these abstract forms by reflecting or meditating, but they are more likely to come about by "process"—letting the paint speak for itself by behaving on its own terms.' The semi-random qualities that emerge from the deliberately uncontrolled phases of the process can then be 'played off' against the premeditated and figurative areas. Here we might conclude that there are no parallels with scientific creativity, for apart from the inclusion of faces, classical pillars, and veined leaves, there seem to be no tight constraints on the process of layering the paint and arranging the representational motifs. Yet Willson is strongly affirmative that there are criteria by which the experimental process of 'art by order and chaos' can be judged to succeed. 'Things begin to work when this "play off" is in a symbiotic and dynamic relationship, and both kinds of imagery are integrated and unified, despite their different points of departure.' Here there are very strong resonances with the self-enhancing or suppressing process of scientific model-making: when the terms in a mathematical theory for nature start to interact, to give rise to a dynamic or an emergent property that their crafter did not put in by hand, then that too is a signal that the path taken promises to be a potentially fruitful one.

Willson was an experienced and consummate artist, yet a delightfully humble man. His own experience by self-admission follows that of Einstein: '99% perspiration and 1% inspiration.' He is coy about using the term 'inspiration,' 'because it has connotations of Romantic soulfulness.' This is not the first time in this exploration where previously downplayed passion and passivity were acknowledged as important alongside cognition, deliberation, and activity. Yet the emotional seems to appear in narratives of creation too frequently for it to be a simple by-product. 'Inspiration' seemed from this conversation to be an important idea in spite of its unfashionableness. Neither could the artistic process be described entirely as active—a 'letting go' as well as a 'long struggle' were together essential aspects of authentic experience.

At the end of this chapter we will revisit Graeme Willson's studio to explore the creative pathway of one particular work of his, but we have now enough evidence for a common experience with the visual in creative imagination of art and science to contemplate the greatest visual metaphor of them all.

The great cosmological model and the visual imagination

The observational science that has made the greatest demands on the visual imagination, and over the longest history, is surely astronomy. From the very outset of human contemplation of the sky, human gazers into the vault have attempted to grasp at the challenges it poses. For, unable to reach into the cosmos and manipulate the stars by experiment, astronomers are at first sight just passive receivers of radiation. To be sure, light from the stars is extremely rich in content—different objects shine with different mixes of wavelengths across the spectrum, and outside the range of visible light they may glow deep into the infrared and scintillate in the ultraviolet. But there is another strong limitation, even for a passive observer. We are not able, for now at least, to change the point of our observation into the cosmos. Information-bearing light falls on us from objects far and near—but all projected onto the apparent sphere of the sky seen from our position within the solar system. We had, until the nineteenth century's new advances in telescopic power, no 'perspective' into the universe. This inability to see depth acts as a strong limitation on what we can immediately deduce from the sky at night.

It is the timeless task of astronomy to pose the question, 'Given the pattern of radiation from the sky, and the changing patterns that it traces out, what is the most likely three-dimensional structure that emits it, and what is the history of that structure?' To take the simplest example, the most cursory glance into a clear dark autumn night notices stars of many different brightnesses. But does *Altair*, the brightest star in the constellation of *Aquila* (the eagle), outshine its sister *Deneb* (in *Cygnus*, the swan), further along the Milky Way to the north, because it is intrinsically more powerful or because it is nearer? To answer that question requires much more than the observations themselves—it demands a mental encounter with the stars through models of their nature. The data are dressed with interpretation from atomic theory, thermodynamics, and gravitational physics. The stars' different spectra can now be highly resolved into the intricate detail of their missing wavelengths. These 'spectral lines' are in turn the tell-tales of the chemical structure of their atmospheres, while the overall distribution of energy in the stellar spectrum from red to blue, speaks of the temperature of their surfaces. Astronomers must 'extramissively' project all this interpretative information to meet the incoming signals. Only then are they able to visualize what these points of light in the sky signify, and perceive the immense and searingly hot but distant 'blue supergiant' that is *Deneb Cygni*, irradiating the galaxy with the glare of fifty thousand suns. The task of astronomy—the reimagination of three-dimensional cosmic structure from its two-dimensional impression upon us—is the immediate scientific correlate of visual art's central act. Astronomy is the science par excellence that combines an extramissive reaching out of mentally created possible worlds with the incoming, intromissive, information from the sky.

The same is true of the temporal changes in the sky, obvious from the earliest times as the regular motions of sun, moon, and planets. Ancient astronomers were also confronted with more irregular alterations: the wispy and haunting comets came and went unpredicted; the *novae*, apparently temporary stars, burst into visibility for a few days or weeks before fading back into obscurity. What was the huge structure surrounding the Earth that we were looking at? The answer that dominated middle-eastern and western thought for centuries was perhaps the most influential, awe-inspiring example of architectural imagination before the modern age. As an imaginative creation of the human mind, the geocentric cosmos is to this day an object of power and beauty.

To identify this ancient cosmos as the first great creation of the scientific imagination might invite the criticism that it was ultimately wrong, that it had to be discarded in the face of better evidence and observation, that it was the product of a pre-scientific age. However, we find in ancient and medieval astronomical writings very much the same motivations, curiosity, and drive to understand the hidden structure of the world as in the astronomy of today. Contemporary scientists recognize their continuity with the long history of their subjects when they engage deeply with it.

Richard Bower, a leading cosmologist of galaxy formation at Durham University in the UK, has worked with a remarkable interdisciplinary project in which scientists and humanities scholars work together on medieval scientific texts.[33] He admits that his educational background had suggested that it would take him only a few days to be able to inform his humanities colleagues that Robert Grosseteste's theory of the cosmos, in his brief treatise *De luce* (*On Light*), was constrained by straight-jacketed thinking, dogmatically confined to blindness in the face of evidence, and wilfully ignorant of the heliocentric structure of the solar system eventually adopted four centuries later. However, as he read the astonishingly imaginative treatise and thought about what one can and cannot know without a telescope, he realized that he was in the presence of as great a mind as any he has conversed with professionally. Furthermore, he began to reimagine how a curious and mathematically inclined mind would work with the impressions of the circling heavens to the human eye alone. Far from an irrational, dogma-driven picture of the world, Bower concluded that the geocentric cosmos, with the Earth at the centre of vast, nested planetary and stellar spheres, would be the correct 'minimal model' to work with, until one had strong

[33] For details of *The Ordered Universe Project*, see https://ordered-universe.com.

evidence that the Earth really does move, by spinning on its axis and whirling annually about the sun.[34]

The ancient and medieval cosmos was more than a three-dimensional solution to the astronomical puzzle of what must be the case, to account for what we see in the night sky. It was a vast mental computer, able to predict future positions of planets and the phenomena of eclipses of sun and moon. More than a static picture, however awesome, it was a fully dynamic image of planetary and stellar motion. In the form of the first-century Alexandrian Ptolemy's *Almagest*, it stood unchallenged—against continuous observation—for over a millennium.

Visual representations in medieval texts on 'the Spheres' may be elegant, and sometimes objects of beauty, but even the most gorgeous do not do justice to the grandeur of the mental image. Colour plate D, from the 1245 *Image du Monde* of French encyclopaedist Gossuin de Metz, is a fine example, but it needs to be read schematically, not geometrically. Gossuin's representation is the medieval cosmological equivalent of the London Underground tube map, where stations are in the right order on each line, but actual distances are compressed arbitrarily. The scale on such diagrams is deceptive: it appears that the entire universe is not very much larger than the extent of the Earth (the central disc with the inverted face in plate D). Immediately above lie the four spheres of the elements themselves—earth, water, air, and fire—the last stands out by virtue of its flame-orange colour. Yet by reading detailed descriptions of the ancient cosmological model by its great proponents, it becomes clear that the celestial spheres are an imaginative creation of immense proportion. To give some idea of the sophistication of ancient positional astronomy, it is clear from medieval texts that stellar positions were measured with sufficient accuracy, and recorded over long enough periods in the ancient world, to notice that the point at

[34] It bears repeating that solid direct evidence for a moving Earth had to wait until Bessell's measurement of the parallax of the star 61 Cygni in 1838.

which the starry spheres seem to rotate (currently close to the north star, *Polaris*) creeps slowly over the centuries through the northern constellations.[35] Measurements of stars' altitude above the southern horizon from different observation points allowed estimates of the Earth's diameter within 10 per cent of its correct value. An accurate knowledge of the size of the globe then permitted the surveyor's art of triangulation to be extended into the heavens, and in particular to the moon.

Such ancient observations required the size of the cosmological model to be truly vast—immense even on the scale of the entire Earth. Cicero, in a fragment of his surviving writing known as 'Scipio's Dream' from his larger but incomplete *De Republica*, writes of an imaginary journey up into the stellar sphere itself.[36] From this vantage point Scipio notices that the stars themselves are huge globes much larger than the Earth, which itself could hardly be made out, so tiny it seemed to be from so distant a vantage point:

> *When I gazed in every direction from that point, all else appeared wonderfully beautiful. There were stars which we never see from the earth, and they were all larger than we have ever imagined. The smallest of them was that farthest from heaven and nearest the earth which shone with a borrowed light. The starry spheres were much larger than the earth; indeed the earth itself seemed to me so small that I was scornful of our empire, which covers only a single point, as it were, upon its surface.*

C. S. Lewis in his remarkable extended cultural essay on the ancient world model, *The Discarded Image*,[37] unearths a lovely passage from a medieval popular text, the *South English Legendary*, which explains that if one were to travel vertically upwards at 'forty mile

[35] For a detailed survey of precession, and rival theories of it in the high middle ages, see C. Philipp E. Nothaft, 'Criticism of trepidation models and advocacy of uniform precession in medieval Latin astronomy.' *Arch. Hist. Exact Sci.* **71**, 211–244 (2017)

[36] Cicero (1928), *De Republica VI.IX.XVI*, *De Legibus*, C. W. Keyes, ed., Loeb Library, p. 269.

[37] C. S. Lewis (1994), *The Discarded Image*. Cambridge: CUP Canto edition.

and yet some del mo' a day,' 8000 years of travel would not reach the sphere of the stars. Of course, these dimensions are still nothing on the scale of the universe we know today, but as Lewis reminds us, 'For thought and imagination, ten million miles and a thousand million are much the same. Both can be conceived (that is, we can do sums with both) and neither can be imagined: *and the more imagination we have the better we shall know this*' (my italics). This model of the entire cosmos, which interpreted every heavenward human glance, at least in Europe and the Middle East, for a millennium and a half, was able to re-create by the imaginative force of human mind, structures and scales unrelated and unencountered by any earthly experience.

The geocentric model was by no means a single unchanging edifice, but permitted considerable variety. Even a cursory reading of the versions advanced by Eudoxus, Aristotle, and Ptolemy reveals disagreement on the number of spheres that they each adopt.[38] There is no divergence on the number of planets requiring a hosted place within a sphere, but for each planet, a number of up to four sub-spheres, each turning on axes mildly displaced from each other, could account more accurately for the finer-grained aspects of their apparent motions, such as the looping retrograde motions of Mars, Jupiter, and Saturn.

The constituency of the spheres formed another running debate. As scholar of medieval science Edward Grant has shown[39] that debate pursued the physical nature of the celestial spheres' materiality—were they fluid or solid? The standard medieval introductory text on the heavens, John Sacrobosco's early thirteenth-century *De Sphera* (*On the Spheres*), seems to decide on their solidity, appealing to the solid geometry of Euclid. Later commentaries on that text, however, such as that of Robertus Anglicus (1271), discuss the issue of whether they rotate in

[38] See e.g. David C. Lindberg (2007), *The Beginnings of Western Science*, 2nd edn. Ch. 2. Chicago: Chicago University Press.

[39] E. Grant (1987), 'Celestial orbs in the Latin Middle Ages'. *Isis*, **78**(2), 152–73.

a continuous or contiguous way, shearing as fluids or slipping as solids. To preserve the imagined influence of the outermost spheres' rotation on the lower ones, through some form of cascading set of forces, he intriguingly opts for the (minority) fluid hypothesis. In any case, when Tycho Brahe determined that the comet of 1571 was moving above the orbit of the moon, *through* the supposedly solid spheres, and contrary in direction to any motion that they would be executing, he was able to declare the hypothesis of solidity roundly refuted:

> *I first showed and clearly established that by the motions of comets [the heaven] is fluid and that the celestial mechanism is not a hard and impervious body filled with various real orbs, as has been believed by many up to this point, but that it is very fluid and simple with the orbits of the planets free and without the efforts and revolutions of any real spheres.*[40]

The point here is, of course, not to debate the truth of all this vast imagined geocentric cosmos, but rather to see it as a great act of visual imagination, creating a cosmos of thought that could continually be brought into contact with the form of observed experience, experimented with, reformed, and reshaped. Ultimately, it was discarded—but even the Copernican Revolution was more a product of the Model's own making than an invader from outside. The hypothesis that the sun was central to the scheme rather than the Earth did not, for Copernicus, discard the notion of solid celestial spheres, the ubiquity of purely circular motion, or indeed any of the intricate machinery of the Aristotelian universe, excepting the identity of its central globe.

Heliocentrism has a quiet alternative presence in ancient Greek thought,[41] and Grosseteste's extraordinary light-generated cosmogeny of his *De luce* of 1224 has been described as 'heliocentric

[40] Tycho Brahe (1922), *De mundi aetherei recentioribus phaenomenis*, in *Tychonis Brahe Dani opera omnia*, I. L. E. Dreyer, ed., Vol. IV. Copenhagen, p. 159. E. Grant, trans. Celestial orbs.

[41] In the third century BC, Aristarchus of Samos proposed a heliocentric model, known to us by report in Archimedes' *The Sand Reckoner.*

in all considerations except the geometrical.'[42] Copernicus, in that sense, completed the shift towards the centrality of the sun over the Earth, a serious leap of the imagination, but one that drew its inspiration from the existing paradigm as much as from new thinking. It was Johannes Kepler's close attention to Tycho Brahe's more accurate observations of planetary motions that brought about the final dissolution of the machinery of crystalline spheres and multiple circular motions, with the elliptic planetary orbits that are recognized, since Newton, to be consequences of the ('inverse-square') structure of the sun's gravitational field.

From the beginning, careful and detailed attention to the night sky has inspired a visual imagination of extraordinary power. There has never been a time when a model of the cosmos has been 'read off,' or deduced, from the sky—rather the universe needs to be imaginatively and inductively re-created. That process, and its historical record, not only speaks to the questions of creativity in science that we are holding open, but also suggests a more generalized conception of what we term 'visual imagination.' A combination of prior material, such as the satisfying notion of perfect circular motion, together with new ideas, such as sub-spheres and epicycles, meets with the 'form' of the heavens as they are observed. The creative process was highly visual in its conception, but we have noted that this is not the same as the 'pictorial.' Depicted images of the Model, such as Colour Plate D, were common, but none did justice to the immensity of the imagined vastness, complexity, and grandeur of the scheme that occupied the minds of their authors. The work by Gossuin de Metz is an example of art, as well as an illustration of science. Its sophisticated use of colour and shade is especially notable—the illusion of three-dimensional spheres receding from the viewer into the centre is remarkable, the asymmetric drabness of the

[42] James McEvoy (2000), *Robert Grosseteste*. Oxford: OUP.

Earth compared with the crystalline symmetry and clarity of the heavens, immediately striking.

The great cosmological model, because it inspires both scientific and artistic imagination, indicates their commonalities as well as their contrasts. There are several objects of this kind that thread through and between the disciplines—the rainbow is another object of shared inspiration, which we will encounter in the final chapter. For a contemporary example of cosmos-generated art, we turn to the *Ordered Universe* medieval science project's artist in residence, Alex Carr, hosted during work by the interdisciplinary team on Grosseteste's treatises *On Light* and *On the Spheres*. The aesthetic and technical challenge of representing the medieval cosmos in a three-dimensional installation grew irresistible, but for a long time without resolution under the weight of its sheer difficulty. The breakthrough that led to the breathtaking *Empyrean*[43] was experimental and conceptual: crystalline spheres can be made of copies of themselves. So, 3980 transparent beads are suspended in regular, nested spherical geometry over a metre in diameter. The whole is highly holographic—each bead, contained within the whole, optically images the entire work itself, when viewed closely. Carr discovered, through lengthy experiment, technical drawing, and calculation, that carefully placed gold highlights on each bead, combined with geometrically placed lighting, would highlight each celestial sphere in turn as the observer walked around the installation. Carr writes:

> *Referencing Robert Grosseteste's treatise* De Luce, *which illustrates how the universe came into being through light, much like our current notion of the Big Bang, I examined the geometry of expanding spherical space. In addition, elements of Grosseteste's* De Sphera *feature, in its attention to observational points in relation to the movements of the celestial bodies.*
>
> *The act of observation and perspective is central in this piece, highlighting how visual perception informs our experience of reality and our place in the cosmos.*

[43] Carr's introduction and photographs of the installation can be seen here http://www.alexandracarr.co.uk/empyrean-1 and in the front cover illustration of this book.

The constraints are multiple: the perspective and perception of the observer, the challenge to capture the notion of dynamic expansion in a static exhibit, faithfulness to the conception of the great model, while allowing innovation and inspiration to play out.

The cosmos, as a fundamental example of the scientific visual imagination, indicates finally the power of conceptual frameworks to drive their own evolution, even to the point of destruction and replacement. *Empyrean*, in its artist's testimony, took on a life of its own as the manipulation of its material elements suggested new forms and resolutions. The Copernican version of the scientific model universe, while truly revolutionary in its reconception of our place in the cosmos, is from one point of view a similar simple exchange of two bodies within an existing framework. The prime difficulty in its adoption was not so much that of conflict with received dogma, but that the visual conception of the world model now required a near-impossible feat of the imagination; for the very familiar Earth, on whose apparent solidity we stand, together with its inhabitants, had now to be exalted to the 'third heaven.' Within a generation, from being only observers of the distant heavens from the vantage point of the basest and lowliest place of all, we became ourselves dwellers of the brightly lit and elevated celestial regions. That was the breath-taking and mind-stretching 'revolution' of Copernicus, Brahe, Kepler, and Galileo.

The visual imagination and astronomy today

What has become of the rich heritage of visual imagination in astronomy, inspired by the fundamental question of what lies behind the sky? Do astronomers today need to draw on the same, or even more highly developed visual creativity, or have the newer methodologies of advanced computing and mathematical modelling also remodelled the imaginative process itself? Conversations with researchers in astronomy today suggest more continuity than contrast. The previous chapter reported Astronomer Royal Martin Rees' comparison of a visual thinker with a symbolic thinker in astronomy, by means of an amusing recollection of a conversation in which they had talked past each other for a while. Present at the same *Imagination Institute* workshop was Dr Ashley Zauderer, an astrophysicist who, when she was a researcher at Harvard, was witness to a remarkable moment of observation. Her example pushes our extended definition of the 'visual' yet further, because the observation was made, not in visible light, but in radio waves. A 'radio telescope' is a very different affair from the lens-or-mirror-in-a-tube variety with which we are familiar. Radio waves are more than a million times longer in wavelength than visible light, so carrying a much lower energy density. For both of these reasons, astronomical receivers of radio-wave radiation need to be extremely large. The larger wavelength requires an equally enhanced aperture if the image resolution is to be at all comparable to optical observations, and the collection area needs to be sufficient to collect detectable radio energy. The sums are initially discouraging, as even moderately sharp radio images seem to require vast antennae tens or even hundreds of kilometres in diameter. Human ingenuity refuses to baulk at such obstacles, however, as the group led by Martin Ryle at Cambridge showed in the post-war years. Such vast apertures could be 'synthesized' by connecting arrays of much smaller telescopes at great distances from each other.

In the spring of 2011, Zauderer was an observer at a descendent of Ryle's early technology—the Very Large Array (VLA) in Socorro, New Mexico, working with a recent upgrade to improve the telescope's sensitivity. For those of us whose experience of astronomy extends no further than star maps and finding familiar constellations, the night sky can seem a completely changeless domain, apart from a few wandering planets. But keener observations reveal many more dramatic changes, especially those resulting from the death throes of stars. Even moderately sized stars such as our sun swell, at the end of their life-cycle, into 'red giants' prone to sudden bright outbursts. Much larger stars can undergo the cataclysmic collapses that drive 'supernova' explosions. For a few brief weeks, as their outer layers are ripped from their cores, the violent death of these stars can outshine all others in their host galaxy put together, in a prodigious release of energy. Chinese astronomers noted, in 1054, that a 'guest star' had appeared in the constellation we call *Taurus*. The records are of such impressive accuracy that we can identify it as the cause of a well-known expanding gas cloud, the 'Crab Nebula.'[44] The remnant—at the ground zero of such explosive stellar deaths—is an unimaginably dense 'neutron star,' essentially a giant atomic nucleus up to a kilometre in diameter. Above a critical mass, even the huge pressures within these objects are unable to resist the crushing force of their own gravity, and final collapse into an example of Chandrasekhar's black hole is the only pathway forwards. Further vast outbursts of energy may occur transiently when other objects, gas, or stars, are captured by a black hole. Such in-fall events release radiation across a huge gamut of wavelengths, even in bursts of 'gamma-rays' that possess the shortest and most energetic wavelengths.

Zauderer's Harvard team were looking to push the limits of what was known about these gamma-ray bursts, including those that may be produced from the merging of two compact objects. Their technique required simultaneous observations at different wavelengths. Alerts would arrive as text messages whenever NASA's Swift satellite detected a burst of high energy photons at gamma-ray wavelengths, so that radio observations could begin immediately. One evening, Zauderer received such a text and waited for the usual radio signals. Her own account expresses the surprise of the unexpected:

> *I had a moment of complete awe when I knew we had observed something very very different . . . when the image first popped up on my screen, I knew something*

[44] For a detailed Hubble image of this object see https://www.nasa.gov/feature/goddard/2017/messier-1-the-crab-nebula.

> *special had happened. Instead of seeing only noise or a very faint image, there was an intensely bright source of radio emission exactly consistent with the other observations . . . multiple gamma-ray triggers, when just one is normal. X-rays from the source continued much longer than any gamma-ray burst. Josh Bloom, a professor at UC Berkeley, was the first to suggest in a telegram a tidal disruption of a star by a supermassive black hole—and the data fit this idea.*
>
> *I will never forget that sense of absolute awe, wonder, and connection to the Universe I felt when I first saw this whopping bright radio source in space that was so unexpected—a lonely star happened to wander too close to a supermassive black hole four billion years ago and we as humans happened to have the right technology and instruments to be looking in the right place to catch this extraordinary event as the photons streamed to Earth.*

This is a beautiful, compelling, and deeply human account of scientific discovery. Several aspects strike a reader immediately as illustrative of science as 'imagination constrained by form': an intricate, developed, and communal attention to the cosmos, together with an interpretative readiness. A desire to understand violent stellar events has brought about a complex extension of our visual receptivity to the sky, encompassing infrared, microwave and radio wavelengths on the low-energy side of our visible spectrum; X-ray and gamma-ray at higher energies. This enlargement of our visual receptive grasp is matched by a similarly extended visual imagination towards the generators of the most violent and polychromatic bursts of energy ever contemplated. The detailed conceptualization of the final cataclysmic explosion of a star is not an unconstrained and vague extrapolation of a firecracker. It is a monstrous and detailed imagination of an immense stellar destruction, releasing more energy at the point of its death than the star has emitted in its entire lifespan. The visual metaphor seems not to be erased or supplanted by the mathematics of calculation, but to guide it, and to be enhanced in turn by it. Zauderer commented on the role that visual imagination plays in her own thinking:

> *In terms of visualization, even though the radio image is two-dimensional, I like to see the geometry in my mind's eye and often imagine what events look like in*

reality because in astronomy, we only see 'in part' and you have to reconstruct the rest in your head. In this case I described, I was simultaneously visualizing both the spectral shape as a consistency check (and then did the calculation), but more immediately I visualized a powerful outflow—a 'movie of what was happening' when I saw the very bright CARMA detection.

Listening to this account of an experienced observer and physicist, one might be deceived into thinking that the spectacle was presented to her on a video screen, unfolding in real time as the 'movie' she writes about, running in her mind. But an observer would have seen nothing of this imagery. It is constructed by the informed imagination, developed within the forms of physics and mathematics to re-create visions far beyond everyday encounter. The interplay of visual imagination and cognitive evaluation is tightly woven.

There is an intriguing reference in this personal account of astrophysical imaginative visualization: that we only see 'in part.' The Biblical allusion to St Paul's celebrated passage on love from his Second Letter to the Corinthians can hardly be missed (chapter.13 v.12: 'For now we see in a mirror dimly, but then face to face. Now I know in part; then I shall know fully, even as I have been fully known.'). This is a world in which vision is partial—we can imagine a world in which we see in full (or, to follow the metaphor, 'face to face'). Until then we must supply by induction what neither sight nor deduction is able to do.

There are strong artistic, as well as theological, parallels to the constructive role of the imagination in visualization. To some extent all visual art makes use of the constructive process of vision, but one movement in art of the modern era demands special discussion here, for it depends utterly on it in unprecedented ways, and so resonates with the scientific imagination particularly strongly.

The excited imagination of the impression

Paris, in 1874, saw a group of near-unknown artists give an exhibition of their own work in a cheap venue far removed from the fashionable round of Salons. Although mostly the subject of derision in the popular and critical Parisian art-press at the time, the report in *Le Siècle* of 29th April strikes a different note:

> *What quick intelligence of the object and what amusing brushwork! True, it is summary, but how just the indications are! . . . The common concept which unites them as a group and gives them the collective strength . . . is a certain general aspect. Once the impression is captured, they declare their role terminated. . . If one wants to characterize them with a single work that explains their efforts, one would have to create the new term of* Impressionists. *They are impressionists in the sense that they render not a landscape but the sensation produced by a landscape.*[45]

For all the signs of incomprehension here, the report signals that a shift in artistic imagination really was afoot, that new ways of thinking about the relations between artist, painting, and viewer, were in the ascendant. Critic Louis Leyroy wrote of Claude Monet's *Impression: Sunrise*, the work which was to give its title to the budding movement:[46] 'Impression! Wallpaper in its embryonic state is more finished!' while celebrated contemporary author and poet Baudelaire wrote to the most senior of the group, Édouard Manet: 'You are only the first in the degeneration of your art.'[47] The well-documented explanation for this outrage, constructed with the benefit of a century and more of art and art history since, is 'the shift from Realism to Impressionism and Early Modernism.'[48] But this moment in the history of art also illuminates the comparison of visual creativity in art and science, because it reminds us that all imagination acts within convention. Manet and the other impressionists had ceased to operate within a web of artistic convention that had simply become so commonplace, so taken for granted, that its assumptions had become invisible.

[45] *La Siècle*, 29 April 1874, in *Impressionism and Post-Impressionism*. trans. L. Nochelin, pp. 329–30.

[46] However, the first exhibition in which the new movement referred to themselves as 'impressionists' was not until 1877.

[47] Quoted in Julian Barnes (2015), *Keeping an Eye Open: Essays on Art*. London: Jonathan Cape.

[48] The edited volume B. Fer, ed. (1993), *Modernity and Modernism: French Painting in the Nineteenth Century*. New Haven, CT: Yale University Press, is a good example.

On the occasion at which the philosopher of science, Thomas Kuhn, turned his attention to the comparison of art and science,[49] he was more at pains to stress their differences than their similarities. For Kuhn, the aesthetic constitutes the *end* in art, but at best the *means* in science. Further, their institutional structures, he claims, legitimately differ: the gallery is the centre of current activity for the artist, the museum a repository of defunct historical ideas, for the scientist. So, in art, Picasso's work does not render Titian's of no current artistic merit, while the geocentric cosmos is an object for the history of science, not for science. For these reasons, he permits that his celebrated (but ill-defined) notion of 'paradigm' might be applied to individual pictures, but not to styles. Had the impressionist movement attracted his attention, it would surely have presented aspects that parallel Kuhn's notion of 'paradigm shifts.'[50] Rather than classical scenes or characters, its exponents painted ordinary people and everyday landscapes. Rather than the accepted techniques of portrayal for shadow, sky, or skin-tone, they experimented with new ways of creating visual impression. Here is writer Julian Barnes on Manet's innovations:

> *He brightened and lightened its palette (where academicians began with dark tones and worked up towards lighter ones, Manet's* peinture claire, *did the opposite); he discarded half-tones and brought in a new transparency (growling at paintings that were 'stews and gravies'); he simplified and emphasized outline; he frequently discarded traditional perspective. . .; he compressed the depth of field, and pushed figures out towards us.*

To the reader familiar with the impenetrably dark 'Manet black' of his shadows, and the stark outlines of figures against them, this reminds us of the shock with which such visual violence would have been received in a world of gentle shadings, half-tones, and

[49] Kuhn, T. (1969). 'Comments on the Relations of Science and Art'. *Comparative Studies in Society and History* 11: 403–12; I am grateful to Steve Fuller for drawing my attention to this work in 'Art and science: representation or expression?' *Interdisciplinary Science Reviews*, **45**(1), pp. 16–22 (2020).

[50] Thomas Kuhn (1966), *The Structure of Scientific Revolutions*. Chicago: Chicago University Press.

chiaroscuro. Paradoxically, although criticism was levelled at such claimed departures from a conventional 'realism,' any recollection of a street scene on a bright summer's day will confirm that an open window offers nothing but the perception of total blackness to the observer.

Another central aspect of impressionism closely parallels the imaginative work of astronomy—the negotiation between a three-dimensional world of depth and a two-dimensional projection of it. Here the same idea arises as the fulcrum of one of the greatest shifts in art of the modern period. As critic Clement Greenberg has commented,[51] the impressionists allowed a viewer's eye to be drawn to the 'flatness' of the painting before, rather than after, becoming aware of the depth of what is portrayed by the flat canvas. He refers to this reordering as the 'integrity of the picture-plane' and a 'dialectical tension' between the two visual representations in the viewer's mind—the representation of the canvas, and the reconstructed representation (via the 'impression' that the picture provides) of the original object. Such a conversation between the visual imagination in two and three dimensions is already familiar to us.

A third and signature move, which touches at the heart of the creation of the visual, was the notion that a painting need not do more than a fraction of the work required to construct a visual impression. Drawing from the energy of 'extramission,' as we have more broadly and psychologically defined it in this chapter, the artist's question becomes not, 'How do I portray this scene?' but, 'What do I paint so that the observer perceives this scene?' The creativities of the artists and of the audience of a painting become collaborative and connected.

The one member of the group exhibiting in 1874, and for the first time under the title 'Impressionists' in 1877, who most

[51] Clement Greenberg (1982), Modernist painting, in F. Frascina and C. Harrison, eds., *Modern Art and Modernism*. London: Harper-Row in association with the Open University.

strikingly represents this discovery of a new source for visual imagination, is Paris-born Claude Monet. Son of a respectable if humble family, he rejected his father's wish to enter the family delivery business in favour of his early love of art. A more supportive mother, who was herself a musician, brought some resolution and the offer of tuition. However, his decision meant that he was not in a position to avoid national service, which he endured in Algeria. Originally a seven-year posting, his ill-health cut the superficially unproductive period to just a single twelve-month, but the experience of the severe light of North Africa had a lasting effect. He was to comment forty years later that Algeria's light and colour 'contained the germ of my future researches.'[52] His choice of the term 'research' (*Fr. récherches*) is significant to our purpose, of course—originality and inspiration are very well and necessary in artistic creation as well as in science, but in both need an execution into form to find their end point. Experiment is as necessary for the best-known artists, it seems, as for lesser. In Monet's case, much of the experimentation was directed at the new degree of tension between the painted surface and the scene represented.

A suggestive parallel between the plane of representation and the natural scene, and the *Imagination Institute* conversation's insight into the pair of 'focal planes' held in tension when constructing a theory in physics, is not merely fanciful. The way that artists on the one hand and scientists on the other talk about the process, the reassembling of ingredients in the representation, and in particular the way that the theory, or painting, is designed to create an impression of the perspective into nature that it represents—these are too close to ignore.

An example will assist. In the Philadelphia Museum of Art hangs a remarkable evening scene, painted by Monet during an extended visit to the French Riviera town of Antibes in 1888. Returning to the south, and to southern light for the first time since his brief military service, he was simultaneously attracted

[52] Jeffrey Meyers, 'Monet in Algeria', *History Today*, April 2015, pp. 19–24.

by its sheer intensity and quality, as well as baffled by how to represent it. He writes to his future wife Alice Hoschede:[53]

> *How beautiful it is here, to be sure, but how difficult to paint! I can see what I want to do quite clearly but I'm not there yet. It's so clear and pure in its pinks and blues that the slightest misjudged stroke looks like a smear of dirt.*

Subsequently there was a breakthrough: at some point Monet writes, 'Eventually my eyes were opened, and I really understood nature. I learned to love at the same time.' One of the results was the glorious *Under the Pine Trees at the End of Day* (Colour Plate E). A near-horizontal sunlight sets the trunks and the undersides of the pine trees' foliage ablaze with dark red tones. The top of the undergrowth scatters the same sunset glow. Horizontally, above both bands of red-dominance, are cooler layers. Stippled dark green treetops lie in shadow but echo the cooler blue-greens of the sky at the top of the painting, and across the centre, lighter blues and pinks in the sky settle onto a mottled sea.

The picture is all motion. As John Zarobell of the Philadelphia Museum puts it, 'staccato spots on the trees keep the viewer's eye moving across the surface of the image.' Exactly right—the evening off-shore breeze has picked up and is tossing the pine treetops to and fro as much as it is ruffling the distant ocean. A viewer knows that, and brings that knowledge to the painting. But the painting's surface does more work still. My own impression when looking at *Under the Pine Trees* is that there really is motion there, and I can't help but wonder if Monet had stumbled consciously or unconsciously on a way of patterning the canvas that exploits, in a gentler way, the illusion of motion that is garishly excited by Akiyoshi Kitaoka in Colour Plate B. To do so, there would have to be an incitement to continuous eye movements, as Zarobell observes, but also a layering of different spatial wavelengths perpendicular to the desired illusion of motion.

[53] Quoted in Richard Kendall, ed. (1989), *Monet by Himself: Paintings, Drawings, Pastels, Letters*. London: MacDonald Orbis, p. 126.

Philadelphia curator of European painting, Jennifer Thompson, accompanied me for a morning's close inspection of the work through her expert eyes. She pointed out that some of the relief of the layered oil paints has been unfortunately flattened during relining before the painting's acquisition. But on close inspection, the suspected ingredient of the visual illusion was visibly present: both the cool upper foliage and the burning upper branches contain persistent diagonals from top-left to bottom right, with a graded illumination (from skylight) in the perpendicular direction. Their form is accentuated by the different patterning of the fluid foreground shadows. It is at a distance from which these structures are just out of obvious resolution, that I find the impression of wind-blown motion is greatest, and that the gusty oscillation of the treetops is indeed towards and away from the top right of the frame, perpendicular to the brushstrokes.

Jennifer pointed out that the other near-contemporary works by Monet that she had hung in the same gallery reflected similar techniques, but arranged in different ways. His 1889 river valley scene, *The Grande Creuse at Pont de Vervy*, employs the same mottled and deep-green and red coloration, but directionally patterned at random, rather than aligned. The diagonal-field motif does appear in *Grande Creuse*, but in the pattern of clouds rather than the trees. Even more obviously deployed on the river banks below his standing *Poplars* of 1891, the motif dominates the central band of the painting in the uniform diagonal leaning of an extended reedbed, but leaves the viewer with a sense of uneasiness. After a while, the reason becomes clear—the reflections of the reeds in the river do not obey the correct optics: they are extensions of the real reeds, not reflections of them. Monet is able to create a psychological state in the observer by making an illusion illusory, by knowing when to break a rule of symmetry.

Returning our gaze to *Under the Pine Trees*, we find ourselves once more taken into that clifftop setting, and even constructing the wider panorama not shown within the frame itself. There are clues that allow us to do this, for example to know that if we were to turn westwards and stare at the sun perched right on the horizon, we would see that it reddens towards its lower edge, but shines more golden yellow at the top of its disc. Monet prods us into this intuition by illuminating the underside of the tree foliage with the deep red, but the tufts of grass on the ground with the gold, in the natural inversion generated by the rays that light them. Off-picture information is not uncommon in impressionism. Pissarro's *Railroad to Dieppe*, for example (also in the Philadelphia Museum), contains a large foreground shadow cast, presumably, by an edifice to the onlooker's immediate right. But Monet's implementation is wonderfully subtle.

Whatever one's personal impression of motion itself, if ever a viewer has stood by a stand of trees on a coastal clifftop at evening light, looking out towards the horizon, then memories of the experience can hardly fail to be excited by *Under the Pine Trees*. In the ancient sense of the word 'theory'[54]—the studied impression of a formal, frequently religious, event by experienced onlookers that we have previously found a helpful way-marker, the painting offers a perfect 'theory' of nature at that moment and place. These are the dynamics of wind, light, and colour that play on our faces, dragging us back from the weariness of a day exposed to the heat of such a climate. The orthogonality of the vectors of wind and light induce a three-dimensional impression, and an immersion into it. By an interesting coincidence, it is possible that this particular picture made an even stronger impression on one of its familiar viewers, for it was once exhibited in the Paris art shop of Theo Van Gogh, the artist Vincent's brother. Vincent wrote much later to Theo from the Mediterranean cost:

> *It's funny that one evening recently at Montmajour I saw a red sunset that sent its rays into the trunks and foliage of pines growing in a mass of rocks, colouring the trunks and foliage with orange fire while other pines in the further distance stood*

[54] See Chapters 1 and 5 on *theoria*.

> *out in Prussian Blue against a soft blue-green sky—cerulean. So it's the effect of Claude Monet. It was superb.*

A good painting successfully transports our imagination to the place and time it depicts. It is a superlative one that, once we have seen it, becomes itself the reference point for our visual memories, when we are present at the place in nature that inspired it. Van Gogh's description is unmistakable and remarkable. A very deep observation of nature has been captured when it serves as the interpreter of later natural impressions. This is why I am tempted to refer to such paintings as 'theories' of their object—going beyond the ancient and applicable resonances carried by the word. The way that 'theory' is used now in science stands comparison in both its construction and its function to such works of art.

An artistic theory of music

For a final visual example of an extended visual metaphor, we return to the Yorkshire studio of Graeme Willson. If the impressionists brought into the greatest light the dialectic tension between the two levels of painted canvas and object, then some of Willson's work interrogates the addition of abstract reference as well as a real object. In this example, he attempted the highly challenging task of 'deliberate synaesthesia'—the visual portrayal of music. For two years, as artist in residence at a series of chamber concerts in the Yorkshire town of Ilkley in the UK, he created a series of dynamic portraits of musicians at work, documenting the series of performances. The final summative work, a large canvas nearly two metres square, attempted something greater, simultaneously abstract and representational.

Quartet (reproduced in Colour Plate F) was inspired by the experience of a performance of a string quartet. Willson wanted to capture the extended moment: the production of music as well as attention to it, and to give an impression of music itself. He also recognized that a musical experience is never disembodied, but takes place within a physical space. At the start of a project such as this, it is never clear to the artist that a satisfactory solution to such a challenge even exists, that the question is well-posed. Just as a scientist, faced with a parallel uncertainty might begin by assembling conceptual ingredients and connecting them with the representational methodology of mathematics, so the artist begins to use the methods and techniques at his disposal, starting at a point of connection.

Surprisingly, the first idea towards making a convincing and contemplative connection between visual and musical art was the shared world of *linguistic* metaphor. Both art and music can present a dominant 'key' or tonality. Both colours and notes can be 'bright' or 'muted.' 'Loud' colours on the one hand, and 'shimmering' string passages on the other, illustrate metaphorically the latent synaesthetic links between the visual and auditory that operate long before the onset of a full-blown synaesthesia. There is danger of banality here, for a superficial translation of musical passage into graphical representation transports us no further than the oscilloscope screen (think of Disney's jejune representation of Bach in *Fantasia*).

Yet *Quartet* works wonderfully. The instruments of the string quartet are all visually present, geometrically arranged (but not in order of pitch either vertically or horizontally), each half-obscured by one of the multiple layers of the painting. Willson explained his desire to show the instruments within the texture of the music, individually identifiable but impossible to isolate. The music finds its dominant 'key' in its pinks, creams, reds, and browns. The picture is therefore rooted within an interval of time, but extends in relationship beyond a single moment, through its briefer modulations into blues, and very occasionally the remote greens. The genre of string chamber music, especially the outer movements of quartets, is frequently a 'busy' one—it is a passage of this nature that the picture captures (think of the finale of Haydn's Op. 72 No. 2 'The Fifths'). There is much fine detail—an impression we are left with after a line of running semiquavers weaving along the weft of the pattern—but also steady, smooth, and sustained key-anchors too, in the vertical bars of colour that constitute the warp of the texture.

No performance is entirely predictable; it is of the essence that live music responds to the chances and emotions of the time and place. Here, Willson deployed techniques of creating tension between order and chaos. As well as controlled figuration, pots of paint are tipped over a tilted canvas and allowed to run and merge. The process of art is always connected to the science of materials at the point of art-making. At this point, for example, it is essential that the front of a viscous liquid advancing down a plane is subject to instabilities that exponentially magnify any perturbations to the line with which it meets the surface.[55] Moments of inattention, of half-grabbed phrasing, an unexpected harsh attack from the viola responded to and shaped, jazz-like, by the professionalism of the 'cellist are possible personal reactions to the random fronts that at first disturb but then enter fully into the pattern of the painting. The architectural forms are taken directly from the roof-mouldings of Ilkley's King's Hall (so our gaze, for those in the know, is mostly upwards). Goethe's epithet, 'architecture is frozen music' is continually in mind, and deployed throughout Willson's performance works. Within the temporal experience of letting the eye and mind drift over a painting like *Quartet* the viewer re-creates a new 'performance' within time, the proportions and keys of represented music and architectural form are unfolded from a frozen moment into a sequence.

Finally, the meditation that the painting stimulates is itself captured on the pensive, concentrating, but relaxed faces semi-submerged in its layers. Whether they are listeners to the concert or statues, they remind us that deliberate thought and emotional affect cannot be simply teased apart. The listening

[55] See e.g. L. Kondic (2003), 'Instabilities in gravity-driven flow of thin fluid films', *SIAM Review*, **45** (**1**), 95–115, and the example of biofilm growth we met in Chapter 2.

faces are caught up in architectural, geometric order, but at the same time adrift in the semi-directed flow of the music and the painting. The artist's account of how the two faces were placed within the work as a whole bears a remarkable similarity to the visual accounts of creating mathematical physics that we came across earlier. Willson first made sketches on paper, allowing him to experiment with their position and orientation on the canvas, much like the rearrangements of terms in a fundamental equation until it reached a balance of sufficiency and parsimony. The faces become the viewers of the painting—observers are drawn into its experience as they re-create it. An expanded title of the painting might even be 'Theory of a Quartet.' Like the practitioners of ancient *theoria*, *Quartet* is a powerful reminder that we participate in much of what we perceive. That we are able to do this 'cross modally'—from sight to sound, and even to a sense of space as well—is one of the painting's messages.

The thoughtful faces within *Quartet* point us towards another connection that we must examine in a study of comparative creativity. We have a representation of music—but what does music itself represent? If it represents nothing at all, then how is it that we interpret, enjoy, find intriguing—even transformational—the experience of listening to it? The musical and the mathematical are frequently linked in commonplace comment. But is there substance behind the claims that 'mathematical minds' are often musical ones as well? And what can be said about meaning, and its creation, in two such abstract forms? Although a visual representation has brought us to questions about the abstract and the auditory, it has suggested that the musical and the mathematical may require us to take our leave of the visual metaphor altogether.

Those questions will take centre stage in Chapter 6, but there is a road we must travel to take us there, whose starting point lies within the ideas that the visual metaphor has already presented, for no example of visual imagination has been static, no concept of the cosmos, no imagination of the microscopic, not even solidified oil paint on canvas. All have surprised us with their dynamics of change—all have a narrative. Created as objects, albeit transparent ones pointing beyond themselves to a world beyond, artistic and scientific pictures also create stories. Their shared narrative imagination is a story in itself, and the one we encounter next.

Experimental Science and the Art of the Novel

Works of fiction resemble those machines which we contrive to illustrate the principles of philosophy, such as globes and orreries.

JOSEPH PRIESTLEY, *WARRINGTON ACADEMY LECTURE* 1761

Art lives upon discussion, upon experiment, upon curiosity, upon variety of attempt, upon the exchange of views and the comparison of standpoints.

HENRY JAMES, *THE ART OF FICTION* 1884

The inevitable collision was predictable from the very start of the debate. Oxford University's interdisciplinary institute for the humanities was hosting a public panel discussion on the topic, 'Is Mathematics Poetry?' Among the contributors were mathematical physicist Roger Penrose[1] and the Nigerian-born novelist and poet Ben Okri. Both are deeply impressive and imaginative people and writers. Okri's *The Famished Road* (which won him the Booker Prize as its youngest-ever recipient) transported me with its magical fragrances from a strange African world when I first read it. Penrose, together with Stephen Hawking, conceived and proved the 'singularity theorem' concluding the existence of the black holes that Chandrasekhar had found so aesthetically appealing. As the Oxford discussion wandered into the topic of writing novels, Penrose began to look increasingly exasperated. Eventually he expressed the objection that must have been in the minds of at least half the audience: 'Look, when writing a story an author

[1] Subsequently Nobel Laureate in Physics, 2020.

The Poetry and Music of Science. Tom McLeish, Oxford University Press.
 DOI: 10.1093/oso/9780192845375.003.0004

has complete freedom in what the characters do—that's the freedom of art; in mathematics, a theorem has to be correct—there is no freedom at all.' Okri's reply was measured, and delivered after a characteristically long pause—'Do you know how many times I had to rewrite my last short story?' The answer was, apparently, more than twenty. His point was that the authors of fiction are not—not nearly—as 'free' to control their characters as their readers might suppose. Okri had to 'experiment' with the writing over and over again until he was satisfied that the narrative and its actions were true to the personalities within it, and that the unfolding story was an authentic and convincing tale. Before those multiple drafts, even the outcome of the tale was unknown to its author. It was as if the novel had a life and logic of its own—it pushed back as Okri wrote. The public discussion after that point continued in a much more thoughtful and comparative, rather than conflictual, mode.

The drawing of parallels between fictional writing and scientific investigation has a long history. Nineteenth-century English novelist Henry James compiled a collection of essays on writing into a well-known volume *The Art of the Novel*.[2] Among his many insights into the creative process in general, James has much to say about the way the novelist formulates ideas. He compares the writer's inventions to the discoveries of 'the navigator, the chemist, the biologist' who do not 'find' things so much as 'come upon' them and recognize them. James would have recognized the 'variety of attempt' in Okri's experimental drafting as parallel to a scientist's many trials in perfecting an experiment. He describes 'observing' his characters rather than 'manipulating' them—here on character development in his novel *The Wings of the Dove*:

> *My business was to watch [the subject's] turns as the fond parent watches a child perched, for its first riding lesson, in the saddle, yet its interest . . . was just in its making, on such a scale, for developments.*

[2] Henry James (1934 [2011]), *The Art of the Novel*. Chicago: University of Chicago Press.

Here is a writer of fiction describing his experience of creation as observation, not invention. He might invent the initial arrangement of characters and setting, but from then on, James picks up his pen, to discover, not to decree, what might unfold. James Thurber once replied to Elliott Nugent, his co-author on *The Male Animal* who had asked him what was going to happen next, that he 'didn't know and couldn't tell him until I'd sat down with a typewriter and found out.'

The analogy between the writing of novels as the arrangement of character and situation on the one hand, and the setting up and observing natural phenomena by scientific experimentation on the other, is too strong to be overlooked in an exploration of scientific creativity. The intricate mutual mapping of experimental science and fictional writing allows literary scholar, Patricia Waugh, to write about novels[3] in a way that reads as a recommendation for statistics texts: 'Carefully read, novels educate their readers in how to read them by offering a workout for Bayesian skills first recognized and then developed in the eighteenth century, the first age of risk.' The sibling nature of experimental science and the novel, similarly elicits from literary critic, Frank Kermode, in his study of fiction *The Sense of an Ending*,[4] a discussion of the vast expansion of time afforded by the advances in geology of the eighteenth century:

> *For literature and its criticism this created problems we have not yet solved, though it is obviously relevant that the novel developed as the time of the world expanded, and that the facts are related.*

Family ties descend to our own times as well. Few novels of the last century have received more and richer comment than Virginia Woolf's *To the Lighthouse*. Woolf's tale is of a family, its narrative telescoped across two events separated by a generation.

[3] P. Waugh (2015), 'The novel as therapy: ministrations of voice in an age of risk', *Journal of the British Academy*, **3**, 35–68.

[4] Frank Kermode (2000), *The Sense of an Ending: Studies in the Theory of Fiction*. Oxford: Oxford University Press.

Woolf draws energy from the unresolved interval between them, observing at every turn, in intense detail. It exposes painfully but endearingly the characters' flaws and tragedies, with the perpetual promise that, one day, they will visit the lighthouse on the horizon. In a later letter, Woolf insisted that 'I meant *nothing* by the lighthouse'.[5] Yet its ambiguous beam, which warns of submerged rocks to those who come too close, seems also to shine back invitingly through the time of the novel from the closing, and final, journey to its first lightfall on the young James, conceiving within him his great longing for the expedition. The novel is structured around a distortion of space and time that could not have generated such vertiginous reference in any other century. The worldlines of the characters entangle with each other in complex and non-linear ways; even after her death, Mrs Ramsay haunts, if not quite literally, their interactions and final resolution of relationships.

To the Lighthouse is also infused with running questions about the human relationship with nature. Through the medium of painting, and its evocation of a particular painter and her art, the themes of visual insight and the challenge of imagery that we encountered in Chapter 3 are woven into the text. Lily, the amateur painter, struggles to represent the spatial relationships between objects in the garden, as she attempts to solve her own relationships as an independent woman in a social setting disturbed by undercurrents of rejection. In both art and life, she needs to plot a trajectory of creation with its hazily defined goals, elicited desire, frustrated labour, and eventual moment of gift.

Sensitivity to the representative power of novel writing and its echoes of the creative task within science allow Pat Waugh to write:[6]

[5] Virginia Woolf, *Letters*, 3, 385.

[6] P. Waugh (2018), 'There'll be no landing at the lighthouse tomorrow': Virginia Woolf and the Godrevy Lighthouse, in V. Strang, T. Edensor, and J. Puckering, eds., *From the Lighthouse, A Collection of Interdisciplinary Essays*. Abingdon: Routledge, p. 211.

> *Woolf's 'I meant nothing' suggests an entangled and emergent understanding of the process of creativity that challenges the idea of art's autonomy and makes it fully part of the entanglement of subjects and objects in life as they enter each other's force fields.*

To the Lighthouse provides Waugh with the material to confront the naive 'idea of art's autonomy,' using a powerful combination of scientific metaphors from the twentieth century. She ends up alongside Woolf's character Lily, exploring the notion of experiment that does justice to her art:

> *[for Woolf], for whom every novel was an existential as well as formal experiment, only a recognition of the unpredictability of things, the complexity of the world, the need to fling oneself forward on the back of risk, might allow the generation of some agential force that might actually change the present and begin to shape an as yet unknown but better future.*

We make experiments, not because we know their outcomes, but because we do not, because complexity emerges from the combination of simple components when they assume a relational life of their own.

Science converses with the narrative arts as much as the visual, and teaches us just as much about the work done by the imagination. We need to follow the clues laid down by Okri, James, Kermode, and Waugh, to find that these immediate parallels are by no means superficial. On the surface lies the ostensible clash of culture between observed science and constructed art, the factual against the invented, the whimsy of imagination against inferred 'laws of nature'; here the well-worn 'Two Cultures' narrative is skin-deep. But beneath lies a deeper commonality in the generalized notion of 'experiment'—the construction of well-defined sub-worlds of uncertain outcome, which can nonetheless teach us about the wider world.

Other conditions seem necessary to both fictional writing and experimental science—a confidence in creative ability, for example, is a shared prerequisite. To set out on a scientific project is a counterintuitive voyage of hope that a complex world can be understood, that simple structures can be perceived through the jungle of observed phenomena. Writers of science as well as writers of fiction (*writing* itself is central to the scientific task) have a readership in mind; both need not only to engage, but to persuade, if only to turn the next page. A scientific account must convince its audience of the reliability of its methods and the likelihood of its conclusion.

Such common narratological necessities in fiction and science draw our attention to another common thread, for authors of both must *infer*: in fiction the likely outcomes of their characters and situations, in science what might be deduced about the 'character' of physical law from observation. The very notion of likelihood, or probability, arose in mathematical thinking in the very same period that saw the appearances of the novel and experimental method in science. This is the background to Waugh's 'Bayesian' comment—it refers to the eighteenth-century cleric Thomas Bayes to whom we attribute a quantitative procedure for inferring the likelihood of an unseen cause from observations of its effects.[7]

We begin an examination of this family likeness between early modern science and the novel with their history—the new forms of writing and scientific method.[8] That will prepare us for a closer reading of Henry James's *Art of Fiction*, which we will open alongside physiologist William Beveridge's parallel *Art of Scientific Investigation*, in a comparison of creative processes. Then, as in the case of visual imagination in the last chapter, some particular times and places will exemplify the deeper and particular entanglement of narrative imagination in writing and in science.

[7] See Ian Hacking (2006), *The Emergence of Probability*, 2nd edition. New York: Cambridge University Press.

[8] This is not to deny the category of the 'novel of late antiquity,' but to recognize that the two forms, as the qualification indicates, differ significantly in their representational and narrative approaches, see Erich Auerbach (1993), *Literary Language and its Public in Late Antiquity and the Middle Ages*, trans. R. Manheim, Chapter 2, 'Latin Prose in the Early Middle Ages'.

A shared early history

The sixteenth and seventeenth centuries witnessed no 'birth' of, or 'revolution' in, science in any meaningful sense, however popular this discontinuous narrative of scientific history has become.[9] A conceptual, and even mathematical, search for an understanding of nature has been a response to human curiosity since writing itself began. Nor did the early modern period witness any sudden discarding of ignorance imposed by precedent 'dark' ages of non-science, as presentist views of history would have us believe. The story of science possesses more continuity than that, as we have already seen in our encounters with the scientific enquiries of thirteenth-century thinker Robert Grosseteste, eleventh-century philosopher Avicenna, and their ancient sources. However, to affirm a long continuity is not to claim that the centuries of Copernicus, Kepler, Galileo, Boyle, Hooke, and Newton brought nothing new to science—that would be an equally absurd claim. The early modern period witnessed possibly the most violent upheaval in humankind's imagined place in the world—the replacement of a geocentric cosmological model, treasured for centuries, by one that reconceived the Earth as an orbiting celestial body like any other. In swift succession, the mathematical project of the ancient and medieval mathematicians was brought to fruition by Newton as quantified laws of motion. The arrival of the 'compound' (containing more than one lens) optical instruments of the microscope and telescope realised the visual metaphor of science as enhanced perception into reality, opening both the heavens and the microscopic world to sight. All this was accompanied by powerful developments in experimental method, however ancient its origins might be.

We are so used to the idea of the scientific experiment in our time that it has sometimes become synonymous with the practice

[9] See, for example, Steven Shapin (1996), *The Scientific Revolution*. Chicago: University of Chicago Press, which notably and memorably begins, 'There was no such thing as the scientific revolution and this is a book about it.'

of science itself. Yet the notion that one can learn about nature by doing more than observing was by no means always an evident path to progress. It was not at all clear that abstracting out of context, simplifying and isolating a few parts or objects, and then manipulating them in unnatural ways, in such a highly artificial procedure, would constitute a reliable source of knowledge about the material world. Induction from the specific to the general was recognized by Aristotle as the only way to proceed in natural philosophy, but additional assumptions are required, which today we now scarcely perceive, that allow us to draw conclusions about the workings of natural processes from studying unnatural arrangements of objects.

One is the *assumption of uniformity*. It allows an abstracted and artificial system to render data that apply faithfully to the more complex settings of the natural world, distant from the laboratory (even distant from planet Earth). It is not immediately obvious that we can learn about the density and convection of huge volumes of air in the upper atmosphere, for example, by studying, as did Robert Boyle, the temperature and pressure of gases within small laboratory vessels.

Equally intrinsic to the experimental method is the ability to *work by analogy*. The insight that one can learn about the optical properties of raindrops by studying the passage of light beams through large water-filled glass spheres, was Theodoric of Freiburg's wondrously imaginative leap in the first decade of the fourteenth century. It led to the first complete theory of the rainbow at the level of geometrical optics.[10] The breakthroughs that permitted the notion of controlled experiments were philosophical rather than practical or cognitive.

That the human creation of special, 'small worlds,' that are the experiments of the sciences, would be in continuity with the rest of the world, is itself a hugely creative and non-intuitive idea.

[10] C. B. Boyer (1959), *The Rainbow: From Myth to Mathematics*. New York: Yoseloff; see Chapter 7.

It is highly counterintuitive to suppose that artificial, simplified, and isolated systems would have anything to say about the natural, complex, and connected world. The contemporary voice that most clearly helps us understand the almost overwhelming case against the new 'experimental philosophy' is that of natural philosopher and poet Margaret Cavendish. Her opposition to experimental method as a viable route to knowledge of the natural world is a repeated theme in her 1668 (2nd edition) *Observations Upon Experimental Philosophy*.[11] The Achilles' heel of experimental *artificiality* is highlighted time and again; a good example is her discussion of various 'sorts of heat and cold':

> *For if men conceive that there is but one heat and cold in nature, they are mistaken—and much more, if they think they can measure all the several sorts of heat and cold in all creatures, by artificial experiments. For as much as a natural man differs from an artificial statue or picture of a man, so much differs a natural effect from the artificial . . . Artificial things are pretty toys to employ idle time.*

Strong words indeed! For Cavendish, experiments are toys for the boys (she is unabashed to gender her argument, reserving the superior 'contemplative philosophy' for women). Her voice belongs at the centre of an early modern comparison of science and fictional literature for another reason—the argument was even more boldly put in arguably the first modern work of science fiction, her fantasy, *Blazing World*.[12] Cavendish's rhetorical moves in both philosophical and fictional form juxtapose the artificiality and over-simplicity of experiment to the rich complexity and multiple forms of nature. Her preference in method is therefore for what she calls 'speculative' or 'contemplative' philosophy. It is all too easy today to overlook the perceived difficulties in acquiring human knowledge or conception of nature that persisted from pre-modern thought in spite of the innovations of

[11] Margaret Cavendish, (1668), *Observations Upon Experimental Philosophy* (2nd edition), London: Anne Maxwell.

[12] Margaret Cavendish, (1666), *The Description of a New World, Called the Blazing World*, London: Anne Maxwell.

the fifteenth and sixteenth centuries. However, there is an echo of continuity here too: Cavendish's seventeenth-century suspicion of the over-simplifications inherent in experiment find resonance in twenty-first-century arguments of the relevance of *in vitro* to *in vivo* methods in biology. To some, the examination of the extraction of just two or three proteins from a cell is essential to grasp the nature of their interaction; to others it is a pointless and misleading simplification of the real object of investigation: the living cell in its full complexity. Though akin to ancient notions of connection between 'microcosm' and 'macrocosm,' the new experimental method, therefore, made bold new moves—the making of an experimental world requires humans to take our creative potential more seriously than in any previous age.

The idea is much larger than science itself; it draws on both philosophically and—perhaps surprisingly for us late-moderns—theologically transformational thinking. The gradual shift from Augustine to Aquinas, from knowledge through revelation to knowledge also through sense, reason, and imagination, brought with it, by 1600, the notion that God is not the only manipulator of nature, but endows humankind with the same ability. This is the theological and philosophical driving force that Francis Bacon codified in his manifesto for experimental science in the early seventeenth century, in his *Novum Organum.* The 'New Organ' proposed that the experimental inductive method essentially extends the human sensory apparatus. Bacon's ideas would influence in turn the early novelists, Defoe and Swift (the latter in famously critical vein). Here he is describing the royal task of discovery that is implied by a stewardship of nature:[13]

> *The glory of God is to conceal a thing, but the glory of the king is to find it out; as if, according to the innocent play of children, the Divine Majesty took delight to hide his works, to the end to have them found out; and as if kings could not obtain a greater honour than to be God's playfellows in that game, considering the great commandment of wits and means, whereby nothing needeth to be hidden from them.*

[13] Francis Bacon (1887), *Works*, Volume III, J. Spedding, R. L. Ellis and D. D. Heath, eds.

As, historian Peter Harrison has shown,[14] there are lessons in this early modern story of the invention of experimental method for the student of scientific creativity: the driving imaginative energy for this counterintuitive leap derived largely (in Bacon, Boyle, and others) from a Christian theology of the Fall and a consequent *faith* in the God-given potential to restore human sense and intellect—a reformed recasting of an earlier medieval 'ladder of restored understanding' that we will encounter again in Chapter 8. The idea that linked both medieval and early modern thinkers was that the recovery of an intimate knowledge of nature, intended for humankind but lost as a consequence of the Fall, implied an act of grace on the part of the Creator, enabling the rebuilding of natural knowledge by small steps, beginning with sense data, rather than ideation. The counterintuitive and theologically driven invention of experimental method is perhaps the starkest example of the way in which the greatest imaginative leaps in science require the highest tensions of interdisciplinary distance to power them.

As the Art of Experiment arose from the spirit of the early modern age, it developed consequences beyond science itself. As we have noted, literary historians are in near-uniform agreement that this same period witnessed the rise of the novel. Of course, fictional narrative itself is as old as the writer we call Homer, and the earlier (Old- and Middle-) English speaking world brings us the rich inheritance of *Beowulf* and *The Canterbury Tales*. But before Daniel Defoe imagined what might befall Crusoe, his flawed hero abandoned on an island after shipwreck and finding footprints in the sand that were not his own, and *wrote it down*, there was never the literary form of a created and experimental sub-world—the stage on which the novel plays out.

There are other reasons for setting the creation of scientific experiment and the creative writing of a novel alongside each other: authors do not simply narrate; rather they arrange conceived

[14] See Peter Harrison (2009), *The Fall of Man and the Foundations of Science*. Cambridge: Cambridge University Press.

characters in experimental configuration and situations in novel relationships that have not historically occurred. They create small alternative worlds, microcosmic mirrors of the universe we inhabit. The world of the novel has its own constraints and rules of operation. It has an unfolding story unknown at the outset that can operate not only aesthetically but also as an experience of learning, of reading the world, and of the empathetic transformation of the self. A mandate to create such living microcosms is a common thread that connects science and fiction in the enlightenment period.

The orbits of the early novel and science

A leading advocate of the historical connections between the novel and early modern science is Stanford University Professor of English, John Bender. Yet in his *Enlightenment Fiction and the Scientific Hypothesis*,[15] he introduces a counter-side to the parallel natures of experiment and fiction with which we began: 'the guarantee of factuality in science increasingly required the presence of its opposite, a manifest yet verisimilar fictionality in the novel.' By this account, the novel arose not as an imaginative parallel to scientific exploration, but as a kind of fictional counterbalance to the factual account of the world emerging from the new experimental method. But even if these two satellites of the enlightenment world gazed down onto it from opposite poles, they revolved in the same orbit. The communities of conversation and dissemination of ideas that sprang up in the coffee houses of London, Birmingham, Bristol, and other foci of trade and traffic sustained new literary and scientific networks. These informal forerunners of the later academies of arts and sciences constituted the street internet for the new cultural forms.

Other shared characteristics blurred the opposition of fiction and scientific fact—for early novelists were hasty to feign, almost

[15] John Bender (1998), *Representations*, **61**, pp. 6–28.

universally and explicitly, a *non-fictionality* of what they were writing. Daniel Defoe declares at the start of *Robinson Crusoe* (first published in 1719) that the account 'is a just history of fact . . . without any appearance of fiction in it.' The pretension of facticity is not just openly declared, but crystallized in a 'hyper-realistic' attention to observed detail. Autobiographical, naval, political, maritime, and geographical detail floods the impressions of a reader from the first page. Such contextual depth hangs perpetually in the background as the story of Crusoe's tentative trials, failures, and successes on his island unrolls in the novel's journalistic style. Let us join Crusoe on his first hunting expeditions, a salvaged rifle and shot in hand:

> *I went out at least once every day with my gun as well to divert myself as to see if I could kill any thing fit for food and as near as I could to acquaint myself with what the island produced. The first time I went out I presently discovered that there were goats upon the island which was a great satisfaction to me, but then it was attended with this misfortune to me; that they were so shy, so subtle, and so swift of foot that it was the most difficult thing in the world to come at them; but I was not discouraged at this, not doubting but I might now and then shoot one as it soon happened; for after I had found their haunts a little, I laid wait in this manner for them; I observed, if they saw me in the valleys, though they were upon the rocks, they would run away as in a terrible fright; but if they were feeding in the valleys, and I was upon the rocks, they took no notice of me; from whence, I concluded, that by the position of their optical organs, their sight was so directed downward, that they did not readily see objects that were above them; so, afterwards, I took this method—I always climbed the rocks first, to get above them, and then had frequently a fair mark.*

Crusoe is more than a survivor here, but also a naturalist. His deduction about the 'optical organs' of the island's goat population is, of course, a key to his survival, but it suggests that Defoe means to say more than that. He endows the 'shy, subtle, and swift' animals a pattern of behaviour opened to Crusoe's grasp after repeated and acute observation. Unstated, but also suggested, is that their natural predators tend to attack from below but not from above. The novel creates a world and an immersed observer, who builds a mental reconstruction of its ecosystem. The depth and subtlety of Defoe's own constructive vision is suggested at every turn. No wonder that John Bender tells us that:

> *The novels of Daniel Defoe, Samuel Richardson, and Henry Fielding . . . pretend to offer densely particular, virtually evidentiary accounts of the physical and mental circumstances that actuate their characters and motivate the causal sequences of their plots.*

In a significant move, author and character engage in complementary aspects of inductive and deductive inference. In the goat-hunting example, Crusoe inductively apprehends the animals' visual perception by repeatedly engaging and observing them. Outside the imagined world of the novel, the logic is entirely reversed—Defoe deduces the behaviour of the goats, of whose eyes and brains he is the creator. In one of the remarkable historical parallels between the emergence of fictional writing like this, and science itself, we find a parallel tension within developing scientific method, in this case between deduction from hypotheses and induction from observation. To take one salient example, in successive editions of *Principia Mathematica*, Newton distances himself from the charge of 'feigning hypotheses' (or traditionally, 'framing hypotheses'). Newton was setting up inductive demonstration against the charge of scholastic proposal of hypothesis and deduction, an accusation he levelled at his Cartesian opponents. His distinct natural philosophy, as he is keen to point out, builds its foundations rather on observational and experimental fact and yields valid generalities about nature:[16]

> *In experimental philosophy we are to look upon propositions inferred by general induction from phenomena as accurately or very nearly true . . . till such time as other phenomena occur by which they may either be made more accurate or liable to exceptions.*

Newton's second and third editions of *Principia* appeared in 1713 and 1726 respectively (the first edition, without the 'General Scholium' in which Newton expanded on the inductive method,

[16] Isaac Newton, *Principia Mathematica* Book III, 400.

had been published in 1687). The first edition of *Robinson Crusoe* dates from 1719. Such insistence on inductive experimental method by the great author of the Universal Theory of Gravity clearly influences the novelists of the early eighteenth century, but Newton had received in his turn an earlier influence from tangled stories from literature and theology told in the previous century, as has become clear to scholarship only very recently.

Newton and Milton—Paradise and procession

Historians have been aware, since Newton was ever their subject, that he maintained lifelong interests in theology and in alchemy, but the significance of these topics has been eclipsed by his scientific and mathematical work. A late-modern narrative that he embodied a secular triumph over the religious, mystical, and scholastic patterns of thought of the late middle ages and the renaissance has coloured our view of the relative importance with which he held these disciplines, even relegating them to late aberrations from an otherwise coldly secular and materialist mind. Fortunately, such a reconstructed view is now fading, to reveal a much more complex and fascinating intellect, roaming in astonishing breadth and depth across Biblical philology, theology, and alchemy throughout his life. The connectivity between different experiences that predispose really novel ideation, suggests that an understanding of his scientific innovation cannot be complete without an account of what other patterns and images were flowing through his mind.

Newton did not believe that the physical 'laws' of motion were adequate to account for all natural phenomena, but that material was ubiquitously animate: '*all matter duly formed is attended with signes of life*,' he writes.[17] This 'vitalist' metaphysics wove a strand of thought through the science, literature, and theology

[17] Query 23 in the 'Draft version of Queries to the *Optics*', Cambridge University Library MS 3970.3, fol. 619r, quoted from http://www.newtonproject.sussex.ac.uk/view/texts/normalized/NATP00055, accessed November 2021.

of Newton's century. If he was its greatest scientist, then his older contemporary, John Milton, must take the crown as its greatest poet. In his defining work, *Paradise Lost*, Milton wrote of the sun:[18]

> *The sun that light imparts to all, receives*
> *From all his alimental recompense*
> *In humid exhalations, and at even*
> *Sups with the ocean'*

As Milton scholar Stephen Fallon has pointed out, this view of a vital exchange between sun and Earth 'harmonizes with and prefigures the thought of his formidable younger contemporary.'[19] For Newton was to write, in the later, *An Hypothesis Concerning the Properties of Light*, of a '*subtil*' fluid that ascends from the Earth into the atmosphere and 'perhaps may the sun imbibe this spirit copiously, to conserve his shining' and 'that this spirit affords or carries with it thither the solary fewel and material principle of light.' We cannot wish to keep the revolutionary new thought of hidden connectivity in lines of gravitational force between the sun and its planets while wishing away other hypothesized connections in the same fertile mind.

Milton's magisterial *Paradise Lost* is by no means a novel, nor yet the intimations of one—it is more a theological poem. Yet its cradle in the seventeenth century, its thematic intertwining of humankind, the cosmos and theological narratives, and its thematic consonance with the century's great natural philosopher all point to connections between science and literature. Milton was acquainted with many of the founders of the Royal Society through its progenitor, the 'Secret College', which met at both Gresham College and at the home of Lady Ranelagh, the sister of Robert Boyle, whom he certainly knew. Milton was asked, though refused (possibly through discomfort at the restoration of the monarchy) to write a poem for the foundation of the Royal Society. His status as first choice to mark the foundation of the

[18] John Milton, *Paradise Lost*. 5.423–6.

[19] Stephen M. Fallon (2016), Milton, Newton and the implications of Arianism, in Blair Hoxby and Ann Baynes Coiro, eds., *Milton in the Long Restoration*. Oxford: Oxford University Press, pp. 319–34.

world's first great national academy of science, suggests that his engagement with the new experimental natural philosophy was recognized and even celebrated.

Fallon is even prepared to call *Paradise Lost* an 'experiment in theology',[20] for he and Newton shared another set of unorthodox theological views: Arianism,[21] was still a somewhat dangerous position to hold even in the later seventeenth century. Striking many a reader today as recondite and hair-splitting, this is the essentially non-Trinitarian belief that the Son is not co-eternal and co-existent with the Father, but proceeds from God as a dependent being, albeit higher than all others. In the Patristic age, it came close to dividing the church completely, and though condemned at the church's first council of Nicea (325 AD), it never entirely disappeared. Milton's long poem contains strong Arian themes and language.[22] He and Newton were primarily motivated to this heterodox view through deep textual, philological, and theological studies of the New Testament: the pattern and form of ideas leap from one medium of creation to another. The theological idea of a creative personal force, emanating, like light, from the ground of all being, suggests a relationship determined by a 'central force'. As anthropologist, Marilyn Strathern, has observed,[23] the late-seventeenth century is also the period in which the 'relation' between two objects, or people, becomes reified on equal, or even more solid, terms than its referents. There is surely a connection with the developing form of the novel here, once more: the small worlds of novels turn as much around their interpersonal relations as their characters. Locke, in his 'Essay Concerning Human Understanding' (1690), exemplifies

[20] Steven Fallon, personal communication, University of Notre Dame du Lac, Indiana.

[21] So named after the fourth-century Alexandrian, Arius, whose views were condemned as heretical at the first ecumenical council at Nicea in AD 325.

[22] Examples abound, but the attributes ascribed in Book 3 to the Father are of greater glory than those of the Son, to whom the angels 'Thee next they sang of all Creation first, Begotten Son, Divine Similitude . . .' (l. 383); the Son is part of the created order, see Michael Bauman (1987), *Milton's Arianism*. Frankfurt, Bern, and New York: Lang.

[23] Marilyn Strathern (2005), *Kinship, Law and the Unexpected*. New York: Cambridge University Press.

Strathern's point by the idea of brotherhood, which becomes a *thing* that can be imagined in some detail between two men of whom one knows nothing. What is true of human relationships may be extended into nature by analogy. Newton could say more about the relationship between sun and Earth than he could say about the sun. When this happens, something like the theory of gravity starts to surface and to become refined in the mind of 'the last sorcerer.'[24]

The art of the probable and the hermeneutic stance of Robert Boyle

The duality of induction and deduction, the crystallization of experimental method, a shift of focus from objects to their relationships, and underlying theological themes suffuse the intellectual background of the seventeenth century. All three had also to draw on the arts of persuasion in order to thrive. For to write effective fiction, disseminate theological interpretation, or to establish new science, requires the ability to *convince*. A community in which it is possible to persuade must possess a shared understanding of 'likelihood', and of its foundation-science of *probability*.

Another hugely creative and influential contemporary mind was the French mathematician, theologian and philosopher Blaise Pascal. A problem posed by a gambling friend, and a subsequent correspondence with mathematician and compatriot, Pierre de Fermat, launched the mathematical theory of probability. The very idea that chance might underlie a quantitative branch of mathematics had been conceived in the previous century by the brilliant Italian eccentric, Gerulaimo Cardano, but the seventeenth-century French developments launched the idea of the science of estimation, with its tool of mathematized probability, into the stream of public consciousness in which writers and readers all swam.

The likelihood of hypotheses given a set of data, and the likelihood of events given preconditions, are part of the necessary

[24] Michael White (1997), *Isaac Newton: The Last Sorcerer*. London: HarperCollins.

mental apparatus of the experimental scientist and the novel writer respectively. But to call on them in the first place points to another fundamental shift in the early modern period whose roots lie in the sixteenth century Reformation. This *hermeneutical stance* describes an attitude to the world on the part of every individual, that moves from a priority of received opinion, to a continual state of reading and interpreting. A central tenet of the reformers from Luther in Germany and Calvin in Switzerland to Latimer in England, was the reading of Scripture in one's own language and its unmediated, personal interpretation. A century on from the stormy ecclesiastical and political storms triggered by the nailing of Luther's ninety-five theses on the door of Wittenburg Castle church, the protestant project of 'universal priesthood' was so deeply written into European culture, that it had begun to colour activities beyond religion. A general encouragement to *interpret* events and observations for oneself, rather than accept received interpretations from the pulpits of the age, be they ecclesiastical or scientific, was a product of the Reformation spirit that broke out of the confines of the parish church into other aspects of public life. The political consequences of the Reformation have been debated and described for centuries, but here we are concerned with its far less frequently discussed reframing of science.

It is a short leap by analogy to apply the hermeneutical stance of personal reading and interpretation of the world to perception of nature. Even if Galileo had not given explicit voice to the medieval notion that God has written 'two books',[25] the first in scripture and the second in the natural world, the extension is self-evident. We would almost expect to find a similar democratic movement towards personal science in the latter half of the seventeenth century, at least within protestant society. The search

[25] Among several early attestations to the concept that has been traced to Augustine, Hugh of St Victor's *Didascalicon (Book 7)* ('*This whole world is like a book written by the finger of God. . .*'), contains an important discussion of the necessity to be able to read its letters in order to interpret its meaning. For an extended discussion of the 'Two Books' metaphor see Peter Harrison (2015), *The Territories of Science and Religion*. Chapter 3. Chicago: Chicago University Press.

is not lengthy, for we need look no further than the outstanding proponent of the experimental method, Robert Boyle.[26]

Boyle was a great intellectual architect of the early Royal Society. As I write this, coincidentally sitting in the Council Chamber of the Society's current London residence, I glance up and see on the opposite wall an elegant portrait of him, wigged and gowned, at an open book, directing the onlooker's gaze to a page of diagrams (Colour Plate G). His bearing evokes the Irish nobility into which he was born, but his gaze is sharp and his manner eager. Although working in the spirit of Frances Bacon's call to experimental science, he was careful not to admit discipleship of any philosopher or school—a personal statement of the responsibility of the individual to win knowledge of the world. He devised, with Robert Hooke, an air pump and carried out extensive experiments published in 1660 as *New Experiments Physico-Mechanical, Touching the Spring of the Air, and its Effects*. In later correspondence he articulated the law connecting the pressure of a gas to its volume that bears his name in Anglophone countries today. He also observed the expansion of freezing water, identified the role of air in propagating sound, and the refraction of light by crystals. He was also instrumental in bringing to birth the new and separate science of chemistry from its progenitor in ancient alchemy. His *The Skeptical Chymist* of 1661 announces in its very title not only the new science, but also the individual attitude that he, and his readers, must bring to it. The book advanced a theory of elements, fundamental constituents of compounds that would not be broken down further, and conjectured their particulate but unresolved nature.

Boyle is also credited with responsibility for the style of writing in which experimental science has been recorded, with modifications, since the seventeenth century. Its attention to detail permitted its readers to repeat experiments without the need

[26] J. Paul Hunter (1990), 'Robert Boyle and the epistemology of the novel', *Eighteenth Century Fiction*, **2**, 275–92.

to conjecture missing elements. At the same time, the new exacting idiom preferred the terminology of simple reportage to overinterpretation. A third innovative aspect was the convention of the passive voice. Taken together, these three attributes of Boyle's style created an invitational relationship with the reader. Not only could they later repeat the experiment themselves, but even at the time of reading were presented with sufficient detail to perform a sort of real-time evaluation themselves. As Bergamo Professor of English Linguistics, Maurizio Gotti puts it,[27]

> *the reader—provided with a detailed account of the events and reassured by the presence of reliable spectators—is nevertheless able to scrutinize the contents of the report, and—by means of this process of virtual witnessing—he can make correct evaluation of them.*

Here is an example of experimental essay writing from Boyle's accounts of his air pump that will serve to illustrate the style:[28]

> *The thing that is wont to be admired, and which may pass for our second Experiment is this, That if, when the Receiver is almost empty, a Bystander be desired to lift up the brass Key (formerly described as a stopple in the brass Cover) he will find it a very difficult thing to do so, if the Vessel be well exhausted; and even when but a moderate quantity of Air has been drawn out, he will, when he has lifted it up a little, so that it is somewhat loose from the sides of the lip or socket, which (with the help of a little oyl) it exactly filled before, he will (I say) find it so difficult to be lifted up, that he will imagine that there is some great weight fastened to the bottom of it.*

Pedestrian in style possibly, eschewing any figurative or metaphorical writing—but it is not dull. Passive and active voices interleave, an aesthetic reaction of admiration is anticipated, and the extraordinary experience of the sheer unexpected weight of the air is described with a vivid immediacy, speckled with the finely pointed detail (we feel the 'oyl' on our fingers as we scrabble to

[27] Maurizio Gotti (2001), 'The experimental essay in early modern English', *European Journal of English Studies*, **5**, 221–39.

[28] W. Johnson et al. (1772), *The Works of the Honorable Robert Boyle, Vol. I, New Experiments Physico-Mechanical Touching the Spring of the Air*. A new edition. London: Printed by Miles Flesher for Richard Davis, bookseller in Oxford, 1682. p. 16.

force the little key open against the newly felt weight of air). This sort of text invites a double imagination in the reader: we are there checking the persuasiveness of Boyle's own past account as well as anticipating the creation of our own partial vacuum. This is technical writing, but not at all unapproachable for the lay reader.

Boyle's reformed Anglican churchmanship would not permit an elitist approach to his natural philosophy, any more than it did his theology. He declined the offer of the provostship of Eton College, as this would have meant taking holy orders. He was convinced that his theological writings would have more force as a lay member of the church. Neither would his science retain any trace of priesthood for Boyle, who added to his scientific research writings what might be described as a manifesto, or even a manual for lay science. With the *Occasional Reflections upon Several Svbiects; Whereto is Premis'd a Discourse about Such Kinds of Thoughts* (1665) he added to a genre of lay writings inaugurated by contemporaries John Flavell and William Gearing, which encouraged a universal habit of detailed observation and meditation on daily experience. The practice was even endowed with a formal name—'Occasional Meditation'—and now had the imprimatur of a famous scientist.

The commonplace experiences of colours and forms in flowers, ripples on a lake, a falling leaf were not only to sharpen the observational powers of its followers, but also to nourish their abilities to interpret. In language that reads as a foreshadowing of Romantic nature poetry (see a brief discussion of Wordsworth and Emerson later in this chapter), Boyle develops the 'Book of Nature' trope, as had Galileo before him, but far closer in his development to the twelfth-century theologian Hugh of St Victor, to suggest how it might be *read*:

> *The World is a Great Book, not so much of Nature as of the God of Nature, . . . crowded with instructive Lessons, if we had but the Skill, and would take the Pains, to extract and pick them out: the Creatures are the true Aegyptian Hieroglyphicks, that under the rude form of Birds, and Beasts etc. conceal the mysterious secrets of Knowledge and of Piety.*

The *Occasional Meditations* contains examples of Boyle's own practice—tellingly we would by no means categorize all of his examples today as narrowly 'scientific'. They do include material from his professional activities, such as 'Looking through a Prismatical or Rectangular Glass,' taken from experiences in his own laboratory, but are more likely to feature records such as 'Upon the Sight of a Windmill Standing Still,' an impressionistic record of a moment during an afternoon walk. Even the scientific meditations refrain from technical or mathematical language. Boyle was committed to establishing a lay scientific community, of which he saw himself a leader, but by no means an unintelligible 'expert'.

'Reading' nature directly, and writing and reading texts about nature, guided by the practice of occasional meditation, became a route to self-improvement, heightened awareness, and a realization of the possible within these new realms of knowledge. Such winds of change in the way individuals engaged with texts were as necessary for the fertile growth of science as they were for the new horizons of the novel. J. Paul Hunter suggests an even closer parallel:[29]

> *Such thinking leads to—and justifies—the subjective interpretation of events that the novel exploits, and the popular proliferation of the idea of reading the universe as text instils in the culture a need for texts that offer similar ambiguous possibilities while themselves remaining stable.*

The provisionally credible interpretation, the hermeneutic stance of the individual reader and writer, and here again, the parallel of the stable, created world of the novel and the law-constrained universe unfolding to the new experimental approaches within science—all these sustain more than simple comparisons. Both the narrative worlds and the underlying writing conventions constitute the cargoes traded between early modern science and the novel. Compare the excerpts from Boyle that we have already surveyed with one more from *Robinson Crusoe*:

[29] J. Paul Hunter, 'Robert Boyle. . .'

> *Finding my first Seed did not grow, which I easily imagin'd was by the Drought, I sought for a moister Piece of Ground to make another Trial in, and I dug up a Piece of Ground near my new Bower, and sow'd the rest of my Seed in February, a little before the Vernal Equinox; and this having the rainy Months of March and April to water it, sprung up very pleasantly, and yielded a very good Crop . . .*
>
> *But by this Experiment I was made Master of my Business, and knew exactly when the proper Season was to sow; and that I might expect two Seed times, and two Harvests every Year.*

With even the briefest nod to the use of the passive, an experimental account is once more presented in reproducible detail. Defoe invokes the hermeneutical approach, inviting his readers into the world of the text to evaluate and to credit. The shared custom of 'collective witnessing' becomes the mutual activity of scientific and fictional readerships alike.

Defoe himself wrote explicitly about science; we have in his rather Boylean account of the well-educated and rounded individual, *The Complete English Gentleman*, a commentary on science that, with a nod to the 'two books' parallel, also echoes the desired effects of Boyle's and Flavell's occasional meditation:

> *Science being a publick blessing to mankind ought to be extended and made as difusiv as possible, and should, as the Scripture sayes of sacred knowledge, spread over the whole earth as the waters cover the sea.*

A later editor of *Robinson Crusoe* took its author's wish to heart: the 1815 edition is an extraordinarily encyclopaedic venture, with as many words of footnotes as in Defoe's text. They cover naval and meteorological terms, classical allusions, detailed notes on ship architecture and engineering, navigation, and much besides.

The authoring of possible and plausible worlds drove the imaginations of early modern scientists and authors alike. Their writing was driven by a new and shared emphasis on acute observation, itself a shared narrative. Theological themes such as individual reformational piety were also constitutive of the common foundations of science and novel writing, and surface frequently. Reflection on the creative process itself, however, is rare in the early modern period. To find examples of the genre of self-examination in fiction and in experimental science we must look over a century beyond the world of Defoe and Boyle, of Milton and Newton, to the nineteenth century. There we encounter another remarkable parallelism.

The arts of fiction and science

Henry James (1843–1916) was a singular nineteenth-century New Englander, of moderate means but wide experience and acquaintance on both sides of the Atlantic. Receiving a very unusual education at the direction of parents who saw that science would be as important as classics in a changing world, he began a course at Harvard Law School, but succumbed to a passion for writers and writing before graduating: Emerson, Gaskell, Dickens, Balzac, Walpole, Zola, Maupassant, Ruskin, and Arnold, among others, decorated a cosmopolitan world of literary invention which formed the background to his work. His older brother, William, became a renowned philosopher and psychologist. Henry's own novels—*Portrait of a Lady* and *The Ambassadors* among them—secured him a reptation for writing that explored the ambiguity and contradictory motives of the people and times he knew. He was therefore particularly well-placed to respond to a nineteenth-century emergence of reflection on the process of writing itself—his 1884 *The Art of the Novel*[30] created a modern genre that continued to develop throughout the following century (James gave it another nudge by entitling a later essay *The Art of Fiction*). This collection of his critical prefaces makes fascinating reading for anyone who reflects on the creative process. James considers the varieties of ideation or initial conception of the 'germ' of a novel, and the 'incubation' or development of the germ into an extensive structure. As the shape fills out, the writer

[30] Henry James (1884 [2011]), *The Art of the Novel*, Chicago: Chicago University Press.

becomes aware of an intensifying illumination of his characters—this is apparent as an undirected and passive activity, the phase in which the story takes on 'a life of its own'. But there is finally an activity of 'verification', of reflection on constraint, of the need to persuade a reader of a credible narrative, which corresponds to the universal creative tension of idea and form.

James is resurrecting a classical title—the first century BCE Roman poet Horace, set out, in his *Ars Poetica*, advice on poetic composition, in self-referentially poetic form. Balancing beauty with instruction, and aware of both the author's and the audience's background of learning, Horace is at once practical and philosophical. James' attempt to breathe new life into the genre of *The Art of* . . . caught on. The first chapter of this book introduced Graham Wallas's 1926 *The Art of Thought*,[31] which will provide the fourfold creative scheme of ideation, incubation, illumination, and verification used in the following. Before a generation had passed, *The Art of Scientific Investigation*[32] had been written by Australian-born veterinary pathologist, William Beveridge. Announcing on its title page that 'Scientific research is not itself a science; it is still an art or craft,' Beveridge's book parallels James's in its methodology of personal reflection. There are processual stages to this *Art* too: preparation, experimentation, chance and hypothesis, imagination and intuition, and just as in *The Art of the Novel*, a reflective stage of verification, reason, and observation. If the novel and experimental science shared the same cradle in the sixteenth and seventeenth centuries, then setting these portraits of the two grown adults of the nineteenth and twentieth in juxtaposition suggests one way to explore why their subsequent life stories seem so rarely to have intersected, and whether their apparent divergence hides a deeper relatedness. In either case, the first creative stage is the conception of the idea itself.

[31] Graham Wallas, *The Art*. . .

[32] William I. B. Beveridge (1950), *The Art of Scientific Investigation*. New York: W. W. Norton & Co. Ltd.

Ideation

James's account presents a wide range of personally experienced seeds of novelistic inspiration. The conception of *The American*'s storyline happened by an abrupt 'accident of thought' some years before he wrote the novel, while the origin of the idea for *The Portrait of a Lady* was 'breathed' upon him, 'prescribed and imposed' by life. Such a direct extrapolation from experience seems also to have generated the seed of *The Princess Casamassima*—he writes that the idea of the novel was the fruit of taking long walks around London, because 'perambulation with one's eyes open' provokes the need to interpret and reproduce. He fed his imagination by prowling the streets, observing their multi-layered life, seeking out minute details to 'add a drop, however small, to the bucket of my impressions'. Resonances with the 'Occasional Meditators' of Boyle and early lay science are unmistakable. James's metaphorical language ranges from the biological to the chemical and physical, and even to the medical. He describes the idea for *The Spoils of Poynton* as a 'single small seed,' a 'precious particle,' or 'fruitful essence.' His imagination 'winced' at the prick of the suggestion, for the virtue of an idea is 'all in its needle-like quality, the power to penetrate as finely as possible.'

In the preface to *The Aspern Papers*, James compares the creative ideation of the novelist to the discoveries of 'the navigator, the chemist, the biologist.' Reporting that 'one never really chooses one's general range of vision—the experience from which ideas and themes and suggestions spring,' he finds feelings and thoughts 'we regard very much as imposed and inevitable.' From passive apprehension it is a small step to recognize a cognate emotional 'passion'—the suffering from insight that caused him once more to 'wince.' James writes from a culture in which emotional expression is guarded, but we detect the affective forces at work not so very far below the surface of his text.

Beveridge discusses the origin of scientific research programmes in very similar language, even adopting James' explicit 'walks about town' as his metaphor for a scientist's search for ideas. For Beveridge, 'the majority of discoveries in biology and medicine have been come upon unexpectedly, . . . especially the most important and revolutionary ones.' Budding scientists must be trained in the powers of observation, and essentially in the 'attitude of mind of being constantly on the look-out for the unexpected.' Had Beveridge taken part in the round-table *Imagination Institute* meeting on scientific imagination of Chapter 2, there would have been wide agreement. Authentic experience is speaking when we read that these 'chance' opportunities of observation come more frequently to those who are actively working in the day-to-day tasks of the lab, and to those who take risks in the spirit of scientific playfulness. To interpret an unanticipated observation as signifier of undiscovered truth, the investigator must have a 'prepared mind,' as free as possible from fixed ideas, and equipped with imagination.

The emotional impetus of the scientific creative drive begins in *The Art of Scientific Investigation* as obscurely as it is in *The Art of the Novel.* But as his account develops, Beveridge finds it more natural to speculate on non-cognitive and non-conscious elements, including those at the early stage of ideation. Curiosity drives the creative work of the scientist. Though curiosity often atrophies after childhood, in the scientist it is transferred onto an 'intellectual plane'—investigators are driven by a strong *desire* 'to seek underlying principles in masses of data not obviously related,' which 'may be regarded as an adult form or sublimation of curiosity.' This scientific inquisitiveness is insatiable, an essential source of the germs of scientific ideas.

Incubation

To continue the organic metaphor, the subsequent phase in the development of ideas, in both James and Beveridge, might be termed 'incubation,' in both consciously active and passive modes. Dormant reflection upon the 'germ' of an idea for a novel sometimes extended for James over many years, during which there was no active attempt at developing characters or plot. He writes, suggestively, of conscious 'thought experiments' that surface amid his non-conscious reflection.

For *The American*, once more, the germ developed through a series of self-posed puzzles: 'What would [a wronged compatriot of an aristocratic society] do' in that particular predicament, and 'How would he right himself, or how, failing a remedy, would he conduct himself under his wrong?' These thought experiments gathered their answers like 'plucking spring flowers.' The story of *The Tragic Muse* seems to have been much less consciously developed. James cannot remember the 'precious first moment of consciousness' of its idea—a story about art. Yet it matured 'with having long waited,' without substantive development for a long time before he began writing. The fruit of such incubation depends upon its author 'cultivating fondly its possible relations and extensions. . . [he] is terribly at the mercy of his mind.' Creation requires conscious nurturing of the seedbed of the mind in which an idea must grow.

The accounts of scientific incubation we have already encountered, from the Newtonian 'by thinking upon it constantly,' to the contemporary reports from the *Imagination Institute*, have mirrored all these experiences, from self-posed puzzles to thought experiments. James also recognizes the strange experience of entertaining a notion that is not consciously understood at all (such as my night of non-comprehension of star polymers). The idea for *The Reverberator* 'remained for a long time thus a mere sketched fingerpost,' and James 'couldn't make out at first exactly what it meant' until 'at last, after several years of oblivion, its connections, its illustrative worth, came quite naturally into view.'

James's accounts of incubation suggest analogies with two types of scientific task. One is immediate and local—the solution of another puzzle: how does an independently minded American woman in nineteenth-century Europe navigate the advances of men whose motives are suspect? What is the structure of a molecule whose emergent properties are known, but not its atomistic form, be it benzene or DNA? The author of *Portrait of a Lady* allowed different ideas to incubate as a scientist would mull over different hypotheses. The other, more holistic and global—the daunting weaving of plot lines, characters, motives, and plausible happenstance—constitutes the activity of a much more expansive and complex system. In the language of Iain McGilchrist's analysis of the left/right lateralized human brain, *The Master and his Emissary*,[33] the first is an analytical example of left-brain analysis, the latter an integrative task requiring the dominance of right-brain activity. The great creations of a Dickens, Brontë, or Tolstoy require not just genius but also time to weave, unravel and re-weave the multiple layers of satisfying complexity. This is not the fictional equivalent of a single puzzle, but more closely analogous to the model-building for a complex multi-level system such as a biological cell, or a self-consistent quantum theory for electromagnetism.

When we turn once more from *The Art of the Novel* to *The Art of Scientific Investigation* the active, and passive, aspects of an incubation period are palpably present. The experiments in thought of the novelist move to the laboratory for the scientist. We need, however, to recall that Beveridge worked in a branch of science for which experiment was the central activity; a theoretical physicist would recognize the idea of a 'thought-experiment' even without translation, and Einstein perfected the art, as we have already seen. The results of James's thought experiments, his 'plucking of spring flowers,' speaks of the particularity of this part of the novelist's labour—it is tempting to call it 'reductionist' in the sense that it focuses on abstracted elements of the complete

[33] Iain McGilchrist (2009), *The Master and his Emissary*. New Haven: Yale University Press, reminds us that both art and science require the integration of left and right hemispheres of the brain, *contra* a simplistic 'Two Cultures' neurological lateralization.

narrative. Beveridge articulates similar abstraction in the laboratory experiment:

> *An experiment usually consists in making an event occur under known conditions where as many extraneous influences as possible are eliminated and close observation is possible so that the relationships between phenomena can be revealed.*

In sum, the principle of research is to isolate the object of interest, 'vary one thing at a time and make a note of all you do'. These are the laboratory flowers that, if well-chosen and arranged, evoke an aesthetic all their own, and together constitute the grasp of pattern over an entire field.

Experiments of both artistic and scientific kinds also work at different scales of significance. The incubation of an idea begins with a sort of preliminary evaluation, an experiment whose purpose is simply to explore whether there is potential for extended labour, not to extract large amounts of information. Such scientific scouting into new territory will reveal whether or not a full-scale test is warranted or if the evidence is 'too slender to justify a large experiment.' Beveridge notes that even rapid and preliminary experiments can claim a sparse aesthetic of their own.

I cannot resist a personal illustration of the pleasure latent in a cursory but informative experiment. I was once unable to persuade any experimental colleagues to attempt a neutron-scattering experiment[34] on carefully synthesized polymer samples that a new theory had suggested might prove highly instructive. Admittedly the complexity of the experiment appeared off-putting—it required a tiny and precious sample to be stretched rapidly while quenching it down through 150 degrees in temperature, followed immediately by its transfer into

[34] Since particles also possess wave-like properties, neutrons that are 'shone' into matter are able to reveal molecular details of its structure much as X-rays can. Their advantage is that, being neutrally charged, they are not disturbed on their way by the electrons that scatter light and would otherwise render the samples opaque. Neutron beams are only obtainable at large central facilities, such as ISIS near Oxford, UK.

a beam of neutrons whose deflections from the polymer had to be measured very carefully. The impasse was broken by a small group of young theoreticians volunteering to try their hand at the experiment—plunging a makeshift rig into a bucket of liquid nitrogen on the site of the UK's scientific neutron source at the Rutherford Appleton Laboratory near Oxford. Using a few spare hours of beam-time gave us barely presentable data, but it did show that the pattern of scattered neutrons was remarkable, begging a more careful study. This was enough to persuade a group of 'proper' experimental physicists to construct a very sophisticated rig, temperature-controlled not by a bucket of coolant but by an exquisite system of automated miniature jets of liquid propane aimed at the sample. This lovely piece of apparatus, and the data it produced after a lengthy programme of experiments,[35] was wonderful to behold. But in a different sense, the quick, dirty, and preliminary 'sketch' of the first experiment could claim a rough elegance all of its own, even without the romantic excitement of new insights won in the small hours of the morning at the scattering facility. The equivalence to the working sketches in overall shape of the novelist is not a fanciful comparison.

Illumination

Illumination speaks of clarity and wholeness of vision, but it also refers to the experience of incoming, rather than self-generated, insight. James presents several autobiographical examples in *The Art of the Novel*, through the language of observation. Once his characters are sufficiently formed, he seems to be able to watch them, rather than to choreograph their movements. He saw the protagonist of *The Princess Casamassima* 'roam and wonder and yearn, saw all the unanswered questions and baffled passions.'

[35] See M. Heinrich, W. Pyckhout-Hintzen, J. Allgaier, D. Richter, E. Straube, D. J. Read, T. C. B. McLeish, D. J. Groves, R. J. Blackwell and A. Wiedenmann (2002), 'Arm relaxation in deformed H-polymers in elongational flow by SANS', *Macromolecules*, **35**, 6650–64.

The visual metaphor of observation breaks down before the creative process is complete: the author becomes aware of his character's state of feeling and becomes emotionally engaged by the experiment's own life.

James gives a counter-example of a self-avowed failure to permit a passive period of contemplation to play out, and a consequent non-illumination, in the composition of *The American.* He admits to forcing his intentions onto the story, so that they 'dug a hole' in which he was 'destined to fall;' his work departed from *terra firma*—he says that it became more 'showy' than 'sound.' Scientists recognize immediately the temptation to force a quick interpretation onto their observations, to 'read' into an experiment with too strong and inflexible an *a priori* framework for the underlying connections and causes. Beveridge achieves as well as any science writer a description of the necessarily subtle balance of preconception and openness. The investigator must have a 'prepared mind', yet be free from fixed ideas to contemplate unexplained observations, so that the initial clue can grow. Once more, attention is directed not so much at the seed, but at the capacity of its incubating environment to incite its germination.

Just as in James's case, the experience of intense concentration of thought, and the 'chance' illumination by insight are described in the same breath. Yet such chance befalls only the prepared eye and mind. The most developed scientific form of such illumination is the *hypothesis.* The *Art of Scientific Investigation* explores the framing of hypotheses as a mental technique, arguing that they can be fruitful even if incorrect, for they prepare the mind to uncover new facts, experiments, or observations. There are lavish historical illustrations, for example the discovery of French physiologist, Claude Bernard (whom we will encounter again in a close reading by novelist Emile Zola), that the rate of blood circulation in animals is controlled locally through the nervous system. An initial hypothesis that severing the nerve leading to a rabbit's ear would shut off the chemical processes that gave off heat within it was confounded: the ear warmed. The hypothesis was partially correct; the nerve severed was, however, a modulator of blood flow not a generator. Bernard realized that he might have made the essential observation long before had his gaze not been exclusively focused on the eyes rather than the ears. Hypotheses are tools rather than ends in themselves. Beveridge delights in an autobiographical note of Darwin on their fluidity:

> *I have steadily endeavoured to keep my mind free so as to give up any hypothesis, however much beloved (and I cannot resist forming one on every subject) as soon as facts are shown to be opposed to it . . . I cannot remember a single first-formed hypothesis which had not after a time to be given up or greatly modified.*

If the first pole of illumination in science is the hypothesis, then the second is the experience of imaginative insight. Beveridge asks 'how ideas originate in the mind and what conditions are favourable for creative mental effort.' The prior stages of a conscious idea and its incubation are preliminaries, but he insists that 'the conjuring up of the idea is not a deliberate, voluntary act. It is something that happens to us rather than something we do.' The moment of ideation comes in a flash, perhaps through a 'connection between several things or ideas' or through a 'great leap forward.' He anticipates the moments that mathematical physicist, Michael Berry, called 'claritons' in the conversation of Chapter 2 and, like Berry, muses on the nature of the mental state most conducive to them—he calls this a 'state of doubt which is the stimulus to thorough enquiry.' In common with all recorded examples of the experience, he insists that 'it is not possible deliberately to create ideas or to control their creation. When a difficulty stimulates the mind, suggested solutions just automatically spring into the consciousness.' All the investigator can do, then, is direct their thoughts to a certain problem and attend to it while suggestions are 'thrown up by the subconscious mind.'

An extended chapter on the imagination is a radical new departure. Now the formerly rational scientist avows that 'all creative thinkers must be dreamers.' He adopts Rosamond Harding's definition of dreaming: 'allowing the will to focus the mind passively on the subject so that it follows the trains of thought as they arise.'[36] Intriguingly, this is also the point at which the animal physiologist starts referring to physicists. Perhaps there, rather than in his own field (which remains to this day much more focused on hypothesis-driven method), he finds other scientific voices that speak of a mysterious experience of creation from the subconscious. Max Planck's words, 'imaginative vision and faith . . . are indispensable—the pure rationalist has no place in science'[37] in his 1932 book, *Where is Science Going?*, perhaps leant Beveridge the authority to venture into the obscure side of scientific imagination.

Yet there are hints in the entangled story of literature and science that intuitive imagination might not after all constitute an utterly impenetrable veil. Indeed, empirical experience points to practices that, through repetition, enhance and predispose it. Beveridge acknowledges, for example, that creative imagination is nourished by the conversations of a community:

> *Productive mental effort is often helped by intellectual intercourse. Discussing a problem allows contributions from those with different backgrounds and a pooling of ideas and information. It is a means of uncovering error, encourages the investigator, and helps her escape from an established pattern of thought which has proved fruitless.*

As part of a rhythm of work, contemplation, conversation, and rest, the temporary abandonment of a problem allows the investigator to approach it in a new light, free from conditioned thinking. The affective power of curiosity can overcome the chasms of counterintuition sometimes necessary to free scientific thinking from its conditioning. Beveridge makes the point delightfully with the story of surgeon, John Hunter's realization that a restriction of blood flow to an organ might actually be

[36] Rosamond E. M. Harding (1940), *An Anatomy of Inspiration.* Oxford: Frank Cass & Co.

[37] Max Planck (1932), *Where is Science Going?* New York: W. W. Norton and Co.

treated counterintuitively by tying off an artery, so at first further restricting the flow, rather than seeking immediately to increase it (this strategy promotes the formation of other vessels—the lesson is not to think of the vascular structure as fixed, but responsive). Initiated by an encounter in Richmond Park with a young red deer, which got him thinking, the Hunter surgical procedure became routine in human patients within a few years.

Curiosity and intuition foster other, still more affective, modes of cognition in this irretrievably non-systematic experience of scientific illumination. Beveridge defines intuition as 'a sudden enlightenment or comprehension of a situation, a clarifying idea which springs into the consciousness, often, though not necessarily, when one is not consciously thinking of that subject.' Though these ideas are 'more dramatic and important progressions of thought,' he conjectures that they likely occur by the same cogitative process as 'the gradual steps in ordinary reasoning.' He adopts the same mode of introspection as Henri Poincaré, writing that the ideas of intuition must 'originate from the subconscious activities of the mind which, when directed at a problem, immediately brings together various ideas that have been associated with that particular subject before.' In a remarkable consonance with the mathematician, he speculates that the subconscious mind might react to new ideas with pleasure and excitement in a mirrored but submerged copy of the emotions of which we are aware, and that this reaction might 'bring the idea into the conscious mind.' Here again is Poincaré's subterranean form of emotion—an affective response of which we remain unaware, but that nonetheless serves to nourish innovative thinking and to winnow fruitful ideas from the fallacious.

Beveridge is well aware of the experience of resting states necessary to the productive intuition of the subconscious. His notion that these hidden processes are best described as a form of aesthetic explains the conditions favourable for intuition, namely 'freedom from other competing problems' and a state of mental 'relaxation.' Hence the title of *The Art of Scientific Imagination*—as

Beveridge identifies 'emotional sensitivity' as an essential quality of an investigator, he makes his central claim that, 'the great scientist must be regarded as a creative artist.'

Verification and the constraint of form

If this book has a central running hypothesis, it is that creativity in both art and science emerges from a tension between imaginative power and the constraint of form, and between non-conscious apprehension of possibilities and conscious analysis. If such a universal dialectic operates, we would expect the authors of works in the *The Art of . . .* genre to invoke it at this point of their parallel accounts. The diffuse wonder of initial illumination is insufficient on its own for either the novelist or the scientist. Both have elaborated, as far as their self-awareness permitted, on the essential roles of the emotions and the non-conscious mind, of the exercise of curiosity, of intense observation, of dialogue, and of the intangible fruitfulness of the fallow periods of mental absence from the conscious creative task. Yet from all this emerges only their raw material; the final work must be subject to the conscious attention to constraints. James and Beveridge would agree with Leonardo's attributed maxim, 'Art lives from constraints and dies from freedom', but the sources of the pertinent constraints in writing and experimental science differ from those of art. A naive anticipation would be that the novelist is constrained, like the poet, from within, while the scientist encounters constraints from the external observed world.

James is unmistakable on the essential enclosure of his themes and subjects by tight form and genre; only in this way could they reach clarity and intelligibility. Creativity is defeated by looseness of form. In the preface to *The Portrait of a Lady*, he compares the process of novel writing to the construction of a monument out of bricks, paying attention to the line, scale, and perspective. He compares himself to a builder, constructing his scheme by driving in supports and laying horizontal beams so that the structure can withstand strain. Only such material arrangement 'opens wide the door to ingenuity'. Commentating on *The Tragic Muse,* he shifts the metaphor from architecture to painting: 'A picture without composition slights its most precious chance for beauty, and is moreover not composed at all unless the painter knows *how* that principle of health and safety, working as an absolutely premeditated art, has prevailed.' Creative energy without constraints becomes a 'large loose baggy monster,' full of arbitrary elements, and ultimately lacking in meaning.

If imagination without form remains impotent, then, equally, form cannot be applied in a vacuum of ideas; the process of creation is never a strict diachronic progression in stages of illumination and verification. James seeks a metaphor for how a writer's creative flow must be held in synchrony with the shaping demands of form, and finds it in the fluidity of a river. In his writing *The Turn of the Screw,* he allowed 'the imagination absolute freedom of hand . . . with no outside control involved,' yet the stream of this improvisation was never permitted to 'break bounds and get into flood' or else it would 'ravage the story.' Ultimately, the imposition of form is temporally indistinguishable from ideation itself—its apprehension may even trigger the flow of ideas. Composition allows creativity to move the work forwards. Echoes of Platonic idea, form, and matter have not died away.

The apprehension of form can of itself constitute an aesthetic—in a work of art whose 'reading' and writing requires a temporal sequence, it may take on the pleasure of a journey. So, James writes that his best and finest ingenuities were always his 'conformities' to the conditions of his work, and he recalls 'the pleasure of feeling his divisions, his proportions and general rhythm, rest on permanent rather than in any degree on momentary proprieties.' The luxury of this aesthetic is greatest 'when we feel the surface, like the thick ice of the skater's pond, bear without cracking the strongest pressure we throw on it.'

Constraints on the scientific imagination do shape from without, as observations or data, but also from within, as with the novelist. Our very paradigms—the immense frameworks of assumption and narrative that Thomas Kuhn identified—are sometimes so large that we fail to see them for what they are,[38] yet they provide formal constraints on our thinking, and even on our observation. The very idea of a 'polymer,' invoked several times in this discussion of ideation in science, furnishes a good example. Until Staudinger's removal of the constraint imposed by the paradigm for chemistry before 1930, that all molecules were composed of a fixed and small number of atoms, it was simply not possible to conceive of any of the macromolecular structures and processes that we have discussed, nor for that matter to embark on a molecular theory of biology. On the other hand, without the footholds and frameworks that our paradigms supply, we have no navigable mental pathway, nor constitutive materials with which to build our models of the world at all. The scientist recognizes James's 'beams and building blocks' with which he builds the novelist's worlds, but also knows that one day she may have to tear the entire edifice down when the building regulations change.

Paradigm shifts rest eventually on observations of the world. Beveridge marshals his material on constraint and form in science under the two headings 'Reason' and 'Observation.' He is remarkably sensitive to the social and human uncertainties surrounding the 'theory-laden' status of observation and experiment that we associate with later twentieth-century philosophy of science. A reflective scientist knows that inner and outer constraints play simultaneously. Arguably Kuhn's paradigm shifts can be understood in terms of this language of inner and outer constraints. Observation is more than just seeing; it requires both sense-perception *and* a mental process that may be partly conscious and partly unconscious. The act of judgement is constantly at work, as Kant perceived, and Michael Polanyi was to explore

[38] Thomas Kuhn (1962), *The Structure of Scientific Revolutions*. Chicago: University of Chicago Press.

later in his *Personal Knowledge*.[39] Although Beveridge does not refer to the psychology of perception, he knows, from his professional experience, about professional experience of the active perceptual modes we encountered when exploring the visual metaphor in science.[40] He categorizes observations into the deliberate, and the passive or unexpected. A scientist should remain open to occurrences outside their selected field of attention. Charles Darwin's son writes of his father:

> *He wished to learn as much as possible from an experiment so he did not confine himself to observing the single point to which the experiment was directed, and his power of seeing a number of things was wonderful . . . There was one quality of mind which seemed to be of special and extreme advantage in leading him to make discoveries, It was the power of never letting exceptions pass unnoticed.*

Possibly this acute attention to peripheral detail, and especially to the constraints that weigh so heavily on existing theories and paradigms (the 'exceptions') was learned through experience of earlier missed opportunities. Darwin himself writes ruefully of a valley explored on an earlier expedition:

> *Neither of us saw a trace of the wonderful glacial phenomena all around us; we did not notice plainly the scored rocks, the perched boulders, the lateral and terminal moraines.*

The sheer exuberance and enjoyment of 'constraint' can redeem all negative connotations of the word. Both James and Beveridge claim, echoing Leonardo, that constraint enhances, rather than diminishes, the creative process. The potency of external form is found in its breaching of the dual inner constraint—the paucity of human reason. Yet it is from such apparently disruptive experiences that the radically new can emerge.

Beveridge draws attention to the paradoxical role played by *limitations* of reason in the scientific process. He quotes from Francis Bacon: 'Men are rather beholden . . . generally to chance, or

[39] Michael Polanyi (1962), *Personal Knowledge*. London: Routledge.

[40] See Chapter 3 *Seeing the Unseen: Art and the Visual Imagination.*

anything else, than to logic, for the invention of arts and sciences,' and from Schiller:

> *The slowness and difficulty with which the human race makes discoveries and its blindness to the most obvious facts, if it happens to be unprepared or unwilling to see them, should suffice to show that there is something gravely wrong about the logician's account of discovery.*

Beveridge stands in the tradition of centuries of thought from Aristotle through Aquinas and Bacon when he identifies the different characters of inductive and deductive reason. Inductive reasoning he charmingly admits to be 'less trustworthy but more productive' than deductive reasoning. Though in certain fields of study—such as mathematics, physics, and chemistry—the initial assumptions or axioms are clearly defined, inductions are nevertheless 'usually arrived at not by the mechanical application of logic, but by intuition, and the course of our thoughts is constantly guided by our personal judgment.' In a second anticipation of Michael Polanyi's masterpiece *Personal Knowledge*,[41] Beveridge points out that the manipulation of experimental science is rarely as precise or controlled as it is formally presented. Personal judgement limits the use of reason in the sciences, while simultaneously extending the range of the imaginable: 'terms often cannot be defined accurately and premises are seldom precise or unconditionally true.' Science moves as much by flow and feel as by formal logic.

James's pleasure at finding his writer's rhythm excites a resonance in the experience of scientific interpretation. Experimental data alone conveys nothing, but by induction is given narrative form by the scientist. Beveridge argues that scientists should *not* 'train themselves to adopt a disinterested attitude towards their work,' for it is better to 'recognize and face the danger' of bias. The pleasure that comes from 'associating ourselves whole-heartedly with our ideas,' is 'one of the chief incentives in science.' He describes an emotion familiar to any story-teller—including the

[41] Michael Polanyi, *Personal Knowledge*.

thrill of walking between the dangers of listening too much and too little to preconceptions (this is the 'bias' in Beveridge and the 'forced intentions' of James). The move from *fact* to *interpretation* even adopts a shifting grammar of tense in *The Art of Scientific Investigation*: induction moves from speaking of the past to vocalizing the possible and probable future. The value of science is its ability to predict (future) from a moment of insight and convergence (present) of a stream of experiential data (past). The constraints on induction turn it into an open plot, for there is a perpetual risk that scientific induction be misled by imperfect data, clouded insight, or alternative interpretations—'generalizations can never be proved.'

The Art of the Novel considers both the inner and outer nature of formal constraints—the shape of the narrative and the properties of its characters channel the possible flow of plot along some contours and not others. But a second layer of form appears in the structure of the text itself—not as clear as in poetry, yet novel writing is never of entirely free verse. Paragraphing, chapter length, and pace are shaping forces as fundamental as the interior logic of the story. Like Boyle three centuries before him, Beveridge advocates a strict organization of scientific written form. Good scientific writing becomes reciprocal with thought: 'careful and correct use of language is a powerful aid to straight thinking' because it requires 'getting our own minds quite clear on what we mean.' *The Art of Scientific Investigation* concludes that the best training for reasoning is 'discipline and training in writing.' Any scientist knows this by experience: an effective working strategy to extricate a research programme from a dead end is often to start writing about it. Semantic logic provides new pathways for ideation and language lubricates thought.

The inner grammar and syntax of writing engages with the outer form of experimental observation in the generation of a narrative of scientific induction, faithful to the world and to its representation. In the same way, a novel must respect its inner structure as much as the outer constraints of authenticity.

This is surely why Ben Okri needed to write twenty drafts before he achieved the coherent narrative that lurked beyond the tip of his pen, but also why the tightly prosaic prescription for scientific writing, adopted since the early years of the Royal Society, might bear reappraisal in the light of the noetic power of other forms no less responsive to constraint.

Entanglements of science and literature

Our juxtaposition of James's *The Art of the Novel* and Beveridge's *The Art of Scientific Investigation* has followed the trajectories of scientific and literary creation as separate—if revealingly parallel—paths. They throw mutual light upon each other in intriguing ways; the shadowy stages in each creative process can illuminate the other. The degree of overlap of their intuitions and their recorded experience is, however, extraordinary. The testimony of literary conception has encouraged us to peer deeper into the origin of ideas in science, while the constraints of scientific paradigm have reflected the imaginative potential in forms of writing. Their parallel juxtaposition has helped us examine literary art and science by the light of the other, but it also invites due caution before claiming inappropriate identities between the two.

The more attentively we listen to the parallel conversations of writing and science, the more we begin to detect the undeniable presence of a real conversation between them. A mutual discourse has unrolled over history, and by a form of interdisciplinary gravity each has affected the course of the other. The rise of the English novel in the first place, its 'scientific' cast in supposed factuality, the simplified and abstracted world that it creates, and the gaze onto the world that early novelists shared with scientists such as Boyle—these all testify to a deeper entanglement between fictional writing and the developments in science.

One treads very carefully through the treacherous territory of *Zeitgeist*. In the twentieth century, overblown claims of connection between relativity theory in physics and relativism in philosophy and literary criticism have left some scholars' metaphorical fingers badly singed. But it would be equally foolish to turn an intentionally deaf ear to the possibility of a true historical dialogue between literature and science. One of the problematic aspects of the example of naive parallels between Einsteinian relativity and the loss of the absolute in literary criticism, or between Darwinian evolution[42] and an arbitrarily imposed sense of social 'fitness' is that it is precisely in the twentieth century that the intensity of conversation between science and the humanities seems to have been throttled down to its lowest ever level. In earlier eras, the cross-disciplinary bandwidth was much greater, so the effectiveness of mutually generative conceptions so much the more plausible. Even in the early nineteenth century, we find the poet Samuel Coleridge and the scientist Humphrey Davy close friends and colleagues, Coleridge encouraging Davy in writing his own poetry, while Davy welcomed his literary friend into the laboratories of the Royal Institution to participate in experiments.[43]

New sources of imagination immediately spring into being through conversations between the authors of science and of literature. The creation of new forms, even new syntax of writing that the first scientific journals stimulated, points to a deeper connection. Furthermore, we have seen how the discipline of writing is itself a road to creative acts in science as much as in the production of literature. *The Art of Scientific Investigation* reports the testimony of scientists that the very act of writing, the formal constraints of grammar and logic through which it leads thought, and the pathways to rich seams of metaphor that it provides, can bring scientific programmes to fruition in ways that purely mathematical logic or bare experimentation cannot. It is a general finding that the broader the base of material

[42] Gillian Beer's masterly 1983 *Darwin's Plots* (Cambridge: CUP) traces the literary influences of the greatest of all nineteenth-century scientific metaphors.

[43] Richard Holmes (2008), *The Age of Wonder*. London: Harper Press.

upon which a conception can draw, the more transformative its ideation can be. One feature of non-conscious imaginative thought is its long-range, and sometimes bizarre, call on distant connections. Yet it is just those very leaps that are needed to disclose new creative pathways of thought that break away from established conformities. It is not ridiculous to suggest that reading Shakespeare can contribute to the solution of scientific puzzles.

The question arises as to whether there are particular periods or movements in history whose flow of ideas break the banks of their disciplinary rivers, creating a floodplain of shared and mutually informing currents. From this perspective, the story already told in this chapter suggests a deeper connective structure beneath the thought of, at least, the European seventeenth century. The shared locus of the early novel and the rise of experimental science points towards the common creation of systems, isolated from the larger world, that nevertheless speak truth about that wider experience. The idea has ancient roots in the Platonic and Aristotelian notion of mimesis,[44] a miniature reality that implies a larger whole. Through mimesis, human beings, in reading the world and writing into it have the capacity themselves to bring into being 'the novel.' An exploration of the future actions of an isolated human being, until the constrained reintroduction of human contact (Defoe), becomes the topic of imaginative discourse in the same way that the isolation of single gases, in order to follow their behaviour, exemplifies the experimental method for reading new knowledge from nature (Boyle). The remaining task for this chapter is to see if there are any connections to discern between the century of inception and the hints of later conversations in the nineteenth- and twentieth-century echoes of *The Art of . . .* parallels.

[44] Erich Auerbach (1953), *Mimesis: The Representation of Reality in Western Literature*. trans. Willard R. Trask. Princeton: Princeton University Press.

Humboldt, Emerson, Wordsworth, and the Romantic scientific aesthetic

Thanks to a number of contemporary biographers, academic and popular, there has been a reawakening of long dormant public appreciation of the Prussian natural philosopher, explorer, and writer Alexander von Humboldt.[45] That name was once as strong an international cipher for 'scientist' in the nineteenth century as Einstein's in the twentieth. Humboldt's passion for discovery led him on two great expeditions: the first as a young man at the end of the eighteenth century, to South America, and the second twenty years later, to the Asian steppes. A consummate observer and recorder, he has more species, geographical features, and places named after him than any human who ever lived, from butterflies to great ocean currents. He scaled lofty Andean mountains, navigated rapids, and pioneered jungle trails and lengthy ocean crossings, but there are three essential elements of his legacy. The first is his possession of multiple scales of mental vision. Like an impressionist painter he was able to combine the attention to the detail of delicate differences between leaf structures or the structure of insect wings, yet in the same work form great conceptions of ecosystems, in vast holistic syntheses. The second is a radically new conception of nature as a single connected system. Humboldt saw, beyond the multiple visual impressions of the Orinoco plain, the interconnected food chains that sustained plant and animal life, and in turn he grasped their dynamic coupling with the local climate. He drew further implications for a web of life on Earth from his observations that the same flora appeared at varying altitudes within different latitudes on the planet. Migration of their seeds, and their subsequent germination, suggested to him the comparability of ecosystems at different latitudes as well as mechanisms for global convection of living matter.

[45] Laura Dassow Walls (2009), *The Passage to Cosmos: Alexander von Humboldt and the Shaping of America*. Chicago: University of Chicago Press; Andrea Wulf (2015), *The Invention of Nature*. New York: Alfred Knopf.

This radically new vision required a radical new form—the third great legacy is Humboldt's incomparable writing; this is why he must appear in a narrative of creative entanglement between literature and science. Although widely known for a magisterial, lifelong (and unfinished) opus, *Kosmos*, a multi-volume compendium of his natural history, Humboldt's twinning of literary style and scientific vision was probably more influential through an earlier work. The 1808 *Ansichten an die Natur* (*Views of Nature*, in a recent translation by Mark Person[46]) was his most polished and widely read work. It is really a collection of essays ('Views') that, taken together, create a satisfying whole. *Ansichten* received multiple editions, and the final 1849 version was translated twice, independently, into English, in addition to the existing Dutch, Spanish, Russian, Polish, Swedish, and French versions, ensuring a long and wide legacy and reception unmatched by a scientific book until Darwin's *Origin of Species*. In *Ansichten*, he consciously set out, as had Boyle long before, to define a new style of writing, an 'aesthetic unity that blended literature and science together,' according to literary critic and Humboldt scholar Laura Dassow Walls.[47] The preface to the first edition conveys its author's tense awareness that the language he uses will be critical for the creative consequences in its readers. This example from the first essay in the book shows how the richness of language and simile suggests the very biological connections that constituted its scientific originality:

> *The surface of the Earth is hardly moistened before the fragrant steppe becomes covered with Kyllinga, with many-panicled Pasculum and a number of diverse grasses. Enticed by the light, herbaceous Mimosas unfold their leaves sunken in slumber and greet the rising sun like the morning song of the birds and the opening blossoms of the water plants. Horses and cattle graze in glad enjoyment of life. The*

[46] Alexander von Humboldt (1808), *Ansichten an die Natur*, Stephen T. Jackson and Laura Dassow Walls, eds., trans. Mark W. Person as *Views of Nature* (2014) Chicago: University of Chicago Press.

[47] Laura Dassow Walls (2017), Ansichten der Natur, mit Wissenschaftlichen Erlauterungen und Zusatze, in *Alexander von Humboldt-Handbuch des Metzler-Verlags*, Ottmar Ette, ed. Berlin, Heidelberg: Springer

> *grass shooting upwards hides the beautifully spotted jaguar. Keeping watch from a safe hiding place and carefully measuring the length of his spring, he snatches passing animals, catlike as the Asiatic tiger.*[48]

In reading this short passage in any language (though the new translation is beautifully and richly generous in its echoes of the original), our minds swim in the early morning of the Llanos, and we are alerted to the possibility of unseen connections between birds and the plants that grow within their habitat. Even the vast intercontinental distances of space and the aeons of evolutionary time are evoked by the observation that American jaguars and Asian tigers share a common feline physiognomy and spring. Humboldt does not simply write about nature, he writes 'nature' itself into discourse. If his language created his science, then his reification of nature also prepared the conceptual ground for others after him.

If the metaphorical fruitfulness of *Views of Nature* seems to have presaged the later developments of evolutionary biology, then Humboldt's conception of the Earth itself as a connected and dynamic whole was arguably a necessary preliminary for the new geology. Plate tectonics—the theory that the Earth's crust is under continuous and segmented motion, with mid-ocean upwelling and intercontinental fault lines of subduction—suffered a difficult birth, taking fifty years of acrimonious dispute before acceptance in the mid-1960s. To change the paradigm from a solid and permanent planet to a liquid and metamorphosing one, was hard enough even with the foresight of Humboldt, who imagined linear sequences of volcanoes connected by deep underground rivers of magma, like planetary blood-vessels. Without his vital metaphor, it would surely have met with even greater resistance.

Humboldt's understanding of the role that humankind plays in nature is both glorious and tragic. On his South America expedition, he encountered river basins where over-grazing and intensive agriculture had so eroded the soil that vast swathes of

[48] Alexander von Humboldt, *Ansichten an die Natur*, p. 38.

land had become bare. The cycle of evaporation and rain had been broken, and the connected system of biological and geological activity sent into irreversible decay. Anthropogenic climate change was already documented, at the regional level, by 1850. Conversely, in regions untouched by human activity, he saw unfettered potential for life to explore variety:

> *All the more freely, then, the forces of Nature manifested themselves upon the steppes in many diverse types of animal life: free and limited only by themselves, like the vegetation in the forests of the Orinoco, where the Hymenaea and the hugetrunked bay tree are never threatened by the hand of Man, but only by the luxuriant crush of entwining growth.*[49]

Humboldt biographer, Andrea Wulf, comments that he was the first to identify the potential for humans to destroy their own habitats. Throughout this labour of observation, reflection, and writing, however, Humboldt experienced the glory of the human recreative mind. He writes of the strangeness of nature, its sometimes terrifying unfamiliarity and difference, and the consequent impulse to discover its deep secrets, even to use them in ameliorating the worst excesses of human exploitation.

Ralph Waldo Emerson, the Romantic essayist and Transcendentalist poet, bears the stamp of Humboldt's sculpted forms of language that 'write nature into being.' Compare this passage from his 1838 address to the Harvard Divinity School with the previous passage from Humboldt:

> *In this refulgent summer, it has been a luxury to draw the breath of life. The grass grows, the buds burst, the meadow is spotted with fire and gold in the tint of flowers. The air is full of birds, and sweet with the breath of the pine, the balm-of-Gilead, and the new hay. Night brings no gloom to the heart with its welcome shade. Through the transparent darkness the stars pour their almost spiritual rays. Man under them seems a young child, and his huge globe a toy. The cool night bathes the world as with a river, and prepares his eyes again for the crimson dawn. The mystery of nature was never displayed more happily. The corn and the wine have been freely dealt to all creatures, and the never-broken silence with which the*

[49] Alexander von Humboldt, *Ansichten an die Natur*, p. 36.

old bounty goes forward, has not yielded yet one word of explanation. One is constrained to respect the perfection of this world, in which our senses converse. How wide; how rich; what invitation from every property it gives to every faculty of man! In its fruitful soils; in its navigable sea; in its mountains of metal and stone; in its forests of all woods; in its animals; in its chemical ingredients; in the powers and path of light, heat, attraction, and life, it is well worth the pith and heart of great men to subdue and enjoy it. The planters, the mechanics, the inventors, the astronomers, the builders of cities, and the captains, history delights to honour.

Humboldtian scientific style is unmistakable alongside Wordsworth's poetic influence, but also the potent simile: the action of a cool night might refresh like a dip into a river. The reader might think of unseen but felt rivers of air, circulating as emergent large-scale structures in the atmosphere. Here is a vision of interconnectivity in the natural world—Emerson's scope encompasses the 'chemical elements' and the 'navigable sea' within the same textual arc.

Emerson also wrote about the relationship between natural philosophers and poets. If his great scientific influence was Humboldt, then the poet he most admired was Wordsworth, to whom he ascribes, 'the bringing of poetry back to Nature . . . undoing the old divorce in which poetry had been famished and false, and Nature had been suspected and pagan.'[50] Emerson's interpretation of the human task was a 'grand narrative' of reading and interpretation of the world. In a threefold task of reading, interpreting, and speaking nature, both scientist and poet draw from imagination. Taking two passages from a later essay, the communicative, prophetic task draws, for Emerson, on the same wellsprings of imagination and inner sight:[51]

Science does not know its debt to imagination. Goethe did not believe that a great naturalist could exist without this faculty. He was himself conscious of its help, which made him a prophet among the doctors. From this vision he gave brave hints

[50] Ralph Waldo Emerson (2010), *Works*, Vols. 8, 66. Boston: Harvard University Press.

[51] Ralph Waldo Emerson (2001), *Emerson's Prose and Poetry*, 1st edition. Saundra Morris and Joel Porte, eds. New York: W. W. Norton & Company.

> *to the zoologist, the botanist, and the optician. When the soul of the poet has come to the ripeness of thought, it detaches from itself and sends away from it its poems or songs, a fearless, sleepless, deathless progeny which is not exposed to the accidents of the wary kingdom of time.*

The Romantic era of nature-writing constitutes one of the great hopeful chapters of history, where imaginative writing and imaginative science seemed on the edge of becoming the warp and weft of a single cultural tapestry. Goethe and Humboldt spent long happy hours together whenever the restless explorer found time to visit the older polymath and poet. The self-conscious preface to *Ansicht an die Natur* expresses Humboldt's desire to write from the soul as well as from the mind, even while acknowledging that a tension between the poetic and the scientific must never mislead, but that powerful writing could interpret and communicate the deep truths of nature more powerfully that a catalogue of facts. Emerson insists on talking of the Poet and the Man of Science in the same breath; the two of them depend on both genius, 'the enchantment of the intellect' and love, the enchantment of the affections.[52] As Walls has pointed out, Emerson—unusually for an American—refused to use the term 'scientist.' The term, coined by William Whewell of Cambridge in the 1830s, was also rejected by Faraday and Maxwell, appearing more commonly in Britain only towards the end of the century. For these capacious minds, the classification seemed to signal the growing fissure between poetry, philosophy, and natural science that they so deplored.

Emerson's choice of 'man of science' rather than 'scientist' contains another clue—for it is also a direct adoption from Wordsworth. At this point, we make contact once again with his preface to the second edition of *Lyrical Ballads* featured in Chapter 1. His distantly glimpsed prospect of a science of 'flesh

[52] Emerson, *Early Lectures*, 57. Laura Dassow Walls, *Emerson's Life in Science*, 62–3, 85.

and blood,' from which a poet could find inspiration, seems puzzling, jarring, or even comic to a twenty-first century reader, now we have the literary and scientific context that allows us a glimpse of what Wordsworth saw on his horizon. As scholar of Romantic literature, Emily Dumler-Winckler has pointed out, both Wordsworth and Emerson saw an unabashed motive for poetry and science as the *pleasure* that comes through knowledge, even hard and painfully won, communicated personally to readers who participate in both struggle and enjoyment. They had also glimpsed in the scientific prose of Humboldt and Goethe a future in which scientific writing might, without compromising the constraint of accuracy or detail, nevertheless unite both cognition and aesthetic in a close partnership. Wordsworth writes:

> *The knowledge both of the Poet and the Man of science is pleasure; but the knowledge of the one cleaves to us as a necessary part of our existence, our natural and unalienable inheritance; the other is a personal and individual acquisition, slow to come to us, and by no habitual and direct sympathy connecting us with our fellow-beings.*

The Romantic period seems to stand as an example, writ large, of the first stage of a common creative process—where a vision of the potential and possible creates a desire to find a pathway towards it. Wordsworth and Emerson perceive a distant view of a partnership between writing and science, if not of equals, then of complementary energies. But their vision bifurcates; they see another scenario in which poetry, prose, and science take very different paths, in which the 'debt to imagination' owed by science is forgotten, and in which the 'direct sympathy' by which science might connect us with 'our fellow-beings' is severed. Towards its darker endpoints are the loveless terrors of Mary Shelley's *Frankenstein*, and an unbalanced deployment of power and mastery over nature, rather than a participative recreation. This is the road that a disquieted Humboldt foresaw, but hoped to avoid by recruiting the power of imaginary writing to the purposes of science. Sadly, the technocratic path, in which science takes its leave of poetry and imaginative writing, would overwhelm his and Wordsworth's vision of a partnership of creative energy. By the end of the nineteenth century, the final fixation of its movements triggered by the divergent vocabulary of 'artist' and 'scientist', had realized their fears, rather than their hopes.

Emile Zola, Claude Bernard, and the 'experimental novel'

Well into the nineteenth-century process of parting ways of science and literature, in 1880, the celebrated French realist novelist, Emile Zola, published an essay, *Le Roman Experimentale* (*The Experimental Novel*). For readers fresh from reading Wordsworth, Humboldt, or Emerson, its thesis comes as a shock. For Zola's notion of 'a literature governed by science' is a far cry from the Romantics' sublime sources of inspiration from nature. As a further example of entanglement of the 'arts' of the novel and of science, it is fascinating, but constitutes a grim way-marker on that trajectory taken historically, so very different from the vision and hopes of Emerson and Goethe.

Zola's inspiration derives from an earlier (1865) book on the application of experimental science to medicine, the physiologist Claude Bernard's (whom we met earlier being surprised by the temperature of a rabbit's ear) *Introduction a l'etude de la medecine experimentale*. Zola read the *Introduction* with immense admiration:

> *Claude Bernard all his life was searchinq and battling to put medicine in a scientific path. In his struggle, we see the first feeble attempts of a science to disengage itself little by little from empiricism, and to gain a foothold in the realm of truth by means of the experimental method.*[53]

He is exultant to find a written account of living organisms, and in particular of the human animal, that rejects ideas of 'vitalism'—that living, and especially conscious, matter contains substance

[53] Emile Zola (1964), 'The experimental novel', in, Maxwell Geismar, ed., *The Naturalist Novel*. Ste. Anne de Bellevue: Harvest House Ltd.

or potential absent from ordinary matter, which then necessarily lifts it out of reach of science. Bernard, who had interestingly harboured early thoughts of becoming a novelist, set out his intentions clearly:[54]

> *It is therefore clear to all unprejudiced minds that medicine is turning toward its permanent scientific path. By the very nature of its evolutionary advance, it is little by little abandoning the region of systems, to assume a more and more analytic form, and thus gradually to join in the method of investigation common to the experimental sciences.*

Zola the novelist, is enamoured greatly of his understanding of experimental method:

> *The end of all experimental method . . . is then identical for living and for inanimate bodies; it consists in finding the relations which unite a phenomenon of any kind to its nearest cause . . . Experimental science has no necessity to worry itself about the 'why' of things; it simply explains the 'how.'*

He differentiates between the 'observer,' whose role is that of pure recorder, one who 'writes under the dictation of nature,' and the 'experimentalist' who 'comes forward to interpret the phenomenon.' This is language we have heard before: Zola's observer could pass for the great-grandchild of Boyle's occasional meditator, and his experimentalist the daughter of Emerson's or Wordsworth's transcendent poet-scientist. But the metaphors are deceptive for there is none of the imaginative or the sublime in Zola; he wants to replace every occurrence of 'doctor' in Bernard with 'novelist.' His analysis is that the novelist has until then acted only as an 'observer,' not as an 'experimentalist.' The transformation from one to the other becomes his project for the 'experimental novel,' deliberately avoiding novelistic imagination rather than enhancing it, in an extreme overweighting of the opposite pole of constraint within their creative tension:

[54] Claude Bernard (1957), *Introduction to the Study of Experimental Medicine*. New York: Dover Publications.

It is scientific imagination, it is experimental reasoning, which combats one by one the hypotheses of the idealists, and which replaces purely imaginary novels by novels of observation and experiment.

Reading Zola's novels themselves, the detailed study of mining families and communities that undergird *Germinale*, or the agricultural life and its intimate liaison with processes of germination, growth, and death in *La Terre*, is fortunately a less depressing experience than reading his recipe for the 'experimental novel,' though their gritty realism resonates with and illustrates his other project. Here the experimental novel is an antithesis of art, at least in so far as 'science' is opposed to 'philosophy.' Scientists, for Bernard and Zola, tackle answerable questions, philosophers only those 'that are in dispute,' acting like musicians who, 'left to themselves, will sing forever and never discover a single truth.' The experimental novel belongs to neither music nor to philosophy—so working through Zola's proposal is confusing, until the realization dawns that our twenty-first century reading is clouded by our current definitions, and that he is really conceiving of the discipline we now term 'social science.'

Zola extrapolates: Bernard has shown how experimental scientific method may be extended from inanimate isolated systems to living individuals; Zola perceives a further stage, possibly at some remote future, in which groups and societies of those individuals can also be studied using scientific method. This is a highly creative step, a vast visualization of an entirely new field of questions, research communities, and methodologies. The strangest aspect of his conception, to us, is that he anticipates that the *novel* will take pride of place in the set of the new science's methodologies. The following passage from section II of the essay resonates most clearly with our contemporary language:

And this is what constitutes the experimental novel: to possess a knowledge of the mechanism of the phenomena inherent in man, to show the machinery of his intellectual and sensory manifestations, under the influences of heredity and environment, such as physiology shall give them to us, and then finally to exhibit man living in

> *social conditions produced by himself, which he modifies daily, and in the heart of which he himself experiences a continual transformation.*

Here is the chain of thinking, starting from Barnard's physiology, keeping a deterministic philosophy of science in mind, and extrapolating method from physical, through medical, and on to the future social sciences. But when applied to mechanizing the human it has dangerous, even terrifying, consequences. Zola sees further still: there are strong intimations of an application of a social science to social engineering, in his words, 'to make ourselves master of these phenomena.'

The origin of social science is usually attributed to Emile Durkheim and Max Weber, who consciously drew, a generation later, on the 'social physics' of August Comte, but here in Zola we have an alternative creative route into an experimentally based science of human communities. The path taken by twentieth-century university departments of social science would, of course, diverge from his vision. The fictional alternative history in which academia followed Zola rather than Durkheim has yet to be written, but it should give us pause. One of Zola's principal motivations for advancing the tool of his trade as the methodology of a new science is the very necessity of form, of authenticity, of constraint that the novelist is subject to, albeit advocated to an extreme and suffocating extent. The need to conform the novel's creative energy to life as it is observed emerges continually, from James's *Art of the Novel* to the writers of 'fiction' in the Oxford debate between novelist and scientist. But the novel is not an instrument of science, even if it bears close resemblance in some aspects. In Zola, we find the novel advocating, even implementing, a mechanistic model for the human—an accusation usually laid at the door of science, but now more clearly a (deplorable) human choice that sits above them both. *The Experimental Novel* is a dark example of how tangles between the novel and science can drag both down to a mechanical conception of the human, as easily as to a shared celebration of creativity and freedom.

Creativity and constraint in the novel of the twentieth century—The Paris Reviews and a Nobel lecture

We began this journey of the relationship between fictional writing and experimental science in our own era, hearing from Ben Okri and reading Virginia Woolf, before digging at the early modern root systems beneath both these fruitful trees of human creativity. The story since then has brought us back to the twentieth century with a question—did the divergence of literary art and science feared by Humboldt or Wordsworth materialize in quite the way that a 'Two Cultures' narrative would report it? Are these twins destined to wander on different paths, or might the family ties that bind them be brought into greater visibility?

A rich resource exists of personal interviews, dating from the 1950s to 2008, between writers and journalists from *The Paris Review*. Now published as a collection, the volumes of *Paris Review Interviews*[55] offer writers' own accounts of their invention, the connection between their fiction and experience, and their views on the novel and on poetry in our time. These intimate conversations might help resolve what has become of the parallel and the entangled trajectories of the stories of science and the novel.

Robert Penn Warren, author of the political, personal, and theological novel of America's 1930s deep south, *All the King's Men*, responds to a question on the role of the novel in experiment and social science, a century after Zola:

> *I think it's purely accidental. For one writer a big dose of such stuff might be fine, for another it might be poison. I've known a great many people, some of them writers, who think of literature as material that you 'work up.' You didn't 'work up' literature. They point at Zola. But Zola didn't do that. . . You see as much as you can, and the events and books that are interesting to you because you're a human being, not because you're trying to be a writer. Then those things may be of some use to you as a writer later on. . .*

[55] Philip Gourevich (ed.) (2006–2009) *The Paris Interviews Vol. I–IV*. New York: Picador.

> *... Some [authors] say their sole interest is experimentation. Well, I think that you learn all you can and try to use it. I don't know what is meant by the word 'experiment'; you ought to be playing for keeps.*

Warren speaks for a much more pragmatic and intuitive approach to writing, in accord with many of the interviewees' invitations to analyse the writing process. Some are clearly reluctant to do so, like French Nobel Laureate Francois Mauriac:

> *When I begin to write I don't stop and wonder if I am interfering too directly in the story, or if I know too much about my characters, or whether or not I ought to judge them. I write with complete naïveté, spontaneously. I've never had any preconceived notion of what I could or could not do.*

Later in the interview he compares writing a novel to 'taking down dictation.' We have encountered such highly experiential [here the French 'experience' does better than the English 'experimental'] framing of writing already in James. It speaks once more of the deep role of the non-conscious mind in creating a story, writing in a much simpler way than Zola's highly constructed experimental novel. Writing can be a journey into unanticipated places, into which the writer takes the first conscious step, but they thereafter become a receptive observer.

Mauriac's experience of writing might appear a very far cry from 'hypothesis-driven' science—after all, a hypothesis is certainly a 'preconceived notion.' But it resonates with a more common and less formalized way of scientific exploration, one that we have encountered many times in the honest accounts of scientists' creative discoveries. 'Spontaneous naivete' describes a late-modern way of doing science much more faithfully than the constructivism of hypothesis-creation, which is often a tidy reconstruction of events after the fact than an accurate record of the path towards discovery.

A classic, though very rare, example of a scientific story-telling of the highest quality can be found in theoretical physicist,

Richard Feynman's Nobel Prize address.[56] Feynman's brazen truth-telling will serve well as a final parallel example in the spirit of our study of James and Beveridge. Almost unique as a published example of the way science happens, rather than a post hoc rationalization of its results, it reads not unlike a short novel. There are characters and plotlines introduced (the tension between advanced and retarded solutions of Maxwell's wave equation, the notion of an electron moving back in time), great loves (advanced formulations of the theory of motion such as the Lagrangian principle of least action[57]) and great loathings (an interesting personal quirk—Feynman disliked energy-based, or 'Hamiltonian' formulations of motion, as much as he loved the Lagrangian), unexpected twists of plot (the discovery that Feynman's methods could calculate, in an evening, a general result from which an immediate special case had previously taken another leading physicist six months to deduce), and a final denouement in which the original cast of characters fades away to make way for the final revelation, yet in which they all find their echoes. E. M. Forster would nod in agreement, 'When all goes well, the original material soon disappears, and a character who belongs to the book and nowhere else emerges.'[58]

The Nobel address speaks to the family relationship between science and the novel at a higher level than formal narrative,

[56] https://www.nobelprize.org/prizes/physics/1965/feynman/lecture/feynman-lecture.html accessed 15th November 2017; all Nobel Prize addresses and speeches can be accessed on this website.

[57] A lovely personal and technical account can be found in *The Feynman Lectures on Physics* Vol. II, lecture 19. The idea is that for any conceivable trajectory (including unphysical ones) of a particle in a forcefield, it is possible to calculate a quantity, called the 'action,' as an integral along the path of the difference between current potential and kinetic energies (that is the 'Lagrangian'). The physical path, obeying the laws of motion, is the one with the least action, a wonderfully elegant result.

[58] E. M. Forster in *The Paris Reviews* I.

for the author is himself both a character in the plot and its writer. His love-affair with one of its characters (the notion of an electron's self-energy) is a candid account of the emotional energy that keeps scientists at work when all seems hopeless:

> *That was the beginning, and the idea seemed so obvious to me and so elegant that I fell deeply in love with it. And, like falling in love with a woman, it is only possible if you do not know much about her, so you cannot see her faults. The faults will become apparent later, but after the love is strong enough to hold you to her. So, I was held to this theory, in spite of all difficulties, by my youthful enthusiasm.*[59]

The way in which Feynman talks about his holding the theory of electrodynamics up to his mind's sight and turning it every possible way is reminiscent of the long contemplation of subject matter of a novel (such as *Wings of a Dove*) by James, or Beveridge's rueful quotation of Schiller on the illogicality of true scientific discovery:

> *To summarize, when I was done with this, . . . I knew many different ways of formulating classical electrodynamics, with many different mathematical forms. I got to know how to express the subject every which way.*

Late-modern novels eschew the happy ending. Reconciliation there might be, but alongside an accommodation to permanent scars and the ache of loss; a homecoming, but with a nostalgia for exile; a resolution, but with loose ends and rough edges. This is the human story of science as well. Feynman muses that although he found an accurate calculational scheme for the interaction of light and matter, the original thorns in the flesh that motivated his search in the first place, the horrendous appearance of infinite quantities, had not been extracted to his satisfaction. Feynman, an honest scientific narrator, is faithful to the eternal hovering, just over the horizon of the mind, of ignorance—the frayed ends of incomplete understanding, and beyond them the trackless territory of unasked questions.

[59] R. P. Feynman, https://www.nobelprize.org/prizes/physics/1965/feynman/lecture/feynman-lecture.html accessed 15th November 2017.

Awareness of such haunting presence of unknown fields of ignorance, to make out the fleeting outlines of answers in the periphery of mental vision is a reason to think and to write, whether our imagination is directed at nature or towards art. We have met with many reasons to believe that the two entangle more than late-modern disciplinary divisiveness would have it. Perhaps the nineteenth-century Romantics' hopeful road ahead, along which science and poetry might journey together, has not petered out entirely, albeit becoming the track less travelled by. A candidate journeyman is Vladimir Nabokov, the twentieth-century Russian-American writer, who possessed very great skill in the creation of novels, short stories, and natural science. Literary scholar and theologian David Hart writes of Nabokov: 'One might almost say that, for him, there really was no ultimate formal distinction to be made between nature and art: practically everything is, if approached with a sufficiently responsive sensibility, a poetic achievement.'[60]

The Russian emigre to America, who wrote in the languages of both his natural and adopted countries, was also a passionate and consummate lepidopterist. His attraction to complex formal structure must also have contributed to his additional success as a serious chess master. His novels and short stories glitter with detail and presence. He had the happy ability to plunge his readers into immersive experiences by a deft pen-stroke—somehow it is the sketchy allusion to gaunt and rain-swept television aerials against a darkening grey sky that conjures up a virtual reality of cold urban bleakness—one that two pages of detailed description could never match. His writing will serve to sum up our journey through and between experimental science and the novel, not only because both co-existed in his mind, but also because he hints, as Feynman does, at horizons of thought that words are unable to bridge. Hart again:

[60] David Bentley Hart (2017), in *A Splendid Wickedness and Other Essays*. Grand Rapids: Eerdmans.

> *[T]here are ghosts hovering just beyond the margins of the page. They are not the phantoms of horror fiction, and they rarely if ever intrude too forcibly upon the action of the tale. Rather, they are subtle beings, possessed of enchanting powers, quiet presences mysteriously and winsomely insinuating themselves into the narrative, and even on occasion into the very form of the text.*

One of the perpetual ghosts is a wordless presence of purpose; writing in the period of which George Steiner can complain of a 'broken contract' between literature and the world,[61] Nabokov chooses enigma, transcendence, and rumour of telos over nihilism or the absurd. His loving celebration of detail and puzzle-making, even in the whirl of insanity and triple-bluff in his extraordinary novel *Pale Fire*, hints at redemption beyond tragedy. He invites us to question and to re-create models of his worlds in the mental images of our own.

Although Nabokov protested any connection between his passionate scientific work on butterflies and his writing, in his (characteristically written, not oral) response to his *Paris Review* interview and elsewhere it is hard to believe so capacious a mind to be as divided as its owner claimed. Even at the methodological level there are signs of flow and counter-flow. He wrote segments of his stories on stacks of filing cards, then arranged and rearranged them until the emergent pattern satisfied. One is reminded of Mendeleev's laying out and rearranging cards marked with the known elements and their properties until the structure of the periodic table clarified. Nabokov's novels and science share symbols too; the gorgeous butterflies of his lepidoptery alight on his literary pages at singular moments, hinting at unspoken presences. At a deeper level, the way he approached his zoological work was profoundly, and anachronistically, phenomenological. Intensely suspicious of the encroaching hegemony of genetics, he always believed that the phenotypes of morphology and behaviour had more to tell us than the sterile skeleton of a genotype. And this not without success-working as curator in the

[61] George Steiner (1989), *Real Presences*. London: Faber and Faber.

Museum of Comparative Zoology at Harvard in the early 1940s, and after painstaking morphological comparison of a group of *Lycaenidae* butterflies called *Polyommatus* blues, he made a radical suggestion that their radiation into the New World took place in five waves of migration across the Bering Straits, between one and eleven million years ago. A genetic analysis by a Harvard paleoecology research group, published in 2011, vindicated Nabokov's hypothesis in detail.[62] His extraordinary eye for structure and patterns, and his wisdom to see that the world is connected and emergent, rather than fragmented and reductionist, gives both his writing and his zoology the holistic depth at which a twentieth-century Humboldt might well have rejoiced.

Scientific discovery and the novel

From the first examples of the mutual orbit of science and the writing of novels, the degree of their entanglement has surprised. Close readings of James's and Beveridge's accounts of the processes experienced by their creators add to historical evidence that the imaginative labours required in both bear not simply general, but detailed comparison. The psychological experiences of conception of a goal, desire to meet it, frustration at failure, incubation of ideas and material, illumination from non-conscious thought, and the dual roles of imagination and constraint in final creation—all these are shared experiences.

There are reasons for such commonality that take us deeper than the methodology of creativity, for novel writing and science are both activities driven by the experience of immersion; they are what humans do in the face of experiencing the world. Both of the great ancient philosophers identified 'wonder' (*thaumazein*) as the beginning of philosophy.[63] I would add, as the beginning

[62] Roger Vila, et al. (2011), 'Phylogeny and palaeoecology of Polyommatus blue butterflies show Beringia was a climate-regulated gateway to the New World', *Proceedings of the Royal Society B, 278*, 2737–44.

[63] Plato (as Socrates) at 155d of the *Theaetetus*; Aristotle, *Metaphysics* 982b.

of writing and of science too—we are meaning-seeking animals immersed in a world of the aleatory and contingent as well as the wonderful and the sublime. Part of our desire to make sense of the world seems to find an outlet in its recreation, or at least in the creation of models of it. An experiment becomes at the same time a window into the world and a local habitation for it. In his trenchant study of the novel's purpose, *The Sense of an Ending*,[64] Frank Kermode insists that the novel has (usually hidden) philosophical force, driven by the tension between the human situation and the contingency of the world. Jean-Paul Sartre's formula is an inclusive one: 'the final aim of art is to reclaim the world by revealing it as it is, but as if it had its source in human liberty.'[65]

We could say the same of science too, understanding that the practice of science in its re-creation of the world is a high expression of human liberty, yet constrained by the observed world. The title of Kermode's theory of fiction awakes another aspect of making sense of the world, namely that it is itself a story with a beginning, middle, and end. The *telos*, the end—in both senses of finality and purpose—run through novel- and science-making.

But in our own age, there seems to be no guarantee that a sense of purpose is not delusion; there nags perpetually, at writer and scientist alike, the suspicion of make-believe, of wish-fulfilment and infidelity. 'How does the novelistic differ from existential fiction?' asks Kermode—we might paraphrase, 'How do experiments differ from the connected complexity of reality?' The theological underpinning of hope, that these reconciliatory tasks between meaning-seeking are not without foundation, was strong enough for the age of Newton and Boyle to develop the experimental project, but in our late-modern era its loss leaves talk of purpose hard to formulate. One consequence of the late-modern cutting adrift of deconstruction, of the separation between word

[64] Frank Kermode, *The Sense of an Ending*.

[65] Quoted by Kermode p.145.

and referent—George Steiner called it the 'Broken Contract'[66]—is the loss of perception that writing and science share anything at all, and a *telos* in particular.

The turning point of the 'story of stories' we have told in this chapter was the Romantic era of Wordsworth, Humboldt, Emerson, Davy, and Coleridge. It was simultaneously the high-water mark of writing that could claim to be at once scientific and poetic, of explicit mutual exchange between natural philosophers and writers, yet also the vantage point from which its exponents saw a parting of the ways as the nineteenth century wore on. The French, after all, still call novels *romans*. The notion that we can 'read' the world (Hugh of St Victor, Francis Bacon, Galileo, and Robert Boyle to name four for whom this metaphor was self-evident) might trace the move from medieval to enlightenment worlds, but the idea that we can *write* it into being is solidly Romantic.

Romanticism might risk the engendering of power and lovelessness, as Mary Shelley warned so terrifyingly, but it also urges an unrestrained submission to the sublime and the wondrous. Facing—janus-like, both inwardly and outwardly—wonder can initiate a divorce between poetry and science, but celebrate their marriage as well.[67] Emerson could write of 're-enchanting nature,' while Keats, the trained doctor turned poet, was despairing that the rainbow lay all unwoven at the touch of 'cold philosophy'. More significantly, recognizing the Romantic can awaken a reader to the role of emotion, affect, and aesthetic in science as much as in art. No parting of cultural ways can deny the emotional response felt at a newly won grasp of the world, of the forging of connections between the local and the universal. After Darwin and the growth of evolutionary biology we know our connection to every other living thing as the vast conception of

[66] George Steiner, *Real Presences*.

[67] Emily Dumler-Winckler, *A Holy Matrimony*.

the tree of life. After Hubble and the cosmology of birth and expansion, the astrophysical narratives of the birth, life, and death of stars, we know that our own life-cycle and atomic makeup participates in theirs.

Returning to Virginia Woolf's *To the Lighthouse*, there is a moment of insight in its eighth chapter that seems to capture the story of writing and science we have told. Lily the artist is putting away her brushes when she notices the older family friend, botanist William Bankes, gazing on the domestic scene of Mrs Ramsay reading to her young son in a window bay:

> *It was love, she thought, pretending to move her canvas, distilled and filtered; love that never attempted to clutch its object; but, like the love which mathematicians bear their symbols, or poets their phrases, was meant to be spread over the world and become part of human gain. So it was indeed. The world by all means should have shared it, could Mr Bankes have said why the woman pleased him so; why the sight of her reading a fairy tale to the boy had upon him precisely the same effect as the solution of a scientific problem, so that he rested in contemplation of it, and felt, as he felt when he had proved something about the digestive system of plants, that barbarity was tamed, the reign of chaos subdued.*

Bankes, the botanist, found no words, but Woolf the novelist did. Calling to mind comprehension of the intimate details of the local, she makes the narrative leap that all writing and all science must do, the inductive exercise of faith that there are connections between the local and the global that we must discover and repair. Writing that, is to write our story too.

But writing is more than stories, as science is more than experiment. To complete an account of the textual mode of creativity in art and science, this chapter's peripheral appearances of poetry must be brought into the centre of attention as complementary to prose, as theoretical science is similarly complementary to experimental. Their equally rich relationship is the focus of the next chapter.

Poetry and Theoretical Science

For a scientist must indeed be freely imaginative and yet skeptical, creative and yet a critic. There is a sense in which he must be free, but another in which his thought must be very preceisely regimented; there is poetry in science, but also a lot of bookkeeping.

PETER MEDAWAR[1]

The surprising kinship between novelistic writing and experimental science, explored in the last chapter, encourages the search for family resemblance between other modes of writing and science. Similarly obscured by modern disciplinary boundary-drawing, they will also require a little cultural archaeology to reveal their historical, formal, and creative entanglements. The literary form suggested by this book's title page is poetry. Admittedly, our initial encounter of poetry with science in the introduction, in the form of Wordsworth's preface to the *Lyrical Ballads*, did not suggest optimism. The stipulation that the discoveries of science would one day 'inspire the poet's art,' but only should they 'become familiar to us as enjoying and suffering beings,' appeared to have been unfulfilled in Wordsworth's time. Yet by the end of that introductory survey, a structural similarity had already appeared to bring some processes of poetry and science into alignment: the creative force of imagination as it meets the constraints of form.

If poetry is the creative meeting of imaginative energies and ideas with the shaping constraint of form, then what could act

[1] *Peter Medawar (1996) The Strange Case of the Spotted Mice and Other Classic Essays on Science* Oxford: Oxford University Press, p. 63.

The Poetry and Music of Science. Tom McLeish, Oxford University Press.
 DOI: 10.1093/oso/9780192845375.003.0005

as a better metaphor for the scientific imperative to describe the world in layered and connected detail as well as grand design? In other words, what could call upon greater imaginative energies than the invitation to reimagine the universe, and what could constitute a tighter constraint of form than to conform that imagination to the universe we observe? Science becomes a metaphor for poetry, or perhaps for a single, polyvocal poem, while poetry patterns science for the same reason. The shared pattern of constrained imagination within a creative process shared by science and the arts suggests more precisely the specific scientific partner to parallel poetry in particular. If *experimental* science communes with the observed and open creativity of the novel, then the formal constraints and imaginative power of poetry point to science's partner to experiment in the notion of *theory*.

Two hidden levels of science and poetry

We have also already met with the notion of *theoria* in Thomas Browne's classical 'true theory of death' on contemplating a skull.[2] A first glance seems to perceive a yawning conceptual chasm between such poetic, priestly, imaginative writing, and current scientific usage of 'theory,' as in the 'theory of evolution,' 'quantum theory,' or 'theoretical polymer physics,' especially when these adopt mathematical forms. Yet the structural similarities outweigh the apparently contrasting content.

First, all theory is representation. Originally the description of ancient priestly rights, *theoria* later becomes the studied mediation exemplified by Browne's skull, with its powerful signifiers of emptiness, sterility, void, and lost life, beyond the object of gaze. Later still 'theory' conceives immense schemes of interrelation and explanation within the material world, such as the 'tree

[2] Chapter 1: Introduction.

of life' that Darwin sketched for his *Origin of Species*—itself a representation in abstract and graphical form of the vast temporal branches of living species on Earth; the idea crystallized in his mind over decades.

Second, the long history over which any great, framing scientific concept takes shape, such as Darwin's grand idea in biology or the expanding universe of cosmology, illuminates another word that Browne selects, and a second common thread linking poetic and scientific theory—'contemplation.' All theory emerges from slow reflection, the mental examination of an object from different perspectives, and the experience of a slowly cohering representation.

Finally, the dynamic of poetry—*poiesis*, as of theory – is directed from the inner imagination outwards towards the world. For Plato, *poiesis* described the creative act that brought nature into being;[3] for Aristotle, as we saw in Chapter 2, it referred to the human reimagining of nature. This is the sense in which theory is the reciprocal partner of observation, creating and expressing by representing the world from within, rather than receiving it from without.

Such a dynamic of expressive, even rhetorical, re-creation of the world appears repeatedly in the reflections of poets themselves. Philip Larkin could write of poetry, 'as a guiding principle, I believe that every poem must be its own freshly-created universe,'[4] and T. S. Eliot frequently mused over the way that poetry moves from particular instances to general truths, writing in his *Poetry and Drama:*[5]

> *It is ultimately the function of art, in imposing a credible order upon ordinary reality, and thereby eliciting some perception of an order in reality, to bring us to a*

[3] See Diotema's proposal in Plato's *Symposium. The Internet Classics Archive: Symposium by Plato,* trans. Benjamin Jowett.

[4] Quoted in Andrew Motion (1993) *Philip Larkin: A Writer's Life*. London: Faber p. 273.

[5] T. S. Eliot (1951), *Poetry and Drama*. Scholar's Select.

condition of serenity, stillness and reconciliation; and then leave us, as Virgil left Dante, to proceed toward a region where that guide can avail us no farther.

The inductive and extrapolative 'procession to further regions' is the function of theory in science. An experiment can illuminate and report on a specific set of events, but, on its own, this is not sufficient to reimagine a cosmos. Scholar David Withun writes of Eliot's discussion of the purpose of poetry:[6]

[It] is able to grant the reader the ability to perceive that reality, in spite of its often chaotic and random appearance, has some underlying unity by which it is bound together. This insight, in turn, provides the terms by which one may make peace with the world.

This echoes once again George Steiner's 'rendering into some measure of communicability the inhuman otherness of matter.' There seem, therefore, to be two levels on which poetry and science in its theoretical mode may entangle, just as between fictional literature and experimental science. Not only is there the possibility of mutual inspiration—of scientific discoveries providing material for creative writing, and of imaginative writing inspiring scientific imagination—but, at a deeper level, there are structural patterns of correspondence to be explored, including a recurrent thread of purpose: to make peace between the human mind and the world. It will be helpful to explore both of these levels, as in the previous chapter, but before doing so, we need to sound out the depth at which they have become buried, in our own times. As in the case of fictional writing, this will require some history.

Two lost cousins

Wordsworth gave a reason for the current disjuncture he perceived between science and poetry: essentially a failure to communicate on the part of the new science. To recall, 'The remotest

[6] In his essay *Making Peace with the World: T. S. Eliot & the Purpose of Poetry*: https://theimaginativeconservative.org/2017/08/poetry-ts-eliot-david-withun.html

discoveries of the Chemist, the Botanist or Mineralogist, will be as proper objects to the Poet's art as any upon which it can be employed,' thus runs the preface to the third edition of the *Lyrical Ballads*—but this vision is conditional—'if the time should ever come when these things should be familiar to us . . . as enjoying and suffering beings.' The rub is that science, together with its treasure chest of metaphor and imagination, had even then become sequestered within the locked safe of its institutional structures: 'Science,' they continue, 'is a personal and individual acquisition, slow to come to us, and by no means habitual and direct sympathy connecting us with our fellow-beings . . . The Poet, in contrast, singing a song in which all humans beings join, rejoices in the presence of truth as our visible friend and hourly companion.' Nearly two centuries later, biologist and science writer Peter Medawar echoes the same analysis: 'no-one questions the inspirational character of musical or poetic invention because . . . something travels; scientific discovery is a private event and the delight that accompanies it does not travel.'[7]

Wordsworth was not alone in identifying a hidden, and contingent, fraternity between poetry and science, while admitting that the two had parted company for a while. An almost contemporary analysis from Goethe in his 1827 *On Morphology*[8] likewise envisions a time when such 'forgotten' connections might be re-established:

> *Nowhere would anyone grant that science and poetry can be united. They forgot that science arose from poetry, and failed to see that a change of times might beneficently reunite the two as friends, at a higher level and to mutual advantage.*

While the English Romantics point to a future hope for a remarriage of science and poetry, Goethe insists on their common

[7] Peter Medawar (1984), *Pluto's Republic*. Oxford: OUP.

[8] Johann Wolfgang von Goethe (1827) *On Morphology*, in Hamburger Ausgabe: Werke Hamburger Ausgabe in 14 Bänden, edited by Erich Trunz (Hamburg: Chr. Wegner, 1948–60; Reprinted, C. H. Beck, 1981). Translation in Christine Lehleiter (2016), Fact and Fiction: Literary and Scientific Cultures in Germany and Britain. Toronto: University of Toronto Press.

ancestry. His notion that their felt separation is a form of *forgetfulness* is fascinating. It patterns that similar cultural forgetfulness which obscured a cousinly relationship between literature and experimental science. Such agreement between early nineteenth century poets in German and English traditions might suggest that, if they saw poetry as the eventual healing of division between science and art, or in more historically faithful language, between reason and imagination, then the falling-out might have begun with poetry in the first place. The introduction noted that painful tradition of poets, from Blake who sees science as 'reasonings like vast serpents,' in opposition to the creative human spirit, to Keats's 'unweaving the rainbow,' and to Poe's 'vulture whose wings are dull realities.' This amnesia of the imagination not only shrouded the powerful kinship of imaginative creation of alternative worlds and the creation of experimental abstractions from ours, but also made science appear, to these poets, as a force of disenchantment, as replacing colour with greyness, and as preying on wonder.

In her powerful book, *Science and Poetry*, philosopher Mary Midgley responds to these opposed Romantic voices of Wordsworth's hope and Keats' despair. In a carefully argued invective against the divorce of subject and object, and against the flattening effect of modern metaphysical reductionism, she describes what she sees when peering into Goethe's future. 'Our visions,' she writes, '– our ways of imaging the world—determine the direction of our thoughts, as well as being the source of our poetry.' She interprets Shelley's claim that poets 'are the unacknowledged legislators of the world' to mean that they are the modern-day *prophets*, not through any power of foretelling, but (and in a more biblical sense) through the forcefulness of their visions. If that is true, then the shrieks of Blake, Keats, and Poe become prophetic words, not against science as such, but against a modern framing of science that had already all but forgotten its imaginative and metaphorical kinship with poetry. The persistent truth is that metaphor, concept, reasoning, formulation, imaginative energy—these needfully pervade all of science.

Swiss scholar of English literature, Irmtraud Huber, is sure that the 'Two Cultures' divide did indeed start with the Romantic falling out of poetry and science:[9]

> *The roots of the idea of the two cultures lie in the much more narrow juxtaposition of science and poetry, rather than artistic culture more generally. There was no conception of an antagonism between science vs. the visual arts or music, for example; and the early novel, with its emphasis on realism, understood itself very much in dialogue with and even in extension to science, rather than in opposition to it.*

She points to evidence in what appears at first sight to be a candidate for a nineteenth century forerunner of this book, namely Robert Hunt's, *The Poetry of Science.*[10] Hunt, however, would not be sympathetic to a view that the creative processes of science and poetry are in any way comparable; on the contrary, he takes them to be antithetical:

> *The fumes of the laboratory, its alkalics and acids, the mechanical appliances of the observatory, its specula and its lenses, do not appear fitted for a place in the painted bowers of the Muses.*

Rather, science is comparable to poetry because it shares the capacity to evoke beauty, wonder, and the sublime, through its discoveries. Like Wordsworth, Hunt is careful to point out that this power is conditional, but unlike the poet, this is not the social condition of a shared familiarity with the objects of science, but a qualitative requirement on its form and results. The aesthetic power of science is latent, for Hunt, in its induction of connective laws, rather than the prior material of listed facts. William Whewell wrote in the same period that scientific facts on their own are like individual 'pearls on a string.'[11] On their

[9] Irmtraud Huber (2020), 'Another *Poetry of Science*: Tom McLeish (2019) in comparison with Robert Hunt (1848), *Interdisciplinary Science Reviews*, **45**(1), 23–29.

[10] Hunt, Robert (1850 [1848]), *The Poetry of Science: Or the Studies of the Physical Phenomena of Nature*. Boston: Gould, Kendall, and Lincoln.

[11] Quoted in Huber (2020).

own they might be pretty, but they lead to nothing beyond. Yet through connection, an emergent object of beauty appears. Whewell describes this process of connection through their underlying patterns as the 'imposition of formal unity.' Science is never 'read off' nature, but 'written into' a picture, a narrative, or an abstract constructed representation of it. Hunt takes this extrapolation further still:

> *In these studies of the effects which are continually presenting themselves to the observing eye, and of the phenomena of causes, as far as they are revealed by Science in its search of the physical earth, it will be shown that beneath the beautiful vesture of the external world there exists, like its quickening soul, a pervading power, assuming the most varied aspects, giving the whole its life and loveliness, and linking every portion of this material mass in a common bond with some great universal principle beyond our knowledge. . . . But if admitted even to a clear perception of the theoretical Power which we regard as regulating the known forces, we must still see an unknown agency beyond us, which can only be referred to the Creator's will.*[12]

Here lies the clue to the separation of ways, for Hunt begins to privilege science, not only with the power to discern the 'phenomena of causes,' but also to reveal the Creator's power. As Huber puts it, 'In Hunt's view science can inspire awe and wonder and humility in the face of the material world even better than poetry. "The phenomena of Reality are more startling than the phantoms of the Ideal," Hunt claims; "Truth is stranger than fiction".'[13]

Revisiting the question of poetry and science, at the heart of the Romantic nineteenth century, will confirm the suspicion raised in the introduction, that the 'conflict narrative' (between science and religion) and the 'Two Cultures' divide are indeed related. The prioritizing of science over poetry that Hunt articulates succeeds in clouding sight of both levels at which they are originally and fundamentally connected. For by replacing poetic vision by

[12] Robert Hunt (1850 [1848]).

[13] Irmtraud Huber (2020).

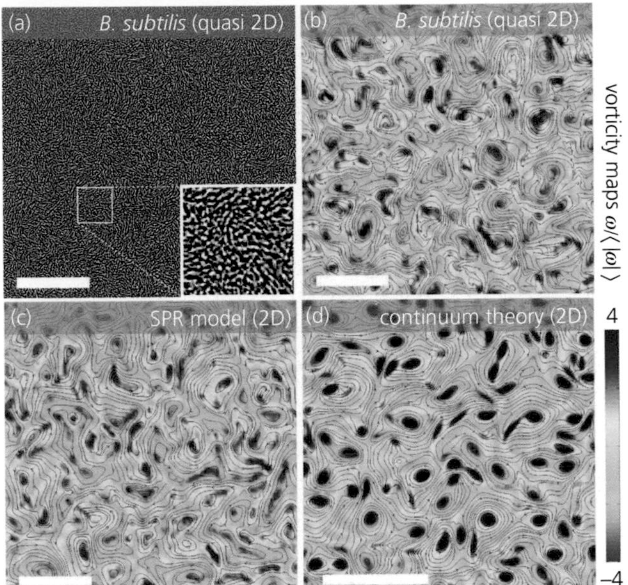

Colour Plate A. Experiments (A—microscopy and B—a plot of velocity contours and vorticity strength in colour scale) and models (C—a mesoscopic model and D—a continuum theory) of a turbulent quasi-2D flow of dense swimming bacteria (*B. subtilis*). Reproduced with permission from Wensink, H. H. *et al.* (2012), 'Mesoscale turbulence in living fluids', *Proceedings of the National Academy of Sciences*, 109, 14308–14313

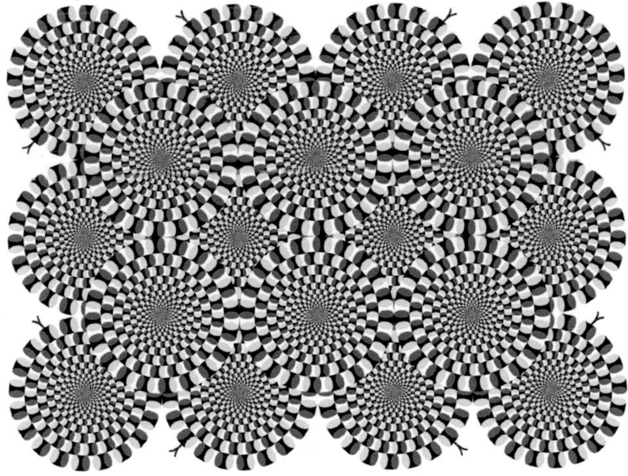

Colour Plate B. Example of the Ashihda and Kitaoka's 'Rotating Snakes Illusion'. Letting one's eyes wander over the image creates for most viewers an illusion of rotational motion in the circles. Reproduced with kind permission of Akiyoshi Kitaoka

(a)

(b)

Colour Plate C. (A) 'Adoration of the Magi', 1303–05 (fresco), Giotto di Bondone (*c.* 1266–1337)/Scrovegni (Arena) Chapel, Padua, Italy/De Agostini Picture Library/A. Dagli Orti/Bridgeman Images. (B) 'Last Judgement', 1303–05 (fresco), Giotto di Bondone (*c.* 1266–1337)/Scrovegni (Arena) Chapel, Padua, Italy/Mondadori Portfolio/Archivio Antonio Quattrone/Bridgeman Images

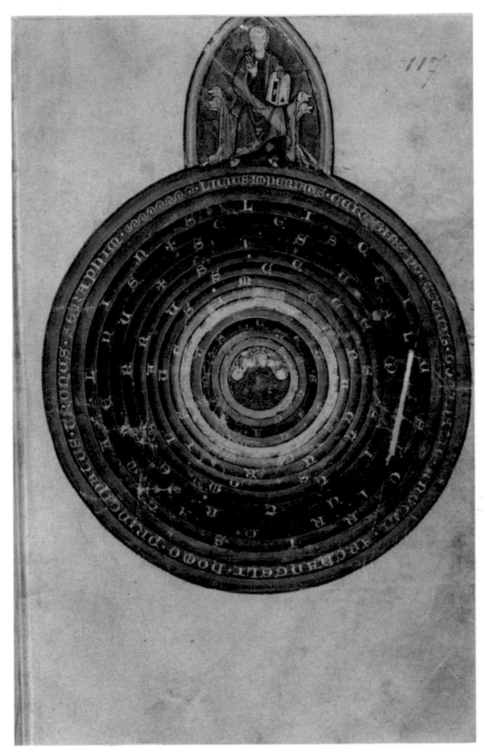

Colour Plate D. 'L'image Du Monde' by Gossuin de Metz, With permission, Bibliothèque Nationale de France

Colour Plate E. Claude Monet; 'Under the Pine Trees at the End of Day' (1888), Philadelphia Museum of Art, 125th Anniversary Acquisition. Gift of F. Otto Haas and partial gift of the reserved life interest of Carole Haas Gravagno, 1993-151-1 with permission of the Philadelphia Museum of Art

Colour Plate F. Graeme Willson (1998); 'Quartet', Private Collection, with permission, Photograph courtesy of Brian Slater

Colour Plate G. 'Robert Boyle' (1627–91), by German painter Johann Kerseboom (d. 1708), with permission of The Royal Society

Colour Plate H. Pablo Picasso (1937); 'Guernica' with permission, Reina Sofía Museum Collection, Madrid

Colour Plate I. The ivory 'Lion Man', Schwabia, *c.* 40 000 BCE, with permission of the Museum of Ulm

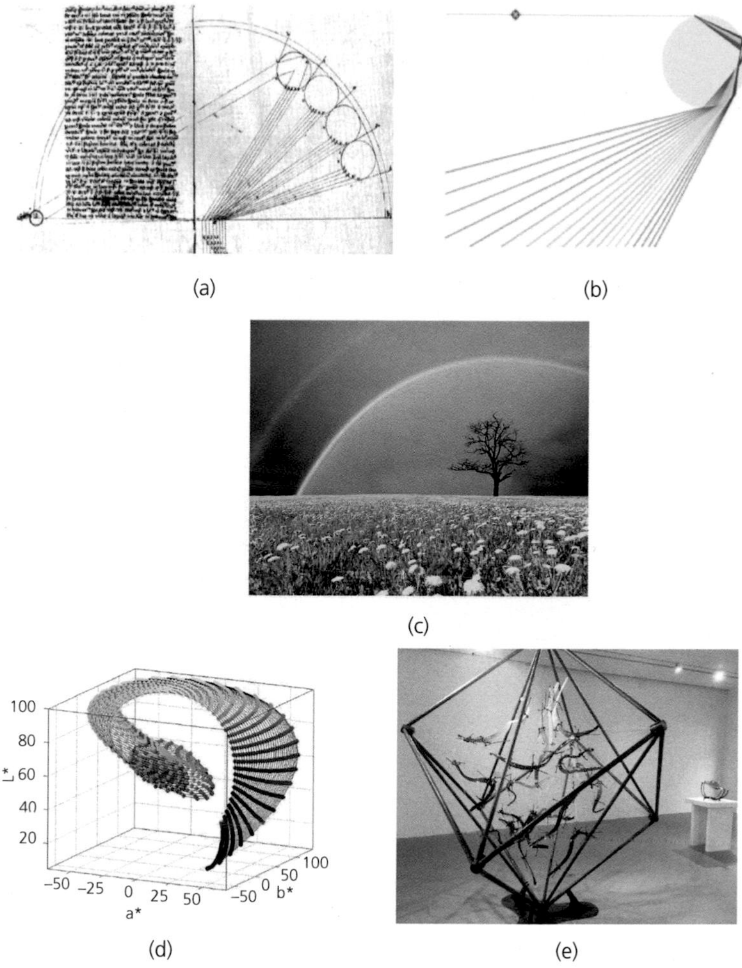

Colour Plate J. (a) Theodoric's diagram of the formation of the rainbow through double-refraction and reflection of sunlight in raindrops towards an observer; MS Basel F. IV. 30, ff. 33v-34r. (b) a modern illustration of the dispersion in angles of refracted and reflected light within a raindrop. (c) A primary and secondary rainbow *https://www.pinterest.com.au/pin/232287293250306356/?autologin=true*. (d) and (e) Art installation by Colin Rennie, Sunderland Art Museum 'Illuminating Colour'. (b) and (d) reproduced with permission of Prof. Hannah Smithson the *objects* of science, attention is drawn away both from the alignment of their creative imaginative *process*, and from the essentially relational work of both art and science. By the same fatal reorientation, the potential for science as a source of poetry is stifled. For Hunt, science is its own poetry.

Though fully manifest by the mid nineteenth century, the seeds of separation between the scientific and poetic imagination were sown in the early modern era—the same that saw the parting of ways between experimental science and fictional writing. Long before Hunt's demotion of poetry, and Keats and Poe's disavowal of science, the institutionalization and professionalization of the sciences was already enshrining the partition through limitation on language. Thomas Sprat, writing what was essentially the manifesto for the young Royal Society in 1667 (under the guise of a 'history')[14] urged his readers to:

> *separate the knowledge of Nature from the colours of Rhetoric, the devices of Fancy or the delightful deceits of Fables.*

From the origins of the scientific institutions, words associated with poetry on the one hand, and with natural science on the other, were carefully disentangled. The divorce of the ideas and actions that they connoted would follow, and their connections buried at such a depth that it would take the acuity of Goethe, Wordsworth, or Lovelace to perceive them two centuries later.

Early modern science-poetry

If it is true that cousinly connections between the creative imaginations of theoretical science and of poetry have been forgotten, then it ought to be possible to find early examples in which their relationship is not obscured. As good a place as any to start is with the remarkable Margaret Cavendish we met in the previous chapter—the same natural philosopher who wrote an

[14] Thomas Sprat (1667), *History of the Royal Society of London, for the Improving of Natural Knowledge*. London: The Royal Society.

early science-fiction novel embedding her subtle critique of experimental method also explored atomic theory through the medium of poetry. More than forty poems, published in 1653, detailed in imaginative form the supposed properties of the microscopic 'atoms' that underlie the properties of the materials they constituted.[15] The poems' images draw from contemporary atomistic philosophy, by which atomic shape at the smallest scale is responsible for emergent hardness, softness, or fluidity. The poems themselves range in length and employ varying forms of rhyming pentameter. A delightful example is *Of Aiery Atomes*:

THE Atomes long, which streaming Aire makes,
Are hollow, from which Forme Aire softnesse takes.
This makes that Aire, and water neer agree,
Because in hollownesse alike they be.
For Aiery Atomes made are like a Pipe,
And watry Atomes, Round, and Cimball like.
Although the one is Long, the other Round;
Yet in the midst, a hollownesse is found.
This makes us thinke, water turnes into Aire,
And Aire often runs into water faire.
And like two Twins, mistaken they are oft;
Because their hollownesse makes them both soft.

The poem employs its own couplet form to link the spatial levels of atomic structure ('hollowness,' 'roundness') and emergent fluid property ('softness,' fluidity—'streaming'). The poem invites reflection upon its own imaginative task—'this makes us thinke,' which is to account for both the similarities and the differences between water and air, as well as to understand how they appear to interconvert. Such observation that 'Aire often

[15] Margaret Cavendish (1953), *A World Made by Atomes*. Electronic edition available at Emory Women Writers' Project. Atlanta: Emory University. http://library2.utm.utoronto.ca/poemsandfancies/2017/06/09/a-world-made-by-atoms/.

runs into water faire,' is, of course, the condensation of water vapour or steam rather than any sort of 'transmutation' of water and air in current scientific terms. Air (all gases and vapours) and water, share the properties of *fluidity* and, compared to solids, low density—it is this combination that Cavendish terms 'softnesse,' and seeks for an explanation in atomic structure. She alights upon the idea of *hollowness*, providing lightness in a natural way, but expressed in different atomic shapes. We would describe her airy atoms as prolate, the watery as oblate. Implied but unstated is that the evaporation of water into vapour corresponds to an ingenious atomic shape-shifting: the cymbal shapes of water atoms must elongate along their axis of symmetry and retract in perpendicular dimensions to adopt the long cylindrical forms of airy atoms.

Cavendish's sources here are the ancient Greek world's notion of atoms, ascribed to the fifth century BCE school of Leucippus and Democritus, transmitted largely through the later, but surviving, work of Lucretius *De Rerum Naturae* (On the Nature of Things), and references to his contemporary Epicurus, read throughout the middle ages and Renaissance, although her elemental categories reflect a Platonic scheme from the *Timaeus*. These ancient philosophers had also envisaged materials composed of atoms possessing different shapes, and noticed the explanatory potential of such an atomic theory. The overarching challenge was to explain the phenomenon of change within constraints. More detailed aspects of material behaviour might also permit atomic explanation—the more 'sluggish' flow of oil when compared to water, for example, might arise from atoms that hook onto each other and cling. There is no evidence, however, suggesting that any of the ancients imagined hollow atoms. Margaret Cavendish is improvising on, and extending, the classical theme here, letting the poetry drive imaginative thought towards an account of 'softness.' It is worth remarking that, although this line of thought has little to do with our current theory of atoms, it does have a legacy at the higher length-scale pertinent to large molecules of complex shape, especially within the molecular-level theory of 'liquid crystals.'[16] To take one example of a result from this level of the 'soft matter science' that we met in the context of polymers in Chapter 3, Cavendish correctly anticipates that randomly arranged rods really do require more volume, and so produce a lower density than randomly packed disks (think of random arrangements of straws, as opposed to coins, in a jar). The other atom poems in the collection move the reader into stranger territory—inwards towards the brain, its thoughts and emotions, and the diseases of the body, and outwards to the hidden natures of the sun and planets.

Through poetry, Cavendish sees, and allows her readers to see, cosmical connections between the human and material world, traced by the universal presence and motion of atoms. Albeit in painterly and metaphorical terms, she comments on why she would choose to couch some scientific thoughts in poetry rather than prose, 'some being done with Oily-colours of Poetry, others with Water-colours of Prose.'[17] Her later works in natural philosophy continue to make, at least occasional, use of poetry as well as prose, in a much more fluid way than does any writer today, but which brings to the surface the distinctive roles each form enjoys in the articulation of creative thought. A watercolour painter has to be very careful—there is no way to hide an over-bold brushstroke. But in the same way that oil-painting can be bolder, so poetry for Margaret Cavendish was the medium in which more speculative ideas could more freely develop.

Cavendish's suggestion of human connection with atoms is all the more remarkable in the context of the seventeenth century, when atoms were more usually identified with the non-human and inhuman. In *Paradise Lost*, Milton diminishes the worth of the Earth in relation to cosmic dimensions by 'atomizing it':[18]

[16] See e.g. Tom McLeish (2020) *Soft Matter – A Very Short Introduction*. Oxford: Oxford University Press.

[17] Margaret Cavendish (1671), *Nature's picture drawn by fancies pencil*, sig. b1r.

[18] John Milton, *Paradise Lost* 8: 87–9.

this Earth a spot, a grain,
An Atom, with the Firmament compar'd
And all her numbered Starrs, that seem to rowl
Spaces incomprehensible

And John Donne sees in his own prophetic poetry the world (or is it our picture of the world?) crumbling into disorder of atoms under the incoherence of that century's 'new philosophy':[19]

new Philosophy cals all in doubt,
The Element of fire is quite put out,
. . .
And freely men confesse, that this world's spent,
When in the Planets, and the Firmament
They seeke so many new; they see that this
Is crumbled out againe to'his Atomis.
'Tis all in pieces, all cohaerence gone.

This brief glimpse into early modern poetic writing on scientific themes, from the period before the strictures of Thomas Sprat's divisive decrees took root, does much more than illustrate how a scientific idea, such as the atom, found a natural place in its poetry. It also exemplifies the co-creative play by which poetic and scientific imagination may inform and inspire each other. Are atoms inevitably the signifiers and embodiments of the world's dissolution and chaos, as Milton and Donne suggest? Or might they undergird the emergence of life, thought, and personhood, as Cavendish's atom poetry envisages? When working at the very limits of scientific understanding, it is not only knowledge that is typically missing, but even the metaphors and categories by which those limits might be extended. As literary scholar, Kevin Killeen, puts it, 'Natural philosophy needed the poetic because there was no other way in which it might address the intangible, paradoxical and frustrating resistance to definition.'[20]

[19] John Donne, *First Anniversarie. An Anatomy of the World.* Lines 205–13.

[20] Kevin Killeen (2021), 'Poetry and Natural Philosophy: The Errant Soul in John Davies, John Donne and Phineas Fletcher' in *Oxford Handbook of Renaissance Poetry*, Jason Scott-Warren and Andrew Zurcher (eds), Oxford: OUP.

The example of atomic theory in the seventeenth century illustrates some of the ways in which poetry and the inner vision of scientific theory may be related. Other examples abound—the equally unseen but unimaginably vast reach of Newtonian gravity also involved poetry in its discursive history—Newton's brilliant friend, poet, and scholar, Elizabeth Tollet, is the prime example.[21] The subsequent insights of Goethe and Wordsworth that there is more to 'poetry and science' than a parting of ways raises one of the important questions of this book as a whole—the role of the *imagination* in the pursuit of scientific knowledge. They also confirm that exploring the often-buried connections between science and poetry might prove fruitful in discovering how the exercise of imagination plays out within theoretical science, in a similar way to the mutual understanding generated by comparing the way that both novelistic writing and experimental science set up and explore 'small worlds' of observation.

Science, poetry, and the imagination

The renunciation of imagination as a route to knowledge, complementary to that of reason, is perhaps the singular most characteristic shift from medieval and renaissance natural philosophy to early modern science. It is captured succinctly in Thomas Sprat's rejection of 'rhetoric,' 'fancy,' and 'fables' in his *History of the Royal Society*, and in John Locke's slightly later *Essays Concerning Human Understanding*, which we encountered in the introduction. Yet Sprat and Locke clearly wrote into a culture within which poetry and science were no strangers. This easy traffic between the arts and sciences of the 'textual creative mode,' in poetry as well as prose, is reminiscent of a historical thread within similar comparisons of the visual creative mode. Examined in Chapter 3,

[21] Tollet and her poetry is discussed in Patricia Fara (2021), *Life After Gravity*. Oxford: Oxford University Press pp. 28–32.

that modern story drew heavily from ideas forged in the explosive medieval Renaissance of learning in the thirteenth century. Early modern developments might pretend to have made a clean departure from their scholastic medieval inheritance, but transitions in the history of ideas are never quite so clean. In the realm of 'imagination' through the textual mode as well, it pays to look into that earlier age of remarkable reinvigoration of sense, natural knowledge, imagination, memory, and understanding.

The creative intellectual world of the thirteenth century Latin West was, as we have seen, invigorated by newly translated science and philosophy from both Ancient Greece and early medieval Islamic commentary on Aristotle and other ancient thinkers. We encountered Robert Grosseteste's description of mental vision (Latin *sollertia*) descending below the apparent surface of materials, as well as Giotto's meaningfully created illusions of just such surfaces. Yet the notion of 'imagination' (Latin *imaginatio*) at that time had a rather different meaning in medieval usage to that of today. Consider, for example, this summary of what we might term a 'theological epistemology' from the early Franciscan thinker St Bonaventure's 1259 *Itinerarium Mentis ad Deum* (the Mind's Road to God):[22]

> *Therefore, according to the six stages of ascension into God, there are six stages of the soul's powers by which we mount from the depths to the heights, from the external to the internal, from the temporal to the eternal—to wit, sense, imagination, reason, intellect, intelligence, and the apex of the mind, the illumination of conscience.*

To a modern reader, Bonaventure seems to be making a purely inner, 'spiritual,' journey, but that would be a projection of a modern mindset. For the early Franciscans, a discovery of God would always also entail a discovery of the divine mind, in which lies an encounter with the world in all its multi-layered physical and material polychromy. Any mention of 'imagination' in the sixteenth and seventeenth centuries would be bound to echo its

[22] English transl. by Alexis Bugnolo, The Franciscan Archive. Source of the English digital text: http://www.ewtn.com/library/SOURCES/ROAD.TXT

use in this high medieval 'ladder of understanding' that Isaac of Stella, in the twelfth century, or Grosseteste and Bonaventure in the thirteenth, would invoke to explain the theological rationale for learning. As we will see in more detail in Chapter 7, for the medieval mind, the study of the material world within a prayerful and disciplined direction of both reason (aspect) and feeling (affect) together could redirect and resurrect the sequence of human appropriation of the world, spoiled and broken since the Fall. Medieval scholars saw a ladder that starts with *sensation*, then via *imagination* and *memory* leads finally to a renewed and healed *understanding*. Their 'ladder' was the background that fuelled early modern creation of the 'small worlds' that experimental method relied on.[23]

So, 'imagination' was, for centuries, that mental place where sense-impressions were first received and curated—'imaged'—not the canvas or tablet on which inner pictures or words were inscribed. It adopted a central role in early experimental and observational science. However, as its meaning became subtly extended in the seventeenth century to allow this same mental locus to receive images and impressions from *within* as well as originally, from *without* through the senses, so through the elevation of rational thought over the forms of creative writing that the Royal Society and other professional scientific bodies outlawed, science slowly forgot its essential debt to the creative imagination.

No one grasps this evolved double meaning of imagination as clearly as the poetic counter-voice, at the turn of the eighteenth to nineteenth century, of Wordsworth's colleague, Samuel Taylor Coleridge. Although it was Coleridge who once insisted (perhaps in the spirit of the times voiced by Keats and Poe) that the opposite of 'poetry' was not 'prose,' but 'science,' by this he meant the dreary assembly of fact and mechanism that science had become under the aegis of its national institutions and their insistence on flattening prose-reporting of empirical data, with no room for the imaginative creation of new inner representations

[23] See Chapter 4 *A shared early history.*

of nature. A closer look, for example, at Coleridge's long collaboration in both poetry and chemistry with Humphrey Davy at the Royal Institution, or his collaboration with William Wordsworth on the *Lyrical Ballads* with its strong invocation of science as a potential source of poetic song, indicates that he believed that poetry and science could rediscover their commonality through a partnership of *theoria* and imagination. At Davy's invitation, Coleridge lectured on *Poetry and the Imagination* at the Royal Institution in 1808, in spite of the scientist's clearly mixed view of the poet's genius which, though possessing 'exalted genius, enlarged views, sensitive heart and enlarged mind,' still wanted, in the scientist's opinion, 'order, precision and regularity.'[24]

Far less well-known than his early poetry, Coleridge's later writings sprang from poetically inspired theological and philosophical reflection. His own experience of the creative imagination was fed by both the science he loved (he read Newton's *Opticks* in its entirety), together with a powerful, even shocking, personal revelation through the contemplation of Moses' encounter with God at the burning bush.[25] He writes in his *Biographia Litteraria* of 1817:[26]

> *The Primary Imagination I hold to be the living power and primary agent of all creation as a repetition in the finite mind of the eternal act of creation in the infinite I AM.*

As poet and Coleridge scholar, Malcolm Guite, comments, 'It is as though the creative word that speaks the cosmos into being echoes back to God from minds made in his image.'[27] Coleridge is not only restoring the imagination to a place in which it rediscovers an essential role in humans' knowledge of the natural world, but is also addressing a related, deep problem of modern philosophy—the divorce, as codified by Kant, between those very

[24] As recorded in Richard Holmes (2008) *The Age of Wonder*. London: Harper Press.

[25] Exodus Chapter 3.

[26] Samuel Taylor Coleridge (1817) *Biographia Litteraria*, ed. J. Engell and W. Jackson Bate (Princeton 1983), Vol II. Ch. 13.

[27] Malcolm Guite (2012) *Faith, Hope and Poetry*. Oxford: Ashgate.

human subjects and the objects that they seek to know. Coleridge answers by the theological insight that humans, created in the image of God ('*in imago Dei*') are ourselves *both* created and observed objects *and* living, creating, and participating subjects. Our very perception of the world is as objects within the world, but made in the image of its original creating subject. Humans are the place and image where the categories of subject and object overlap; and part of the structure of that image is the creating ability itself.

Coleridge is here writing, not even of the imagination that science or poetry requires, of hidden inner structure to nature (that related human endeavour is the 'Secondary Imagination'), but of 'mere' sensory perception itself. This is his 'Primary Imagination'—and is, as we saw, the medieval meaning of *imaginatio*, whose power draws from the projected energies of Creation itself and comes from without. Note the sheer power that Coleridge ascribes to either form of 'imagination.' For the weaker, late-modern association of the word to those fanciful reattachments of fragmented pieces of memory, Coleridge reserves the word 'Fancy.' Once this is understood, the connectivity between the cousinly, secondary imaginations from within, of both science and poetry, is laid bare through their shared reflection of the primary imagination. The greatest of all early modern astronomers, Johannes Kepler, would have understood—he who contemplated the humble glory of 'thinking God's thoughts after Him.'

Coleridge's high vision of the imagination within science is a rare one in his century, but not unique. The inventor of 'fantasy literature,' lauded by C. S. Lewis and J. R. R. Tolkien, is himself now not very much read. Yet George MacDonald's literary production, including fictional works such as *Lilith*, opened possibilities for the literary creation of worlds that enabled others to call them up into the forms of Narnia and Middle Earth, that have not yet seen an equal. Like Coleridge, MacDonald also wrote in philosophical/theological mode, similarly less well-known than his artistic writing. It is worth quoting an exchange from his 1867 essay, *The Imagination, its Function and its Culture* in full. It starts with an imagined dialogue with a disciple of Thomas Sprat:[28]

> *But the facts of Nature are to be discovered only by observation and experiment! True. But how does the man of science come to think of his experiments? Does observation reach to the non-present, the possible, the yet unconceived? Even if it showed you the experiments which ought to be made, will observation reveal to you the experiments which might be made? And who can tell of which kind is the one that carries in its bosom the secret of the law you seek? We yield you your facts. The laws we claim for the prophetic imagination. "He hath set the world in man's heart," not in his understanding. And the heart must open the door to the understanding. It is the far-seeing imagination which beholds what might be a form of things, and says to the intellect: "Try whether that may not be the form of these things;". . . Nay, the poetic relations themselves in the phenomenon may suggest to the imagination the law that rules its scientific life. Yea, more than this: we dare to claim for the true, childlike, humble imagination, such an inward oneness with the laws of the universe that it possesses in itself an insight into the very nature of things.*

The unmistakable resonances with Grosseteste's sanctified gaze beneath the surface of the world, and the insight from Coleridge that we possess 'the world in man's heart' because we are each a 'little I AM,' parallel the juxtaposition of 'poetic relations' with 'the scientific life.' MacDonald continues, 'to inquire into what God has made is the main function of the imagination . . . The man has but to light the lamp within the form, his imagination is the light, it is not the form.' In MacDonald, we encounter once more, but illuminated now by theological insight, poetry becoming not only a great metaphor for, but a generator of, science. Both shape the power, or 'light,' of imagination by the creative constraints of 'form.' In poetry, the form is literary, in science simply the form provided by the world as we observe it. Both draw together the double dynamic—the inward and the outward flow—of imagination.

[28] George MacDonald (1893), "The Imagination, its Function and its Culture" in *A Dish of Orts* (Lexington, 2015).

As Coleridge and MacDonald hint, as the early modern examples of science-poetry illustrate, and as Wordsworth foretold, there is a closer connection between science and poetry than the merely metaphorical. The north-east of England's most visionary twentieth century philosopher, Mary Midgley, chose *Science and Poetry*[29] as the title for a book which, although it does not discuss much poetry, nevertheless sees the nexus of poetry and science as a necessary road to bridging the science and arts, imagination and reason. Midgley's ambitious book also suggests that another argument is connected: the recovery of human freedom from determinism, and minds from reductionism. In particular, *Science and Poetry* explores the 'dependence of detailed thought on entirely non-detailed visions.' This captures precisely the first stage of 'creation narratives' that we have already seen, as common in artistic creation as in scientific, in which a distant, defocused, half-conceived vision of a picture, theory, hypothesis, novel . . . is glimpsed, but without at first either a firm structure or a clear pathway to its realization. It is the imaginative conception of this apparition in the form of *theory*, and its generation of the desire to discover it in its fullness and entirety, that Midgley terms 'poetry' for the sake of her thesis. She continues:

> *What makes theories persuasive in the first place is some other quality in their vision, something in them which answers to a wider need. There is always an imaginative appeal involved as well as an intellectual thirst for understanding.*

Science and Poetry also tackles, as did Coleridge, the related dualism of subject and object, proposing that there is a right way—but also a wrong way—of attempting to unite them. The mistaken route is to make something called 'consciousness' an isolatable, objective puzzle. In this endless self-referential and circular labyrinth the subject becomes its own solipsistic object:

> *To suppose that we have a problem about the existence of other minds is to be in trouble already because it is to have started in the wrong place—Descartes's*

[29] Mary Midgley (2001), *Science and Poetry*. Abingdon: Routledge.

wrong place. If we once sit down in that place we shall never get rid of the problem (Bertrand Russell, who was wedded to this starting point, never did get rid of it). This approach conceives of minds—or consciousness—unrealistically as self-contained, isolated both from each other and from the world around them. It is terminally solipsistic.

Midgley's vision bursts through Descartes's isolationism that insists on suppressing the essentially relational task of all art and science. The task is a healing of a set of broken relationships to each other and to the natural world itself. The 'art' that George Steiner wanted to wake 'into some measure of communicability, the shear inhuman otherness of matter' can never hope to do so if its 'imagination' is caught in a solipsistic loop of self-reference. It must be, as Steiner writes elsewhere in his weighty little book,[30] 'a wager on transcendence.' Imagination's source, as Coleridge perceived, is external, but, as MacDonald clarified, it shines through us to illuminate the world *for* us, and for each other's consciousness, by reflection.

Malcolm Guite has written powerfully on the topic of 'reimagining imagination.' Although his *Faith, Hope and Poetry; Theology and the Poetic Imagination*[31] does not signpost an immediate relevance to science in its title, it is interesting that this poet, scholar, and priest finds himself referring frequently to science as he writes on poetry, imagination, and theology. His declared task is to rouse from a long cultural amnesia an understanding of imagination as a route to knowledge in partnership with reason. Guite has no illusion over the magnitude, nor the essential importance, of this task, and articulates supremely well the challenge of centuries of modern assumptions that confuse 'imagination' with mere 'fancy' (in Coleridge's terminology), and so bar it from any power to acquire knowledge. For Guite, as it was for MacDonald and Coleridge before him, the insights of Christian theology and experience become essential to understand both

[30] George Steiner, *Real Presences.*

[31] Malcolm Guite, *Faith, Hope and Poetry.*

the problem and the task. From Augustine to Bacon, reason is supposed less 'fallen,' less damaged, or less prone to perversion than 'imagination,' yet 'these two ways of knowing are mutually enfolded and depend on one another.' The key idea, which also echoes Midgley and Coleridge, is this:

> *If part of the Imago Dei is itself our creative imagination then we should expect the action of the Word, indwelling and redeeming fallen humanity, to begin in, and work outward through, the human imagination. If this is so then we should be able to discern the presence of that Word in the works of art which are the fruit of our imagination.*

Furthermore, Guite knows that this must be true of science as well:

> *I want to support [Mary Midgley's] thesis that the poetic imagination is fully engaged in scientific endeavour and also that poetry is capable of refining and expressing the doubt, as well as the faith, that is part of the dynamic of both science and theology.*

Poetry, science, and theology combine in the perspective, or the projection of gaze, onto and into the world.[32] We look upon the world as an animated image and with the same imagination of the gaze of love that is bestowed by its first Creator. Our poetry, finding form for expression, and our science, exploring in the imagination of theory the form of observational constraint, are related acts of 'waking into some measure of communicability, the shear inhuman otherness of matter.'

There is another resonance found here, for Steiner's 'waking' in relation to both science and poetry. One of the great powers of science is its ability to distinguish between the familiar and the understood. To take up Keats's challenge: just because we are familiar with the rainbow in a sunlit rain-shower does not mean that we *understand* it, nor even that we dissect it in the poet's negative sense, so that is lies in so many dismembered, unwoven, and

[32] See also in Tom McLeish, *Faith and Wisdom in Science*, Chapter 7.

wonder-less pieces at our feet. It is a glory of science that the delicate interlacing of light and water, geometry and atmospherics, retinal cells and the brain's processing of their signals, all combine to yield the perception of the bow (we will have cause to revisit the rainbow in all these ways in the final chapter). We know and see through the familiar into a richer appreciation of its underlying structure. Compare this effect of the theoretical gaze on nature with Coleridge's explanation of the aim of the *Lyrical Ballads*:[33]

> *by awakening the mind's attention to the lethargy of custom, and directing it to the loneliness and the wonders of the world before us; an inexhaustible treasure, but for which, in consequence of the film of familiarity and selfish solicitude, we have eyes, yet see not, ears that hear not, and hearts that neither feel nor understand.*

For Coleridge, the same blindness, or better the 'sleep,' of familiarity—'custom'—to the world's inexhaustible wonders, is cleared and woken by the light of poetry in an uncannily similar way to science's lifting of familiarity's veil.

An explicit expression of this commonality of gaze in science and poetry, theologically conceived, surfaces in Coleridge's celebrated long poem, 'The Ancient Mariner.' Guite comments on the moment of redemption towards the end of the poem's narrative when the Mariner gazes down at a shoal of writhing water-snakes illuminated by reflected moonlight, and realizes their happiness and beauty. 'It is though, by seeing these creatures in moonlight that he is given, however briefly, some notion of how God sees them.'[34] That idea, that we can and must learn to look upon nature from a higher perspective (within the theological frame of Guite's analysis this is none other than the perspective of nature's Creator), turning that into a creator's perspective is a very ancient and poetic notion.

[33] Samuel Taylor Coleridge, *Biographia Litteraria*.
[34] Malcolm Guite, *Faith, Hope and Poetry*.

The theological shape of scientific imagination

At every turn, attempts to run a connecting wire between the energies of science and poetry seem to activate ideas within the tradition of theology. It is worth emphasizing that to work with ideas from a theological tradition is not to imply any necessity for a personal, or confessional, commitment to that discipline (although in Guite's case it is clear that energy and insight are drawn from that experience as well). Rather, it may simply be the recognition that the discipline of theology has developed particular avenues of thought, narrative structures, or analytical tools that work effectively in a related field. This happy academic 'borrowing' is not uncommon in interdisciplinary work—we have already come across examples in meetings of biology with physics in Chapter 3. Another example drawing on theology will arise in the final chapter, in an approach to the purpose of creativity in art and science. At this point, we notice that the theological metaphors of creation, 'I AM' and revelation, have emerged from the narratives around 'imagination' that resonate when poetry and science are juxtaposed.

Biblical texts from the wisdom tradition have already helped to illustrate the long tradition of imaginative reflection on nature in the ancient world. It is now perhaps less of a surprise that these texts take the form of *poetry*. This core-textual tradition suggests the next trail to follow for the source of the light that theological ideas throw on the nexus of poetry and science. Here is Berkeley Hebrew scholar, Robert Alter, reflecting on the nature of poetry itself, from the perspective of one who has spent a career immersed in Hebrew wisdom literature:

> *Poetry . . . is not just a set of techniques for saying impressively what could be said otherwise. Rather, it is a particular way of imagining the world—it has its own logic, its own ways of making connections and engendering implications, . . . its distinctive semantic thrusts that follow the momentum of its formal dispositions and habits of expression.*

Suppose, as an exercise, one replaces 'poetry' with 'science' in this remark, especially the work of theoretical science with its 'formal dispositions' of diagrams, image, and mathematical symbolism. Those formal tools also 'engender implications' for the imagination. Alter is witness to the same outward, inductive, and re-creative movement at the heart of *poiesis* in Biblical poetry that Elliott and Larkin described in the poetry of our own time.

To illustrate how Alter's poetic structure plays out in the form of scientific theory, Margaret Cavendish's atomic quest presents itself, now in its modern guise. The early twentieth century breakthroughs in atomic theory that go by the name of 'quantum mechanics' start with Louis De Broglie's reimagining the materiality of an electron as a wave, rather than a point-like particle. Then the 'habits of (mathematical) expression' carry that idea, with their own logic, into Erwin Schrödinger's famous 'wave equation.' Reading Schrödinger's account of how he arrived at his celebrated result maps closely onto accounts we possess of poets' creative processes. In the same way that words are exchanged and moved around, stanzas shifted within the engagement of imaginative energy within formal shape, so the physicist records his formal experiments with different mathematical structures. In particular, the order of the time-derivative was a key step;[35] replacing a second-order term (usual for a wave equation) by a first-order term with an imaginary coefficient, proved a leap of theoretical *poiesis* that drew back a veil on the strangeness of the atomic world. The move revealed the atom's almost musical vibrations and harmony, and in due course its challenge to reconfigure what the 'real' might mean.

The biblical poetry of the *Book of Job* will reappear in more detail in the final chapter, but a glimpse of what Alter refers to as the highest of all Hebrew poetry—the 'Lord's Answer' in

[35] For readers unfamiliar with calculus, as an example, for 'first order time-derivative' read 'velocity' and for 'second-order' read 'acceleration.'

Chapters 38–42—will serve to illustrate how his reading converges on the theo-poetical perspective of Coleridge, MacDonald, and Guite. After an extended silence, in response to the suffering Job's complaints that he is being punished unjustly and that, furthermore, the entire cosmos is in uncontrolled chaos, the voice of God finally speaks from a storm. We encountered some of the cosmically-oriented questions from this extraordinary epiphany in the introduction, but here is a little more to extend the style and content, exploring the meteorological realm:

Have you entered the storehouses of the snow? Or have you seen the arsenals of the hail, . . .
Where is the realm where heat is created, which the sirocco spreads across the earth?
Who cuts a channel for the torrent of rain, a path for the thunderbolt?

We are presented once more, in this highest and best of all Hebrew Biblical poetry, with a double vision of divine and human imaginative gaze onto the natural world. For when Yahweh finally answers Job's anguished demands for an answer to the uncontrolled and unjust world as it appears to him, the righteous suffering human is taken on a questioning exploration of the heavenly, watery, and earthly structures of that very cosmos. These are questions, sprung from an imagination confronted by the tensions of nature's order and chaos, that can find resolution only through deep observation and contemplation. That very transformation of chaotic energy, channelled but not suppressed into order, is reflected by the form of the poetry itself, in which the ill-controlled energies of the indignant Job, and the creative energy of God, are shaped into the exquisite Hebrew verse.

At larger textual scales, the 'Lord's Answer' to Job, also constitutes a response to many earlier sections of the book. In some ways, it connects the entire sequence of discourses between Job and his friends (Alter's 'own ways of making connections'), for whom natural objects (rocks, plants, trees, stars, milk, winds, floods, . . .) are a continuous source of metaphors for the human condition. But in form and content it once more aligns the gaze of the human, Job, with that of the Creator, onto the natural world. As Alter puts it:

> *Through this pushing of poetic expression toward its own upper limits, the concluding speech helps us see the panorama of creation, as perhaps we could do only through poetry, with the eyes of God.*

A moment's reflection on the theories of biological physics that we met with in Chapter 3 provide us with parallels of perception into the tension between order and chaos that we (and Job) have to understand as necessary in a life-supporting world. The self-assembly of biological cells' lipid membranes, displaying spontaneous order among a sea of thermal chaos that turns out to be necessary to their formation, or the random vibrations of protein molecules that carry information rather than erase it, parallel the Joban discourse of how seemingly chaotic floods are channelled into watercourses. Like Coleridge's *Ancient Mariner* turning from the initial strangeness and fear of the roiling underwater snakes to find in them symbols of healing, humans can face the inhuman materiality of the world through the scientific imagination, and turn from its infinite spaces without horror, but with a redeemed reverence and respect, and an understanding that leads homewards. The apparently threatening inhuman forces of nature that confront us in our immaturity become understood and reconciled when we build the 'poetic' forms of a scientific theory of nature to meet them.

An important aspect of the poetic not yet considered is rooted in the ancient tradition of the spoken poetic word. A poem's full resonance of rhyme and meter is appreciated only when it is recited. At first glance the spoken, the rhetorical, aspect of poetry would appear to have an even weaker mapping onto scientific theorizing than its structural or imaginative content. But in the same way that we have discovered that their deep commonalities only seem far-fetched because they have been buried, there is a considerable element of 'performance' in even mathematical reasoning, as we will explore further in the next chapter. But there are theological ideas behind rhetoric that break surface in the current discussion. The outwardly directed movement of creative imagination of *poiesis* has long been associated with speech, or Word, since the writer of the Gospel according to St John brought the ancient Greek ordering principle of *logos* and the Hebrew account of creation together in the sublime 'In the beginning was the Word . . .'[36]

American orthodox theologian David Bentley Hart is especially attuned to the persistence of the rhetorical throughout human participation in creation. Discussing the theme of 'Creation' in his response to post-modernism, *The Beauty of the Infinite*,[37] Hart points out that the created 'other' (other, that is, than either God or human) is 'known as other not in the silence of immediacy or identity, nor in the darkness of infinite alterity but in the free and boundlessly beautiful rhetoric of a shared infinite.' He continues: 'The rhetoric of the other *evokes my representations*' (my italics). This seems to capture my personal experience as a scientist, in words I could never have found for myself. Scientists who spend time thinking theoretically about the systems they seek to understand know the 'in-betweenness' of a 'representation of the other' evoked, for example, in the terms of theoretical physics, by nature's 'infinite and beautiful rhetoric.' Admittedly this describes a 'very good day in the lab,' but it is more than a philosophical take—it's a description of the experience of doing science that clearly resonates with the poetic and rhetorical. If at this point the argument would benefit from a worked example in scientific *poiesis*, to illustrate what nature's 'infinite and beautiful rhetoric' might mean, then the theoretical physics of symmetry will surely serve.

[36] The opening line of the Gospel according to St John, in the King James English translation.

[37] David Bentley Hart (2003) *The Beauty of the Infinite*. Grand Rapids: Eerdmans p. 300.

A poiesis of theory—Emmy Noether and the symmetry of the world

Einstein was once moved to write, in an obituary, 'Pure mathematics is, in its way, the poetry of logical ideas.' He was writing in eulogy of the person possessing arguably the deepest theoretical capacity of any scientist that the twentieth century saw, including his own—the German theoretical physicist and pure mathematician Emmy Noether. If ever there was a scientific imagination that mirrored those aspects of poetry, and that captured the sense by which it reaches beyond prose, as well as poetry's entanglement with scientific theory through the ancient ideas of *theoria* and *poiesis*, then Noether has to stand as the example by which these connections must be judged. Her deep, beautiful, and foundational contributions to the theoretical physics of our own time seem to capture the external perspective onto creation, the evocation of rhetoric and a formal elegance, the way of imagining the world, and a fluidity of previously unseen connection. These are the hallmarks of the poetic that have emerged in this survey in writers from Cavendish to Coleridge.

Amalie Emmy Noether was born Erlangen, Germany in 1882, the oldest child of four born to Amalia and Max, and their only daughter. Her father was a professor of mathematics at the city's university, but the daughter was far to outshine him in proficiency, and that in a time when it was near impossible for women to achieve university appointments. An early love of languages and literature suggested an initial path towards teaching French and English, but the allure of mathematics drew Noether increasingly to its formal and abstract aspects. Her doctorate in 1907 explored, in bewildering detail, the 331 'invariants' of abstract structures in algebraic geometry known as 'bi-quadratic forms.' Though shamefully refused the *Habilitation* (the post-doctoral qualification necessary in Germany to be employed as a university teacher) on account of her gender, she nevertheless agreed to teach at Erlangen's mathematical institute without pay, a situation that lasted until 1919, by which time her extraordinary discoveries could not be overlooked by even the discriminatory prejudice of the University of Göttingen at the time. The cross-currents of such contradictory experiences must have been bewildering: she seems to have been accorded the greatest respect by leading theoreticians and mathematicians such as Klein, Hilbert, and Einstein, yet refused recognition by lesser, but appointment-electing, academics.

By this time, Noether had been drawn into the swirling, gestating physics and mathematics of general relativity, Einstein's theory of gravity. Touched-on in Chapter 3 in visual mode, this is the beautiful geometric insight that gravity can be conceived of, not as Newton's mysterious 'force at a distance,' but as an intrinsic curvature of space and time. Einstein had embarked on extending his 1905 special theory of relativity (applicable only within gravity-free space) as soon as the following year, propelled by one of the famous thought experiments that illustrate so well his visual mode of scientific imagination. Yet it was to be a full decade before he was able to write the fundamental equations that properly coupled the curvature of space-time to the mass and energy that generates it. This second creative road for Einstein was much more tortuous and painful than the first, requiring, among other things, the acquisition of a body of formal mathematics that he clearly found very difficult to master. There remained a strange puzzle, however: that most precious of all laws of physics, the 'conservation of energy,' seemed not to survive into the new formulation in a recognizable form. Beneath this problem lay the deeper and unresolved issue of how conserved quantities of physics arise in the first place. Originally proposed by the brilliant French translator and commentator on Newton's *Principia*, Émily du Châtelet,[38] the issue of conservation of energy proved a spark to Emmy Noether's imagination. Her earlier 'invariants' may have seemed a dry, formal chapter in

[38] *Principes mathématiques de la philosophie naturelle par feue* Madame la Marquise du Châtelet (1st edition, 1756; 2nd edition, 1759). Paris: Desaint & Saillant

algebra, but an 'invariant' quantity in *time* is what is meant by a *conserved* quantity in physics, and the dual vision of intense experience of detail together with her unsurpassed panoramic vision of mathematical structure of physical theories, gave her the form and the content for a work of consummate beauty and depth.

Noether's outward movement of *poiesis* creates and opens up, from the narrow doorway of the energy-conservation puzzle, a vast landscape of possibilities in physics and a clear vision of the origin of conserved quantities. Energy is not the sole example; momentum—the product of mass and velocity of a moving body—is another. Its conservation is recognizable in the head-on collision of billiard balls, as one comes to a halt, bequeathing its motion to another. Conservation of the 'angular momentum' of spinning objects by contrast, maintains constant the product of their mass, speed, and distance from the axis of spin. The faster rotation of a ballerina executing a pirouette as she draws in her arms and legs, maintains a conserved angular momentum. In a non-mechanical example, all interactions and transformations of matter ever observed conserve the total electric charge: it can only be transferred from one body, or particle, to another. Without conserved quantities, the world would dissolve into chaos—they constitute the formal constraints for all dynamics of change. Knowledge of them had built gradually since du Châtelet's proposal of the energy law, derived from Newton's laws of motion, but there was no understanding of how they arose, of what fundamental aspects of nature determine which quantities are conserved, and which not.

The key to finding the origin of conserved quantities lay in Noether's earlier work on invariants, the synonyms of conserved quantities, but more clearly connected with the mathematical and aesthetical concept of *symmetry*. By its definition, any symmetry implies and requires an invariant. A square figure possesses symmetry under rotation, for example. Any rotation by a quarter-turn will leave its geometry exactly the same as before. The invariant in this case is the geometry. The square is said to have a 'fourfold rotational symmetry.' An equilateral triangle, similarly, possesses 'threefold rotational symmetry.' A circle may be rotated through any angle to leave its geometric form invariant—it possesses a 'continuous rotational symmetry.' Other forms are symmetric with respect to translation, rather than rotation. Consider an infinite checkerboard pattern of black and white squares. If it is shifted along the direction of its rows by two squares, the pattern is precisely restored—or invariant. The pattern has a translational symmetry (a discrete one in terms of the measure of the pattern).

In the same way that geometric shapes may have symmetries, physical laws may also be symmetric under displacements and rotations. The way that the universe behaves physically is not altered, as far as any measurement has been able to discern, when any measurement apparatus is moved ('translated') from one point in space to another. The laws of physics can therefore be said to possess a (continuous) translational symmetry. Nor does the universe behave differently when looking in different directions—like the circle (or more completely, a sphere), physics is also rotationally symmetric. Less obvious at first, but just as powerful a symmetry, is the temporal invariance of physical laws—again as far as even the most careful measurements have shown, there is no change in how the universe operates in its forces and interactions over time. So, if physical laws show such symmetries in translations and rotations through space and time—in the same way that geometric shapes do—Noether was able to ask what would be the *physical* correspondences to the *geometric* invariance of the shapes. What are the particular invariants that partner with those *physical* symmetries? She was able to explore the question not only intuitively as here, but also mathematically, since the mathematical descriptions of gravity and of electromagnetism were both already known. She arrived at a simple and beautiful conclusion: for every symmetry of physics there was paired a conserved quantity. From translational symmetry through space arose the conservation of momentum, from translation symmetry through time arose the conservation of energy, and from symmetry through rotation arose the conservation of angular momentum.

There was even more to this theoretical construction that germinates new divergent ideas, and a far bigger landscape than the one first glimpsed. Such budding and propagation of new ideas works in the same way that the formal and rhetorical act of beginning a poem evokes more, even of David Bentley Hart's 'other,' that lies beyond the original knowledge of the writer. The modern British poet and critic Stephen Spender wrote a candid essay on his personal experience of the poet's art, *The Making of a Poem*, elements of which resonate with the experience of crafting theoretical science in a similar way to those of Robert Alter:

> *[T]he poet is aware of all the implications and possible developments of his idea, just as one might say that a plant was not concentrating on developing mechanically in one direction, but in many directions, towards the warmth and light with its leaves, and towards the water with its roots, all at the same time.*

Spender points out that this self-germinating growth has not only such imaginative expression, but also a formal one:

> *Now the line . . . is a way of thinking imaginatively. If the line embodies some of the ideas which I have related above, these ideas must be further made clear in other lines. That is the terrifying challenge of poetry. Can I think out the logic of the images?*

Noether's work of (in this sense poetical and) theoretical creativity did indeed think out the logic of its first image of symmetry connected with conserved quantity. It carried through the idea to other, non-spatio-temporal, conserved quantities such as electric charge. These more general symmetries turn out to be related to less obvious but fundamental symmetries in the physical laws for the electromagnetic field (in the case of electric charge). Such 'gauge symmetries'[39] became foundational to the quantum field

[39] It is possible to think of each point in space carrying the dial of a 'gauge' whose needles can point to any value on the circles of their dials. If the physics of the space is deemed to be symmetric under any choice of the dial readings, then that space possesses 'gauge symmetry.' From Noether's theorem then arises the conservation of electric charge.

theories for light and elementary particles that emerged as the bedrock of theoretical physics in the latter half of the twentieth century. Not only do these new symmetries imply, through Noether's Theorem, new conserved quantities of 'charge,' but the entire physical behaviour of the fields generated by these charges. In the simplest case, all of electricity and magnetism bursts into existence, once the symmetry holds. The germination of theoretical imagination of this power calls into being an image of the universe in all its structure and dynamics. The technical details would become the realm of only very highly trained mathematical physicists, but the glorious aspect of 'Noether's Theorem' is that its deepest statement is open to contemplation by anyone: that for every symmetry in physics there arises one of the conserved quantities that act as the shaping, formal constraints on the future evolution of the universe.

Emmy Noether not only theorized poetically, in the metaphorical and parallel ways in which these creative forms align, but she also inspired a remarkable poem. American poet, and writer on the connections between mathematics and poetry, JoAnne Growney, was moved in 1964, on encountering a mention of her at an exhibition under the misapplied title '*Men of Modern Mathematics,*' to write a biographical and conceptual poem about her. It includes the lines:

> *Direct and courageous, lacking self-concern,*
> *elegant of mind, a poet of logical ideas.*
> . . ..
> *I followed you and saw you choose*
> *between mathematics and other romance.*
> *For women only, this exclusive standard.*
>
> *I heard fathers say, "Dance with Emmy—*
> *just once, early in the evening. Old Max*
> *is my friend; his daughter likes to dance."*
> *If a woman's dance is mathematics,*
> *she dances alone.*

. . .
Students said, "She's hard to follow, bores me."
A few stood firm and built new algebras
on her exacting formulations.
. . .
She's a pacifist, a woman.
She's a woman and a Jew.
Her abstract thinking
is female and abstruse.

Today, history books proclaim that Noether
is the greatest mathematician
her sex has produced. They say she was good
for a woman.

The poem condenses Emmy Noether's mathematics into the context of her struggles against the overwhelming and unjust headwinds that women in science suffered throughout the twentieth century, and many of which remain unresolved in our own day. It also signifies that recent decades have seen a flowering of science-poetry. If modernism triggered a stifling of the poetic voice in science, perhaps post-modernism's greatest gift is its return.

Science-poetry and poet-scientists

Any proof of any claim that *poiesis* and *theoria* retain a cousinly, cultural, and creative relationship must surely be found, not only in a resonance of their creative processes, but also in an explicit poetic contemplation of scientific imagination, along the lines of the *Lyrical Ballads'* vision. Science has indeed proved an increasing inspiration to poets, to the point where serious anthologies of science poetry have appeared,[40] as has an online

[40] For example, Maurice Riordan and Jon Turney (eds.) (2000) *A Quark for Mister Mark*. London: Faber and Faber.

journal, *Consilience*, the first dedicated to science poetry.[41] Since 2017, pan-disciplinarian Maria Popova has organized an annual festival of science in poetry, *The Universe in Verse*.[42] Furthermore, if the connection between the scientific and poetic imaginations lies as deep as this chapter has traced, then examples of science and poetry emerging from the same minds might be expected. This final section hears from both science-inspired poets and from poet-scientists.

Contemporary Welsh poet John Barnie challenges any persistent claim that science has not, in contrast to Wordsworth's hope and vision, inspired poetry. He describes an early educational experience exclusively focused in the humanities, which coloured a first impression of science towards the technical, uninspiring, and dehumanizing narratives of the Romantics. Yet an inner recognition that there had to be more to science than the instrumental and mechanical, drove a search for the power latent in some of its 'great ideas.' Inspired by the life-sciences in particular, and by science-writing from Darwin to Dawkins, he writes autobiographically:[43]

> *I have internalised aspects of evolutionary biology and paleo-anthropology to the extent that they become part of my mentality and bubble up from the place where poems are formed in the shape of images and themes without my having to think about them.*

The experience of writing poetry, inspired by years of deep dives into the life-sciences, he describes as 'fly-fishing'—a contemplative casting of the conscious poetic mind onto the waters of its subconscious, with an alert attention to what surfaces, and to how it can be shaped. An opportunity to draw from the mental spring-water of paleo-biology arose in the form of an invitation

[41] https://www.consilience-journal.com The journal invites submissions around specific themes; an example on geoscience can be found here: https://www.consilience-journal.com/si/geoscience

[42] https://www.brainpickings.org/the-universe-in-verse

[43] John Barnie (2020), 'Some thoughts on *The Poetry and Music of Science*,' *Interdisciplinary Science Reviews*, **45**, 46–50.

to be poet in residence during the Oxford Museum of Natural History 'Visions of Nature' year in 2016. After talking with the museum's scientists one day as they carefully cleaved, sectioned, and scanned 'nodules' of fossil-containing rock from the Silurian period—the only way of revealing the creatures hidden within—he found the poem *The Sea Spider Speaks* appeared fully formed in his mind:[44]

The Sea Spider Speaks
(Rescued from a lagerstätte)

You took your time but at least you're here
shut in a ball of stone for millions of years
you've no idea;
I don't mind being sliced by the micron,
exposed in three-D,
made to do back-flips on a computer,
I know I'm just pixels, but everything changes
and it's a better fate than yours who'll be forgotten after death;
when in the Med, say 'hi' to the pychnogonids
I'd strum you a tune, but my banjo's lost its strings.

The temporal contrast is stark: of deep-time, sequestered in the Silurian sands, set against and seen through the animated reconstruction of the spider, in false-colour, rotating on the computer screen as software reassembles the coded three-dimensional creature from the past. The notion of 'giving the world a voice,' the implicit theme of the poem, struck this reader as a task required of science too, as its human task must also contemplate the value of life, death, loss, and the constitutive connection of living organisms to the ordinary matter of the world. In musical metaphor for stuff coming alive, the poem itself animates in the deliberate insertion of the technical and rhythmical *pychnogonids*. Scientists recognize such poetic 'fly-fishing' in our own thinking: the lure to attract poetic fragments remains on the visible (conscious) surface; they must swim up, unseen, to meet it. Some of Barnie's fragments were formed from scientific ideas, but his experience

[44] Reproduced by kind permission of the Oxford Natural History Museum.

of writing poetry also resonates strongly with the testimony of Beveridge to the creative process (in science) and Poincaré (in mathematics) recorded in the previous chapter.

Turning from a contemporary poet inspired by the life-sciences to one who finds her chief source in quantum physics, American Mary Peelen began writing poetry after taking degrees in mathematics, theology, and creative writing. She describes poetry being as 'flexible and strong' as mathematics itself—for her a necessary unifier and connector of ideas. Reminiscent of William Whewell's pearl necklace of ideas, 'reality exists in connections' is her reason for pursuit of both poetry and mathematics in the construction of connections to the transcendent aspects of the world introduced by physics. Her favourite, and almost exclusively employed, form comprises ten terse couplets. Originally set by her first poem, '*x*,' the remainder of its anthology, *Quantum Heresies*,[45] followed suit. A supreme example of the connection of ideas in her poetry is the astrophysics-inspired *Supernova*:[46]

> Dying is an art, *said Sylvia Plath,*
> *dark energy providing the opposite of gravity.*
> *A future sun will rise up in all its glory*
> *so red and ravenous it devours the daytime sky,*
> *matter ripping itself into sound and light*
> *in one last explosion uncontainable as art itself.*
> *Heaven performs a billion spectacular finales,*
> *it's up to us to conjure the rest.*
> *We'd all start with divinity and work backwards*
> *If we could manage the math*
> *but even Lady Lazarus burned her miraculous hair*
> *in the calculus of resurrection.*
> *Here at the table, event horizon flickering pink,*
> *we begin with the absolute:*

[45] Mary Peelen (2019) *Quantum Heresies*. Glenview: Glass Lyre Press. I am indebted to Kaley Casenhiser (Yale University) for a rich discussion of this poem.

[46] Reproduced by kind permission of Mary Peelen.

the emperor of ice-cream, Mrs. Ramsey's charm,
and light, of course,
the way it always travels at light speed.
Everything else is contingency –
cutlery glinting like a phantom,
peaches in a milk white bowl, figs going bloody blue.

Starting and ending with references to death and decay, the poem sets the reader's mind humming with connection. Referring to Sylvia Plath's poem on the focused fate of an individual woman in the holocaust, in turn referring to the account of the raising of Lazarus in John's Gospel, which itself refers to the resurrection—all this weaves a web of reference that is illuminated by the ultimate glory of stellar death in the astronomical supernova. This ultimate bright final flash of a distant galaxy's star is typically the only event we record from it; the rest of its life-cycle must be inferred through astrophysics. So Peelen provides a hypnotic image of the immediately preceding stage of red-giant in our own sun's future 'ravenous' consumption of our own sky from horizon to horizon. Her parallel acquaintance with theology leads the poet to ask if one might infer a theological calculus of meaning in a similar way. Yet even the apparent absolute of a 'still life' (the unspoken pun on resurrection at the end is unmissable), a reference to a poem about ice-cream by Peelen's own great influence, Wallace Stevens, and the solidity of Virginia Woolf's central character in *To the Lighthouse*, are divested of their apparent materiality, and instead all communicated impressionistically by physics's only true absolute, the speed of light. The conversation of science with art is fluid and mutual. The connections once made are unforgettable.

If Mary Peelen shows how poetry connects the substance of science with the wider web of human experience, then the north-east English poet, Katrina Porteous, is a contemporary inheritor of Coleridge in indicating how the imaginative process of knowing is a shared one. Her anthology *Edge*[47] draws from her own deep dive into the science she writes about, and of working alongside the communities of scientists who listen, observe, and imagine. In Porteous' case the collaborations extended to musician Peter Zinovieff, whose music forms part of the poems' public performances. The role that poetry plays within the larger work of sharing science, as well as in its conception, is articulated in her introduction to *Edge*:

> *So these poems have an epistemological element: they are not just about* what *we know, but about* how *we know. The relation between scientific empiricism and poetic idealism fascinates me. In each case, I found myself writing my way towards a crude understanding of the subject, just as I would if I were grappling with a difficult human experience. The struggle to understand was also the effort to forge a language in which to interrogate the subject. Since understanding was never more than proximate, the aim of each piece is not to convey accurate scientific information, but rather to translate the experience of trying to understand. As in any poem, my intention is not to explain anything, only to evoke some things. Much of this is achieved through metaphor, and through the physicality of sound.*

It is the haptic, immersed, visualized, sensed, and heard qualities of the material world that, when responded to between Wordsworth's poles of suffering and joy of the human condition, become the science of poetry.

Personal, methodological, and aesthetic comparison of poetry and science can perhaps be best testified to by those who are both scientists and poets. A recent (multi-)biographical study on this topic draws our attention to the personally felt connections on which his subjects can uniquely report.[48] Ada Lovelace, mathematician, daughter of Lord Byron, and herself a poet, writes of scientists as 'those who soar on the fair white wings of imagination.' She also confesses that her best personal preparation for composing poetry would be a week of mathematical work. A

[47] Katrina Porteous (2019) *Edge*. Hexham: Bloodaxe Books.

[48] Sam Illingworth (2019) *A Sonnet to Science*. Manchester: Manchester University Press.

scientist-poet with a less romantic approach was the Czech immunologist, Miroslav Holub, who wrote in his native language, but who supervised very closely the English translations of his poetry. The turbulent twentieth century political history of his home country seems to be reflected in his fluctuating attitudes to the relationship between science and poetry. In a 1994 interview[49] he announced, 'I don't think science can inform poetry . . . science presents a firm ground for all personal feelings.' Holub tended to prefer the 'hard edge' of science, including its constraint on science-poetry, to 'versifications of science data.' Yet in a 2006 collection of essays on contemporary science poetry he admits, albeit grudgingly, to a relationship with the scientific imagination: 'The best one can get from poetry in a scientific career is a kind of vivid imagination which must stay at all times under the strict control of available knowledge.'[50] He could be poetically lyrical in the communication of scientific experiences, describing for example, a diseased tissue in an early poem *In the Microscope*:

Here too are dreaming landscapes,
lunar, derelict.
Here too are the masses,
tillers of the soil.
And cells, fighters,
Who lay down their lives
For all the world. . .

Holub restricted his poetic deployment of technical terms to the point of censure in his early work, yet avers that 'today, I would have fibrin fibres, interleukins, cellular stress proteins, and oxygen radicals included, because they are essential for the battlefield, because they are the driving force behind the apparently still

[49] Honeycutt, 'Interview with Czech Poet Miroslav Holub.'

[50] Miroslav Holub (2006), 'Rampage, or Science in Poetry' in *Contemporary Poetry and Contemporary Science* ed. Robert Crawford. Oxford: Oxford University Press.

life, because they are better known today.'[51] I am not quite as sure as Holub that these terms would really meet Wordsworth's criterion of 'familiar to us,' but there is, he admits, another reason for their introduction—they have value as 'dark images or sounds in an otherwise clear development of the poem.' As Gillian Beer observes in her afterword to the essay collection, 'Holub here recognizes, if a little uneasily, that terms can shed their professional stability and re-emerge as dark matter, not to be described, simply present.'[52]

For a stunningly succinct image of 'dark matter,' we can do no better than to turn to a short poem of that title, by poet-astronomer Rebecca Elson:[53]

Above a pond,
An unseen filament
Of spider's floss
Suspends a slowly
Spinning leaf.

We cannot detect dark matter directly, but infer its existence and even its form by its gravitational effect on luminous, visible stars and galaxies. Elson, who tragically died in 1999 of non-Hodgkin lymphoma at the age of 39, is best known for her work on the luminosity and structure of rich star clusters in our Milky Way galaxy and its satellites, much of it making use of data from the (then newly deployed) Hubble Space Telescope.[54] She wrote poetry throughout her scientific life, but not infrequently faced

[51] Miroslav Holub (2006).

[52] Gillian Beer (2006), 'Afterword' in *Contemporary Poetry and Contemporary Science* ed. Robert Crawford. Oxford: Oxford University Press.

[53] Poems by Rebecca Elson reproduced by kind permission of the publisher, from Rebecca Elson (2018), *A Responsibility to Awe*. Manchester: Carcanet Press Ltd.

[54] For example, Rebecca A. W. Elson, 'The binary star population of the young cluster NGC 1818 in the Large Magellanic Cloud,' *Monthly Notices of the Royal Astronomical Society*, **300**(3):857–62 (1998).

criticism from colleagues that this 'side interest' was a waste of time, as well as suffering, like Noether, the debilitating sexism still endemic to academia. Undeterred, she continued to write wonderful poetry, convinced that it contributed to her professional development and curiosity. The metaphorical link between intergalactic dark matter filaments and spider silk is an instructive example—fresh potential forms for dark matter leap into question. In any case, after reading her work one is left with the impression that here is a human being for whom the questions, 'Why do you write poetry?' or 'Why do you do science?' could only be answered by admitting that they were both ineradicable and interconnected parts of her makeup. Elson's poems certainly reveal and articulate the inside of her science thinking and feeling in much the same way but, like Mary Peelen's, they also reach out and entwine around other distant ideas, bringing them close as if through the optics of a giant telescope. Her posthumous anthology, *A Responsibility to Awe*, edited by her husband and close friends, contains complete poems, but also unfinished work from her notebooks. A thought-provoking juxtaposed pair, titled *Isaac and Eve*, is found among that second group:

Before the Fall	*After the Fall*
Of the apple	*Of the apple*
Mutual attraction	*Mutual attraction*
Was not fully understood	*Was better understood*

Elson ties the apocryphal fall of Newton's apple, stimulating in story the conception of his theory of gravity, to the Edenic forbidden fruit from the tree of the knowledge of good and bad, from which biblical Adam and Eve ate. Yet the notion of a fall from grace in the poem, like starlight in the Hubble Space Telescope, reflects twice: first to consider that 'mutual attraction' requires understanding in both human and physical spheres, and second to the consequent suggestion that the early modern era of Newton might itself have witnessed a Fall in the moral sense. As this very connection of science and poetry has highlighted, Newton's century did indeed sow seeds of a failing, if not a falling, of imagination in service of science, which would later grow into its exile from the scientific community.

But Elson, the other scientist-poets, and the continuous tradition of the entanglement between scientific and poetic imagination witnessed by Cavendish, Donne, and Milton, through Goethe and Coleridge to Peelen, Porteous, Barnie, Holub, and an increasing host of others, promise that the exile might have been partial, and could be temporary. If these poetic voices demonstrate the *theoria* present in poetry, then the creative, rhetorical, reimagined cosmical structures of Noether, Newton, and du Châtelet point to the *poiesis* in the work of scientific theory.

Poetry has taken this tour of the scientific and artistic imagination as far as words can go. Through its pointing to mathematical language on the one hand, and to the rhythm and harmonies of music on the other, it has already prepared the way for the next chapter, which explores these examples of the wordless mode of imagination.

6

Music and Mathematics—Creating the Sublime

And since the arrangement of time and the composition and harmony of the lower world and of all things composed of four elements come from celestial motions, and, moreover, since it is necessary to find the harmony of causes in their effects, the art of music also extends to knowing the proportions of times and the arrangement of the elements, and even the composition of all the elements themselves, by the composition of the lower world.

Robert Grosseteste, *De Artis Liberalibus*

Once in a while, a television or film documentary manages to capture a moment so singular and profound that it deserves the equivalent of leather-binding and a place on the shelves of the British Library or the Bodleian. Fortunately, the BBC in the UK has begun to build some form of equivalent in its permanently accessible online archive. In consequence, anyone can now replay the 1996 *Horizon* programme produced by Simon Singh,[1] featuring the demur and gentle Oxford (formerly Princeton and Cambridge) mathematician Andrew Wiles's long and ultimately successful search for a proof for Pierre de Fermat's celebrated 'last theorem.' No one who has seen the opening sequence of that programme will ever forget it. In the very first frames, Singh brings us the point in their long hours of interviews at which Wiles recounts the moment he finally saw how, after several months

[1] Available at https://www.bbc.co.uk/iplayer/episode/b0074rxx/horizon-1995 1996-fermats-last-theorem (accessed November 2021).

The Poetry and Music of Science. Tom McLeish, Oxford University Press.
 DOI: 10.1093/oso/9780192845375.003.0006

of worry and effort, living with the knowledge that his first attempted proof was faulty, he would be able to bridge its gaping flaw: 'It was so beautiful . . .' he just manages to gasp, before wrenching his gaze away from the camera as his voice chokes up and his eyes moisten at the memory of this treasured epiphany. Singh freezes the frame to let it sink into his viewers' own memories, though he hardly needs to. We are at his mercy for the next fifty-nine minutes, though burdened with a question—how can a piece of mathematics be so beautiful, so able to touch the nerves of aesthetic response, that one sobs with pure joy at recalling it? How might we onlookers, trailing light years behind Wiles in intellectual distance, comprehend the creative progress, let alone the erudite thinking that won such a prize?

My own memory is jogged to recall another moment of surprising joy. Our family lived in the north English city of Sheffield for a few years in the 1990s, while I was undertaking my first academic post in the university's physics department. At that time, as at present, the old steel town had a vibrant musical life, and was fortunate to host the internationally renowned Lindsey String Quartet. Although customarily formal in other settings, the quartet maintained an extraordinarily collaborative rapport with their local audience. This approach to their local performance community was very special to my wife and to me, as it made our introduction to string chamber music all the more engaging. The four performers would talk rather informally through the structure of the works to be played before each concert, and chat over coffee with their regular audience in the interval. Such a collegial approach generated some remarkable examples of the way that musicians and listeners collaborate during performances in subtle but tangible ways. Occasionally, the musical experience would touch a hidden nerve that, though collectively felt, was almost beyond comprehension. One evening's concert concluded with Shostakovich's eighth string quartet—his most intensely personal and soul-searching chamber work that threads the haunting and eponymous D-S-C-H theme throughout its five interlinked movements.[2] The quartet was written during an intense period of personal suffering, physical and emotional, as the composer had just begun to suffer from a chronic muscle weakness, while inwardly and mentally had succumbed to intolerable pressure to join the communist party.[3] There are various meanings read into the work: the score carries the dedication 'to the victims of fascism and the war,' but Shostakovich's children have suggested that the piece is essentially self-referential (and self-pitying). Yet as the performance progressed, it began to take on other layers of meaning. Players and audience shared this awareness, although none was able afterwards to say how. The felt expression of loss, longing, pain, and darkness was palpable; yet at the same time a distant hope of light began to rise among the shifts of key and voice. At the end of the final chord no one moved. Tears streamed down the faces of the musicians, who must have played the work dry-eyed dozens of times before. Eventually they processed out in silence. Only our delayed applause summoned the performers back into the auditorium when we had all calmed down a little. This doesn't happen very often at concerts, even extremely good ones. It is unforgettable when it does. How does a twenty-minute piece of four-part string writing, its development of themes, rhythm, pitch, and modulation manage to carry such emotive meaning? How does its beauty pierce and its profundity inspire awe? With what prelinguistic human longings, hopes, and pain does it manage to connect?

These two examples of emotional response to a deep aesthetic carry their own questions and power, but all the more so because they are tantalizingly related. As we have recalled, music and mathematics are frequently brought into comparison.

[2] An abbreviation for D. Shostakovich's name into musical notes is possible in the German notation (and transliteration of his surname, which contains the C) where H represents B, and S (from Es), E-flat.

[3] Laurel Fay (1999), *Shostakovich: A Life*. Oxford: Oxford University Press.

There is even an academic journal whose focus is on the relationship.[4] After many conversations triggered by knowledge of individuals who simultaneously possess mathematical (or more generally scientific) and musical gifts, I have, however, not found any robust statistical study demonstrating correlation of mathematical and musical ability. There are a few celebrated short-term studies comparing the performance in mathematical reasoning tests between groups of children exposed, or not, to prior musical training. The significant correlations found may reflect neurological development that supports a mathematical and musical connection at a deep level, or they may arise from the more general training of sustained concentration that musical listening cultivates.[5] To say more would require the lifetime longitudinal studies that become prohibitive under today's short-term demands on academic research. To be sure, Einstein enjoyed amateur violin playing, while Borodin earned a living, not from composing the Russian impressionistic music for which he is best known, but as a research chemist. Mathematicians have commented anecdotally on the prevalence of piano-playing among their colleagues.[6] Perhaps Russian twentieth-century composer Igor Stravinsky gave the most precise voice to the musical and mathematical meme when he said,

> *Musical form is close to mathematics—not perhaps to mathematics itself, but certainly to something like mathematical thinking and relationship.*

This sounds a more plausible relationship to explore than a direct but superficial overlaying of the two, and it generates a more

[4] *The Journal of Mathematics and Music*, https://www.tandfonline.com/toc/ tmam 20/current.

[5] Martin F. Gardiner, Alan Fox, Faith Knowles, and Donna Jeffrey (1996), 'Learning improved by arts training,' *Nature*, **381**, 284

[6] For example, Cambridge Mathematician Tim Gower's article in *The Independent* of 5th July 2011 https://www.independent.co.uk/arts-entertainment /classical/features/the-enduring-myth-of-music-and-maths-2307387.html accessed 8/7/2017

powerful question—how is music 'something like' mathematical relationship? A second question brings this chapter as the next step from the last, for the progression from experimental science to its theoretical constructs motivates a discussion of mathematics, as the wordless but animated forms of those theories. In a similar way, a journey in creative language from fictional prose to poetry also points beyond words to artistic forms that are similarly transcendent. Does this comparison also suggest an entangled relation of music to mathematics that bears comparison to those of fiction to experiment, and poetry to theory? These questions will be our guide through the exploration of the third shared mode of creative imagination, for when we leave visual images and words behind, music and mathematics remain as examples of creation in the world of the abstract.

The numerical threads of music

Whatever anecdotal biography might tell us, a structural comparison of music and mathematics turns up sufficient commonalities to suggest interpretations of Stravinsky's remark, and to motivate a deeper look. To begin at the very fundamental level, when we ask about the constituent 'particles' of music and mathematics, we find the structure of number at the atomistic level of both. Mathematics starts with counting, so does music—if a counting with feeling. Music's 'bars' or 'measures' contain in their time signatures the finite, circular number systems known to mathematics as the 'cyclic groups,' notated as Z_n where n is the number of elements in the group. Counting in the cyclic groups is rather easy: in Z_2 it goes 1, 2, 1, 2, 1, 2, . . ., counting in Z_3: 1, 2, 3, 1, 2, 3, 1, . . . Compare the numerators in the musical notation of 'time signatures,' $\left(\frac{2}{4}, \frac{3}{4} \text{and} \frac{4}{4}\right)$ with the suffices of the corresponding group theory notations Z_2, Z_3, and Z_4 and you get the point. Added together and 'unwrapped' from the circular counting of single bars into extended musical sequences, musical beats stretch along the infinite number line. We almost hear a conductor counting beats during rehearsal.

In the fifth century BCE, the Pythagoreans noticed another way in which numbers could be said to generate music. For, as any violin player knows, a plucked string generates a series of pleasingly related tones, known as 'harmonics,' when it is stopped at exactly half, a third, a quarter, a fifth, and so on, of its original length. Each shorter length, a whole number fraction of the first 'open string,' creates notes higher than its 'fundamental' tone by exactly an octave (the half-length), then additionally by a fifth (the third length), another octave (the quarter length), a further major third (the fifth length), and so on. If the rhythmic structure of music embodies mathematical *addition* in different number-worlds, then its harmonic structure reflects the *multiplicative* properties of numbers though ratios: it is literally 'rational.' There is growing evidence from the archaeology of prehistoric wind instruments that the harmonic series has framed human musicmaking since long before writing.[7]

A third dimension of musical perception, beyond rhythm and pitch, maps onto audible volume. Musical loudness rides on another, more subtle, mathematical structure: the intriguing connection between addition and multiplication. The notion of a 'logarithm' is precisely that: every number has a twin—called its logarithm—chosen so that the multiplicative product of any two numbers is itself always twinned with the *sum* of their two logarithms. Any operation of multiplication on the (positive) number line is shadowed by an operation of addition on another number line in which the logarithms dwell. A common example is found in the 'index' notation for powers of ten. Even those with a cursory acquaintance with mathematics know that we can write 100 as 10^2 and 1000 as 10^3. Multiply 100 by 1000 and we have 100,000, or 10^5. Rather than perform the multiplication we might just as

[7] Chris Scarre and Graeme Lawson (2006), *Archaeoacoustics*. McDonald Institute for Archaeological Research and Oxbow Books.

well have added the indices (or equivalently the number of decimal zeros) 2 + 3 = 5. Furthermore, psychological research on perception of all kinds of sense-data, including brightness of light, have shown that our sensations respond in proportion to the logarithms of signals, rather than to the signals directly. In the case of sound, careful experiments on human subjects have shown that equally spaced steps in the perceived volume of a sound correspond to equal *multiples* of acoustic energy. So, increasing the sonic energy of a tone by ten, then by a hundred times, will be reported as two equal steps up in volume. In the crescendo from *ppp* to *fff* we are listening to logarithms.

Whatever element we pick from a close inspection of the musical tapestry—pitch, rhythm, volume—we find numbers adding, multiplying, taking logarithms, or even circling around in spaces of 'finite arithmetic.' But now let us stand back from the warp and weft of the bar- and pitch-lines of a single musical measure, and view the structure of a piece of music as a whole. When a musical work is 'viewed from a distance,' other levels of structure appear, with their own numerical or geometrical relation. We will need to say more about musical form later in this chapter, but whether we are listening to improvised jazz, to a sonata, a symphony, or to the multiple complexities of a fugue, the music makes sense because of its self-referential links across many scales of temporal distance. A short musical figure is repeated a moment later, or transformed in key. Harmonic ideas pervade entire symphonies, creating the impression of a musical journey to strange places signified by successive modulation of the music's harmonic centre to 'remote keys,' followed by a return 'home.' Music that does not return to its home key, or does so only ambiguously, leaves the listener in a strange limbo of unease.

When I asked a professional oboe player whether musical performers recognize a connection between music and mathematics, she replied straightaway, 'When I am playing, I have numbers in my head all the time.' It might be objected that such a performer's viewpoint suggests a superficial divergence between music- and mathematics-making—for while musical creation requires a community of composer and performers, there seems at first sight no analogue to 'performance' in mathematics. New theorems in algebraic geometry or number theory must arise within individual minds, after all. There are no public performances of mathematical results, or iTunes downloads of proofs. Yet on closer acquaintance, it is not so obvious that there is no analogy to be drawn. The story of Andrew Wiles's momentous achievement in proving Fermat's last theorem points to forms of mathematical performance at several turns in its romantic plot. As he approached the final form of his proof, after seven years of solitary work, he needed the assistance of colleagues to critique, guide, and shape the final stages. The challenge was to arrange this without advertising that he was on the trail of so great a quarry. So, he offered a series of Princeton graduate seminars within the general field of the tools he was using. Making sure that the right colleague attended the seminars, he rehearsed week by week the elements of the lengthy construction. For those who have attended mathematical research seminars, the analogy with musical performance is not a far-fetched one. In both cases, the material might be the same on two different occasions, but the emphasis, connectivity, and style can differ. Above all, these 'performances' possess an undeniable *aesthetic*. They can be more or less elegant, beautiful. Mathematics can be presented orally in a dull and pedestrian way, or in the hands of great communicators be delivered with pace, elegance, and drama.

If this analogy of performance in mathematics appears tenuous, or somewhat exclusive, then Wiles's story exemplifies a more democratic way that mathematics can be performed. For it has inspired the permanently available *Horizon* documentary whose first moments I have already related, and subsequently a best-selling book by the programme's producer Simon Singh. Singh is one of those rare technically and linguistically adroit authors, able to convey the structural beauty of mathematics to readers without advanced mathematical training. He and they surely become the analogy of the performers and audience respectively of a musical masterpiece, which once written becomes a cornerstone of cultural achievement, while being re-presented to each generation that follows.

This chapter's task is not an exhaustive categorization of the mappings and connections between mathematics and music, but to ask whether juxtaposing the two can shed light on the process of creation in both of them. Their shared fundamental construction-material of numbers, ratios, and patterns is a superficial beginning, but leads to further avenues we must explore. One has already arisen in a first encounter with musical structure at different temporal scales. Its language called on notes, bar-lines, and staves—a visual and notational expression of music. Building, shaping, and working with creative notation is a process that theoretical physicists have described as essential to the way that many of them work. The methodologies of musical and mathematical creation both touch on visual representation, although neither ultimately refers to anything in the visual world. We will need to listen to people who build with these ideas to create the new—some with famous names and others not so well-known, yet whose everyday experience is to connect the symbolic in music and mathematics with the performed, heard, and thought.

Beyond method lies the energy of the creative purpose, and under that the meaning of the created object itself. Here lies perhaps the deepest shared aspect of the two arts—their ability to excite both our intellect, and aesthetic sense testifies to the power of meaning. Yet it eludes language and depiction alike to describe the meaning of music apart from explicitly programmatic pieces, or of mathematics a part from that which is designed ('programmatically') to model physical systems. Set Victor Hugo's comment on music:

> *Music expresses that which cannot be put into words and that which cannot remain silent.*[8]

alongside mathematician Richard Courant's

> *Mathematics as an expression of the human mind reflects the active will, the contemplative reason, and the desire for aesthetic perfection. Its basic elements are logic and intuition, analysis and construction, generality and individuality.*[9]

These are both wordless and image-free disciplines, yet they resonate with fullness of meaning, contemplation, desire, and aesthetic at the heart of what it means to be human. Behind Hugo's words, and explicitly in Courant's, is the inextricable presence of the emotions, once intellect and aesthetic are combined in creation. As a renowned orchestral conductor once put it, 'Everyone knows that music means something, but no one can say what it is.' As with other signposts within the hidden thought-world of human creativity in art and science alike, this remark points beyond itself. For there is yet another approach to the shared life of mathematics and music, just as there is for experimental science and fiction—in their historical inheritance. This chapter began with another quotation from the thirteenth century commentary on the liberal arts that has already suggested an alternative way of categorizing our disciplines. When we leave the medieval thought-world behind, we lose sight of an essential stream of contemplation that fed the foundation of our contemporary culture. We also forget its embodied family relationship of music and mathematics. Once again, a retrospective glance at the medieval foundations of our contemporary worlds of thought will create a more detached perspective onto the connections that suggest themselves today. We begin with a return to the contemplative context of the thirteenth century, and its cosmic model of heavenly spheres that presented its visual splendour in Chapter 2, but here paying more attention to their harmony.

[8] 'Ce qu'on ne peut dire et ce qu'on ne peut taire, la musique l'exprime.' Victor Hugo (1864) *William Shakespeare,* Part I, Book II, Ch. IV.

[9] Richard Courant (1983), *The Australian Mathematics Teacher*, **39–40**, 3.

Music—the medieval mathematical art

In the medieval curriculum of the seven liberal arts, itself inherited from antiquity, music held pride of place in the 'quadrivium'—the quartet of mathematical arts.[10] It stood alongside the three other advanced disciplines of astronomy, geometry, and arithmetic. Ancient and medieval thinkers chose not to group music alongside what we might today term the 'humanities,' the language-based arts of rhetoric, grammar, and logic that formed the elementary curriculum known as the 'trivium.' The quote at the head of this chapter explains why the thirteenth-century polymath, and our regular correspondent from the thirteenth century, Robert Grosseteste, thinks this must be so. Music orders time, domesticating without trivializing it, and by an extended cosmic metaphor, begins to accommodate by the ordering force of number the strangeness of space as well. It consists of, and conveys, pattern and regularity. The question of the meaning of music itself seems not to arise for this medieval thinker as it does for us—the idea of music seems more akin to meaning itself. Music is that which connects—it responds to the harmonious relationship of the four elements, themselves linked, in that thought-world, to the regular motions of the heavens. This old construction of relations between the heavenly and earthly fell under the study of astrology, which was *not*, we must be clear, primarily concerned with foretelling the future. Medieval astrology was rather an attempt to grasp the unity and harmony of the cosmos in the face of an apparent division into the two worlds of the heavens and the Earth, tracing the influences of the one on the other.

Reading further in Grosseteste's *On the Liberal Arts* clarifies the way that thirteenth-century philosophy thought of music as a

[10] For further reading on the medieval curriculum see, *for example*, Edward Grant (1996), *The Foundations of Modern Science in the Middle Ages*. Cambridge: Cambridge University Press.

sort of disciplinary subcutaneous connective tissue, rather than as a patch on the surface of the curriculum:[11]

> *But when we focus not on what is brought about through bodily motion, but on the regulation in the very motion itself, music is the rectifying [art]; for she teaches in which proportions of motion harmony may be found. The proportions of motion, then, are studied according to a twofold divisibility of motion. For motion is divisible according to a divisibility of time, and according to this divisibility a motion is said to be double in respect to another that is measured according to a double length of time, in the way that a long syllable is the double of a short [syllable]; but motion is also divisible and proportional according to the divisibility and proportionality of space, and in this way a motion may be said to be double in respect to another motion that traverses double the space in the same time. There are therefore five proportions, of which three are of the smallest multiples, and two of the largest superparticulars; for these are between the maximal and minimal divisions in motion according to duration or speed, or according to both. These, I say, are found in motions as a perfect guide.*

This is of course a highly technical piece of writing, and it is easy to forget, after the first sentence, that it is talking about music, for its tenor seems to turn towards ideas we associate with mathematics and with the physics of motion. But music is the world in which proportion, ratio, motion, multiples, and divisibility all reside. Music 'teaches us in which proportions of motion harmony may be found.' Elsewhere in the text, we find that Grosseteste knows that music and motion are connected by the phenomenon of vibration—a struck gong or a plucked string, when closely observed, displays an oscillatory motion by which their elements depart from and return to their natural place of rest many times before settling into stationary equilibrium. It is only while the vibration is visible that a sound is heard. Pre-modern thinkers could

[11] Giles E. M. Gasper, Cecilia Panti, Tom C. B. McLeish, and Hannah E. Smithson, eds. (2019), *The Scientific Works of Robert Grosseteste Vol. 1 Knowing and Speaking: Robert Grosseteste's De artibus liberalibus 'On the Liberal Arts' and De generatione sonorum 'On the Generation of Sounds'*. Oxford: Oxford University Press, trans. therein Sigbjørn O. Sønnesyn.

conceive of a chain of motion by which the string sets up a sympathetic vibration in the surrounding air, which then propagates to our ears. Motion could be perceived as sound.

Motion is also the generator of mathematics. There are two reasons that Grosseteste develops a discussion of *proportion* in his foundational text on the curriculum. One is simply that motion implies velocity, and velocity in turn generates the notion of a *quotient*.[12] A speed is a distance covered divided by the time taken. In all his writings Grosseteste was fascinated by the way that physical phenomena call on mathematical description—or even generate mathematical ideas. He says that one motion can be 'double' another in two ways, in that it can cover twice the distance, or it can persist for twice the time. In the simple (quotient) equation that condenses the words of the last sentence, velocity (V) is defined as the distance travelled (L) divided by the time taken (T), so $V = L/T$. We can double the distance covered L or the time taken T to generate two non-equivalent new motions from an original one.

But what are the strange 'five proportions,' and the even stranger 'two superparticulars'? Here Grosseteste is returning to the 'atoms' of music and mathematics, the counting numbers 1, 2, 3. The whole numbers 2, 3, and 4 are the 'smallest multiples.' A 'superparticular' is just a ratio (the word is transliterated from medieval Latin); new numbers not in the list of simple multiples can be generated by taking quotients of them, the first two being 3/2 and 4/3 (4/2 simply repeats the multiple 2 that we have already listed). Here we confront a more ancient source for this text—for the connection between music and ratio had been recorded long before by the ancient school of Pythagoras.[13] It is hard to know what can be attributed to Pythagoras himself, since none of his own writings survive, and we have only indirect testimony from several Neoplatonist authors, such as the fourth century

[12] A quotient is a simple division: a/b is the quotient of a and b.

[13] The more immediate sources for Grosseteste include the works entitled *De Musica* by the scholars of late antiquity, Boethius and Augustine.

Iamblichus.[14] Yet it is clear that, in any case, the pattern connecting ratio and pitch had been inherited from Greek thought predating Plato. Returning to the plucked string, reducing its length to half of the original will sound a note exactly an octave above the note of the full string. Now stop the length down at one-third of the original length and the note is a 'perfect fifth' above that. (Think of that first dramatic rising interval in Richard Strauss's *Also Sprach Zarathustra* linked forever in Western culture to space exploration by its use in the film *2001: A Space Odyssey*, and derivatively by NASA's Apollo moon-landing programme, or, if you prefer, the first two bugle notes of *The Last Post*.)

Harmony has a human meaning as well as a technical one—there is a medical resonance in music that we have lost in our contemporary thought-world, but which is vital to the history of musical ideas. Behind inherited Pythagorean notions, and explicit in medieval writers such as Grosseteste, there is a sense of established fact in discussing the healing properties of music. Before the advent of the germ theory of disease, illnesses were attributed to imbalances between the four bodily 'humours.' Healing was achieved by restoring balance—by recovering correct harmony and proportion. If music is the liberal art that embodies these two qualities, then it becomes the natural servant of the physician. The idea ought not to be dismissed too lightly, for although music will not, as far as we understand, be effective in combating illness arising from infections, there are other maladies just as incapacitating that affect mental health. 'Music therapy' is an established and effective technique (we will later hear from a practitioner who also composes music). But to grasp why a mathematical art might be invoked to heal, and why mathematics and music reveal such deep connections, we must encounter another ancient text that casts a very long shadow indeed over subsequent

[14] Iamblichus, *De Vita Pythagorica* (*On the Pythagorean Life*), c. 300 AD *Iamblichus, Life of Pythagoras*, trans. Kenneth Sylvan Guthrie (1920).

centuries' musical thought. For the great North-African theologian, philosopher, and bishop, Augustine wrote his magisterial *De Musica* (On Music) from the Christian and intellectual community of fourth-century Hippo—a text that not only survived, but has been read avidly by scholars of every age since.

Augustine on music

The biography of the most influential Christian thinker of the Latin west in the patristic era is as well-known as it is romantic. Born to a well-to-do North-African family living in the fourth century Roman Empire, the admixture of opposing tensions that were to mark his life and thought began their operation early in the devout Christian faith of his mother and the paganism of his father. As a boy, he excelled in Latin grammar and language, and fell under the spell of its literature. Yet he reported his Greek teacher to be a cruel man, beating his pupils—the strongheaded Augustine rebelled by refusing to learn the language. Arguably this simple choice of linguistic path towards the western, rather than the eastern, tradition of thought for Augustine was to be decisive in the division of Christendom half a millennium later, a personal and cultural 'butterfly effect' of immense reach. As a young man, Augustine's lifestyle was famously hedonistic. The parade of vaunted lovers within his Carthage set created another strange dichotomy alongside his espousal of Manichaeism, an early sect of emotional religious fervour. With a mind and soul deeply divided, it is perhaps not surprising that, following his remarkable conversion to Christianity, assisted by the teaching of Ambrose, Bishop of Milan, his great works *Confessions*, *City of God*, and *On Music* itself, carried a strong undercurrent of reconciliation, theological and personal.

The *De Musica* as we have received it (through nearly eighty manuscripts from as early as the eighth and ninth centuries) is constituted in six books, of which the last is the most significant. The first five are effectively preparatory reading for it—Augustine himself seems almost to dismiss them as trivial at the start of the sixth, perhaps because they were written before his conversion, the sixth after. It is written as a dialogue, a pupil standing in for the reader, responding to the master's searching questions. From the very outset these take up the theme of perception of sound, and where its locus is to be found—in the vibrations of the air? in the ear? in the producing voice? in the memory of the recipient? Music takes the stage as the connection between speaker or singer, listener, and the natural air between them. The text poses a problem for the translator in how one should render the Latin *numerales*, which can stand for either 'numbers' or 'rhythms'—a blurring of music and mathematics is enshrined in the Latin *lingua franca* of western thought for over a millennium. The very ubiquity of this idea throughout *De Musica VI* is an example of how a different language-world can help to open up unfamiliar thought-worlds. Sometimes a greater variety in vocabulary leads us to distinctions that are blurred in our own language. A celebrated example is the necessary mapping of the 'Four Loves'[15] of ancient and koine Greek to our single English word. But unequal correspondence can work the other way around as well. A single word used in Latin for two in English forces us to recognize connections to which we are linguistically blind. So, in Latin it is possible to think of musical rhythm, not simply as open to a description by number, but as an embodiment of number itself.

Music, for Augustine, becomes both a window into the mind, and a source of direction towards what is beneficial, and away from the harmful. It is his vehicle for probing the questions we still have today on the locus of free judgement and of the possibility of causal action. Its pursuit of an understanding of what we term 'perception' excites remarkably modern questions. The discussion of Chapter 3 encountered the surprisingly active

[15] C. S. Lewis (1960), *The Four Loves*. London: Geoffrey Bless.

modes in which humans create visual perception of the world—*De Musica* works through the auditory equivalent. Although the question of active perception is familiar to us, the terms in which Augustine poses it are nevertheless strange, in that they are theological. Can the mind (or as Augustine is usually translated, 'soul') really be a simple passive receiver of music, acted-upon rather than acting? Surely if so, this would relegate the soul to a lower status than that of the ear. Yet does this not seem to jar against the relative value of eternal things over the temporal? 'Therefore, it is not strange that the soul, which acts in the mortal flesh, perceives the reaction of bodies,'[16] teases the master of the pupil. The paradox is that of apparently passive reception of musical stimulus, by a human soul that ought to be perpetually active, always taking the initiative in its hierarchical superiority over inanimate matter. Augustine solves it elegantly—he conceives of the soul perpetually generating the *numerales*, the mathematical 'sounding numbers/rhythms' of music, and emitting them in a constant stream from the mind to the ears. It is when these numbers, generated internally, resonate with the same numbers arriving at the ear from the outside world, that the perception of sound occurs at the corresponding pitch of the matching numbers. A swift recollection of our summary of the intromissive and extramissive theories of vision, at the start of Chapter 2, will immediately identify a common pattern. For Augustine, the mind is as perpetually generative of musical numbers as it is of the potential rays of visual perception supposed by Plato in the *Timaeus*.

A consequence of this ancient line of thought is relevant to our search for the source and story of human creativity in art and science, for it identifies the mind (soul) as a perpetual creating source of numbers and tones, as it is of pictures and words. Just as in the

[16] Aurelius Augustinus *De musica* liber VI A Critical Edition with a Translation and an Introduction M. Jacobsson (2002), Stockholm: Almqvist & Wiksell International, IV.7 p. 23.

case of visual perception, the resonances with findings of modern science are remarkable. Not only is the brain continuously alive with the pattern-making of memory: visual, auditory, narrative, and tactile, but recent research indicates that the ear itself is an active, rather than a passive organ. Its individual 'hair-cells,' tuned to pick up particular frequencies of incoming sounds, are almost at the point of bio-mechanically oscillating at those frequencies themselves.[17] This active tuning produces the exquisite sensitivity to very soft sounds that must have saved many ancestral hunter-gatherers' lives. As a consequence, the mammalian ear is actually able to make sound as well as to detect it.

Music is not simply a form of creative art; as a correlative of the evolved human sense of hearing, its own story parallels those of its creators. Small wonder that the period (c. 40,000 BCE) in which cave-art and jewellery first leave archaeological record, preserves early bone flutes that sound the same harmonic series with which we build music today. Like art, science, and story, music has its own developing narrative of discovery and creation. Since music, like mathematics, is wordless and picture-less, it is perhaps more challenging to write about it than for literary forms of art. Relationships between the domains of the novel and the 'art of scientific investigation,' were illuminated by reading Henry James, an exponent of the nineteenth-century novel alongside parallel reflections by William Beveridge, a twentieth-century scientist. It would be equally instructive to find masters of music and mathematics who also wrote about their creation during that same pivotal period for science, fiction, and art, because the nineteenth century also somehow sowed the seeds of their division that we are seeking to understand and to mend. Fortunately, there are candidates for such articulate commentators. We will take a close comparative reading of a great musician

[17] For background see, for example, P. Martin, A. J. Hudspeth, and F. Jülicher (2001), 'Comparison of a hair bundle's spontaneous oscillations with its response to mechanical stimulation reveals the underlying active process,' *PNAS*, **98**, 14380–5.

from the nineteenth century, and a great mathematician from the early twentieth who, likewise, not only represented great creativity themselves, but also wrote in depth about the creative process.

Robert Schumann—creative tension, form, and genre

If asked to name the most prolific of composers in the western tradition, many would suggest Mozart, or after recalling the number of his symphonies, perhaps Haydn. Yet the music we have from the great Romantic composer, Robert Schumann, comes from little more than the decade 1841–51. In his several year-long bursts of prolific activity, he outshone in sheer rate of radiant musical output even Mozart or Haydn. The *Liederjahr* (Year of Song) of 1840–1, following his and his equally gifted wife, Clara's long-delayed marriage, saw the creation of over 140 works, and transformed the genre of solo poetic song forever. The following year he took up the formidable challenge of symphonic writing in the shadow of Beethoven; then in 1842 he found new pathways for the more concentrated form of chamber music, including the invention of a new genre—the piano quintet.[18] Yet the politically turbulent year of 1849 excelled all others: three instrumental chamber works, thirteen choral or vocal ensemble pieces, four further collections of lieder, and among other works the unique bolt from a blue musical sky that is the *Konzertstück* for four horns and orchestra.

Schumann makes an inviting subject for a comparative study of creativity, for other reasons than his own prolific and innovative output. For two of our constant themes, whether exemplifying the radically new in art, science, or literature, are exemplified

[18] The instruments of a standard string quartet together with a piano, in which all partners are treated as equal in the roles of melody and accompaniment.

strongly in his life and work. First is the idea that to generate the radically new requires the forging of connections between distant streams of thought. Schumann worked throughout his life between disparate sources of ideas, principally the literary and the musical, enabling him to innovate beyond the incremental. Second is the power of creative duality, be its poles those of desire and difficulty, imagination and constraint, or the flood and drought of ideas—such inner tensions seem as conducive to ideation as the reading from two or more disparate external sources. Schumann's biography is replete with examples of dualities. Among his many influences, for example, the two that stand out could not seem more removed from each other in character. It was to the formal, structured, mathematical music of J. S. Bach that he turned to again and again in intensive study throughout his life. Yet in literature he immersed himself in much the same manner in reading the work of the romantic, unstructured, emotional, and explorative writing of Jean-Paul Richter. Such dual passion for literature and music became a constitutive creative force for him, to the extent that his biographer John Daverio declares him to be the first composer to think of writing music as a sort of literature.[19] Daverio is not thinking of their obvious intersection in the genre of song, but of a shared texture of meaning and narrative in all music and writing.

French philosopher Rolande Barthes wrote a perceptive essay, *Loving Schumann*, in which he searches for the underlying cause for the unique visceral resonances of his music. The duality of art and its relation to the world is his starting point:[20]

> *For Schumann the world is not unreal, reality is not null and void. His music, by its titles, sometimes by certain discreet effects of description, continuously refers to concrete things: seasons, times of the day, landscapes, festivals, professions. But this reality is threatened with disarticulation, dissociation, with movements not violent (nothing harsh) but brief and, one might say, ceaselessly 'mutant': nothing lasts*

[19] John Daverio (1997), *Robert Schumann, Herald of a 'New Poetic Age'*. Oxford: Oxford University Press.

[20] Rolande Barthe (1985, English translation), Loving Schumann, in *The Responsibility of Forms*. Berkeley: University of California Press.

> *long, each movement interrupts the next: this is the realm of the intermezzo, a rather dizzying notion when it extends to all of music, and when the matrix is experienced only as an exhausting (if graceful) sequence of interstices.*

Barthes develops a paradoxical thesis that Schumann responds to the dualisms around him with an insistence on unity—a sort of musical unison that does not erase harmony but keeps returning to a musical centre, something he refers to as the composer's 'singularity.'

There are, as for any artist, important clues to be had from biography. Life events created other tensions between opposing forces without as well as within. Schumann's great inner lure to a musical career was frustrated by an overbearing mother and guardian who insisted on a legal training. The struggle to break free from that expectation, eventually achieved in 1830 after a move from Leipzig to Heidelberg, was his first act of self-liberation, but by no means the last. For his original wish to be a concert pianist was frustrated by a tragically self-inflicted hand-injury when still a young man. In any case, it is clear that Clara outshone him as a performer, as she did all other pianists of their generation. Their marriage was won only after an even greater struggle against the stony and protracted refusal of her father, Frederick Wieck—once Schumann's piano teacher—to entertain the engagement. Ironically, only Robert's recourse to the assistance of his once-rejected legal profession made it possible for them to marry in 1840.

The marriage was clearly a passionate and mutually supportive one, though not itself without the huge stresses that arise from two careers in the public eye, a large family, and the perennial financial worries of those in the musical profession. Yet a shared love of music fuelled their life together in ways that they recorded in their shared *Tagebuch*, completed daily throughout the 1840s. The couple made it their shared discipline, for example, to pursue the extended study of the highly structured fugues and forms of Bach, while Schumann's own music continuously sought to push at the boundaries of structure, and to innovate new genres.

Alongside many other influences, and as a constant but ambiguous companion, he seems to have suffered from a severely split personality. Whether he was bipolar in the strict psychological sense is now called into question, though once accepted as uncontroversial. It is true that the last two years of his life were spent in the asylum at Endenich, an outcome that he had feared for much of it, but it is wrong to project that final mental illness back into his earlier life and work. That he suffered several extended periods of very dark depression and others of ebullient joy, is not, however, in doubt. He managed to draw on the energies of both in his musicmaking. Even when he wrote about music, it was under two pseudonyms: 'Florestan' and 'Eusebius,' the first romantic and given to transports of emotion, the second more staid, analytical, and dry.

As well as composing, Schumann wrote about music powerfully and perceptively. This is the final qualification that recommends him, paradoxically, to our close study in a chapter on the creation of the wordless and abstract. In the 1830s, before composing in earnest himself, he founded in Leipzig a new musical journal, *Die Neuer Zeitschrift für Musik.* Florestan and Eusebius, among other pseudonyms, were joined by a close group of musical writers, eventually including Clara (whose own pseudonym, *Chiara,* occurs as a frequent by-line). The like-minded little band commented on new music as it was published through the years of late Schubert, early Mendelssohn, Chopin, and Liszt. These great composers, whose names we know well today, constituted of course only a minority of those writing at the time. The pages of *Die Neuer Zeitschrift* were more likely to contain reviews of new piano pieces by Hertz, Hiller, or Lachner, or of symphonies by Müller, than by works more familiar today by Berlioz, Chopin, or by Schumann himself. But his sometimes florid but always keenly targeted analysis leaves a reader also an admirer, for Schumann's criticism was rarely misplaced. He has an almost prophetic ear for the timeless (perhaps his most celebrated comment is his first report on Chopin—'Hats off gentlemen, a genius!'), yet reserves his lengthy point-by-point criticisms in general for the worthy but mediocre. Occasionally, he comments explicitly on the quality of longevity in musical reception, as in an 1835 piece on new piano sonatas by Mendelssohn and Schubert:[21]

> *Only that which has intelligence and poetry vibrates on into the future, and the slower and longer the vibration, the deeper and stronger the strings that were struck.*

Such physical metaphors occur often in his most perceptive and deepest remarks; it is also lovely to read of poetry and the physics of vibration combined so effortlessly into a statement that says that the best music is like literature. Within his measured criticism of less profound compositions are set nonetheless some very finely polished observations of musical wisdom that reflect an intense awareness of how music is constructed. An example is this excerpt from an 1836 review of two piano concerti by the London-based pianist and composer Moscheles:[22]

> *A genuinely musical art form always has a focal point towards which all else gravitates, on which all imaginative impulses concentrate. Many composers place it in the middle (like Mozart), others reserve it for nearer the close (like Beethoven). Wherever it lies, the effect of any composition is dependent upon its dynamic influence. If one has been listening, tense and absorbed, there should come a point where, for the first time, one breathes freely; the summit has been reached, and the view is bright and peaceful—ahead and behind.*

It's a striking passage, if only once more for the repeated physical metaphors (gravity, impulse, dynamics, light). Schumann seems himself to be articulating Barthes' notion of singularity. It also touches an aspect of musical and mathematical creativity that alerted us at the start of this chapter—the performative relationship between artist and audience, or 'creation in community.' The 'summit on which ones breathes freely' is recognizable as the point in any creative process where the fundamental difficulties are finally removed, the dead-ends navigated. The end is by no

[21] Robert Schumann (1965), *Schumann on Music; A Selection from the Writings*, trans. and ed. Henry Pleasants. New York: Dover p. 88.

[22] Robert Schumann, *Schumann on Music*, p. 108.

means achieved, but the path towards it is finally perceptible—and this because of a central idea that clears the way and provisions the journey. It is the experience captured by mathematician Henri Poincaré in a celebrated moment of clarity within a rare example of mathematical introspection:[23]

> *For fifteen days I tried to demonstrate that any function could exist, similar to what I have called Fuchsian Functions, but I was very ignorant, and every day I sat at my desk. I spent an hour or two, I tried many combinations and I came to no results. One evening, contrary to my habit I took black coffee, I could not sleep; ideas rose in crowds; I felt them come against me, until two of them were clinging to form a stable combination. In the morning I had established the existence of a class of Fuchsian Functions, those derived from the hyper-geometric series; I had only to write the results, which took me a few hours.*

Poincaré narrates his own 'reached summit' (the 'morning') and 'focal point' (the class of functions). The cognitive and emotional points of comparison between these musical and mathematical testimonies are clear—the contrast is that Schumann is here referring to the experience of the *listener* to the music, not its composer. It is as if the audience were asked to retrace the creative steps of the composer, perhaps in a shadowed and guided form, but one that nonetheless requires focused concentration and an active anticipation of direction and goal. But if music (or mathematics) need not refer to anything but itself, and need not be *about* anything, it is hard to understand how such an intense communication between originator and listener could arise. Artist and viewer can both point to the canvas and what is represented there, in the richly layered imaginative work of painting and its reception, but for the picture-less and wordless, the absence of referent is one of the mysteries that has brought the two into comparison. The young Schumann suggested that the aesthetic power of music lay precisely in this absence. The resonance to this apophatic

[23] Henri Poincaré (1915), Mathematical Creation, in *The Foundations of Science*, trans. G. B. Halsted. Lancaster, Pennsylvania: The Science Press.

tradition, originally conceived in theology,[24] is unmistakable in this musing passage from an 1835 *Neuer Zeitschrift* article on Hiller's opus 15 études:

> *In no other field of criticism is it so difficult to offer proof as in music. Science can argue with mathematics and logic. To poetry belongs the golden, decisive word. Other arts have accepted nature herself as arbiter, from whom they have borrowed their forms. Music is the orphan whose father and mother no one can determine. And it may well be that precisely in this mystery lies the source of its beauty.*

Just as the role of the theologically apophatic is to direct from *what is not* in order to define, if only in shadow, outline, or relief, *what is*, so the mystery of music's origins developed, within Schumann's thinking, beyond the utterly unsolvable. By 1835, he can say more. Hector Berlioz's *Symphonie Fantastique* had recently appeared, and although Schumann never attended a performance, he had Liszt's 1834 piano reduction before him when he wrote the most detailed and the longest critical piece ever to appear in the *Neuer Zeitschrift*. Berlioz's orchestral *tour de force* is highly 'programmatic'—it tells an impressionistic story of episodes in the life, dreams, and nightmares of an artist. Its symphonic form breaks moulds in many ways; for example, there are five movements rather than three, each carrying characteristic motifs that are maintained in stark simplicity, rather than developed.

Schumann recognizes that the response of such novelty arises in part from the burden that any symphonist in the mid-nineteenth century must bear, writing in the shadow of genius: 'With Beethoven's Ninth Symphony, the greatest of all purely instrumental works in respect of sheer size, it seemed that the ultimate had been reached in terms of both proportions and objectives.' Of course, Schumann understood only all too well the almost intolerable task of writing major orchestral works after

[24] For a recent, imaginative, reception, and reimagination of the apophatic tradition of theology from Gregory of Nyssa and Dionysius the Areopagite to Judith Butler, see Catherine Keller (2014), *Cloud of the Impossible: Negative Theology and Planetary Entanglement*, New York: Columbia University Press.

Beethoven. A symphony in G written in 1832–3, replete with Beethovenian influences, went unfinished and unpublished, and he was not to find his fully fledged symphonic voice until 1841. Berlioz's approach to innovating within the 'shadow' alerts the critic to what *can* be said about source for musical ideas:

> *Alongside the purely musical fantasy there is often, all unwitting, an idea at work; side by side with the ear the eye, and this ever-active organ retains, amidst the sound, certain contours and outlines which, as the music itself takes shape, crystallize and develop into distinctive images.*

He reflects that Beethoven himself had been seized by thoughts of immortality and fallen heroes, and transformed them into music, or rather let those thoughts inform the developing 'music-related' and 'tonally produced' elements of composition. He alludes to Mendelssohn's inspiration from Shakespeare in writing the overture *A Midsummer Night's Dream*. In an extreme and personal example, he claims an experienced connection between a piece of music and a visual and historical setting, assuring any incredulous reader that he and a friend were both independently put in mind of the same period of medieval Seville by a sonata of Schubert's.

We are a long way from the 'mystery' of musical source here, and now in the realm of concrete sources of inspiration, but nonetheless aware that the threads that connect the visual, experiential, and historical to music are long and tenuous. Not unexpectedly, there are strong similarities in Schumann's description of the multisensory and multiply connected shape of the musical imagination with the shape of poetic imagination explored in the previous chapter. Musical 'forms' and 'images' are not the fairy scenes in the forest court, nor the Spanish city, nor images of heroic figures, but dwell intact in their own world of ideas. They may nonetheless carry a 'mapping' onto patterns in the world of non-musical experience. Once more a reader with a little mathematical background will raise an eyebrow, for the notion of mutual mapping of objects and the relations between them, from two entirely different 'worlds,' is ubiquitous in mathematics as the 'homeomorphism.' The identifications do not have to be exact, like a perfect photograph of a real scene; they may even collapse complexity in one domain onto simplicity in the other, but any patterns in a homeomorphism that remain will be faithful in the one domain to the other.

Andrew Wiles, the mathematician who opened this chapter, describes the process of noticing such structures using metaphorical language strikingly similar Schumann's: 'suddenly you see the beauty of this landscape and you just feel it's been there all along.'[25] The larger project within mathematics, greater even than finding proofs of theorems such as Fermat's, is essentially about constructing homeomorphisms. In its most extreme form, the 'Langlands Programme' looks for the undergirding patterns of connection between different branches of mathematics itself, including those that find roots in number theory on the one hand and geometry on the other. Russian-American mathematician Edward Frenkel puts it simply:[26]

> *The Langlands Program is now a vast subject. There is a large community of people working on it in different fields: number theory, harmonic analysis, geometry, representation theory, mathematical physics. Although they work with very different objects, they are all observing similar phenomena.*

It is significant that the motivation for Frenkel's book, and the reason for his introduction of Langlands' panoptic vision to a lay readership, is that he wants a non-mathematical readership to experience the aesthetic loveliness of deep mathematics. There are musical versions of the spirit of Langlands programme; arguably each composition makes a structural and relational link between two or more elements, but Schumann's ability to bring the new and unknown out of connecting and transforming the known but unrelated, is of the first order.

[25] Andrew Wiles, 'What does it feel like to do Maths?' *Plus Magazine*, 1 December, 2016.

[26] Edward Frenkel (2013), *Love and Math: The Heart of Hidden Reality*. Basic Books.

Although Schumann does not have a psychological, let alone mathematical, vocabulary, the mental fuses that are lit by external scenes or narratives, and eventually lead to the explosive creation of new music, are also hidden in his non-conscious creative mind. One indicator of the mental depth at which musical influences lie is their apparent defiance of the simple logical analysis that belongs at shallower levels. In the Berlioz example, Schumann is surprised at the stark simplicity of harmony, the breaking of formal rules of musical expression, that nevertheless 'works' in communicating between composer and listener. The level at which a satisfying resolution is given and taken is not purely cognitive. Schumann writes of both conscious and non-conscious levels of compositional creation. The articulated discussions of composition in his letters and diaries lean more towards a consciously meditated compositional technique, especially in his later years. A diary entry for 1846 even suggests that a shift from improvisational (more subconscious) to deliberate invention represents a maturation:[27]

> *I used to write most, practically all of my shorter pieces in inspiration; many compositions with unbelievable swiftness, for instance, my First Symphony in Bb major in four days . . . only from the year 1845 on, when I began to invent and work out everything in my head, did a completely new manner of composing begin to develop.*

By 'in my head,' he means in contrast to composing at the keyboard. But we should not jump to the conclusion that a deliberate mental construction meant a suppression of the inspiration that comes through processes hidden from conscious cognition. In a later (1852) letter to Carl von Bruyk, he advocates purely mental composition for precisely the opposite reason:[28]

> *Accustom yourself to conceiving music freely from your imagination, without the aid of a piano; only in this way are the inner wellsprings revealed, thus appearing in ever greater clarity and purity.*

[27] R. Schumann, *Tagebücher* 2, p. 402.

[28] R. Schumann, *Briefe, Neuer Folger*, p. 356.

Tension between the conscious deliberate sculpting of form on the one hand, and openness to the 'upwelling' of ideas from mental regions hidden from conscious thought on the other, has become a familiar theme across the spectrum of arts and sciences, but it seems especially familiar to musicians and mathematicians alike. Making one more textual leap from Schumann to Wiles, here is the mathematician in a similar advisory mode:

> *Then you have to stop, let your mind relax a bit and then come back to it. Somehow your subconscious is making connections and you start again, maybe the next afternoon, the next day, the next week even and sometimes it just comes back. Sometimes I put something down for a few months, I come back and it's obvious. I can't explain why. But you have to have the faith that that will come back.*[29]

Schumann gives another reason for taking leave of the piano. He wants to break away not only from writing at the keyboard, but from writing *for* the keyboard—here the dialectic is not between the conscious and unconscious modes of composition, but signifies the old creative tension between imagination and form. Two excerpts from letters in 1839 illustrate the emotional appeal of a larger form. To his composition teacher, Heinrich Dorn, he wrote:[30]

> *Sometimes I would like to smash my piano, it has become too narrow for my thoughts. It is true that I have so little experience of the orchestra, but I hope to arrive at this in due course.*

Later that year he wrote to his (then fiancée) Clara Wieck, relating the experience of a rehearsal of Schubert's Great C-major Symphony under the baton of Mendelssohn. Schumann had rather a high stake in the work, as he himself had discovered the manuscript in Vienna on a visit to Schubert's brother, and suggested that Mendelssohn take it on.

[29] Andrew Wiles, 'What does it feel like. . .'
[30] Written 14th April 1839.

> *Clara, today I was in seventh heaven.*[31] *There was a rehearsal of a symphony by Franz Schubert. If only you had been present! It is indescribable, the instruments are men's and angel's voices, and everything is so full of life and spirit, and an orchestration defying Beethoven—and that length, that heavenly length! Like a novel in four volumes, longer than the Choral Symphony. I was utterly happy, and would have wished for nothing but that you were my wife, and that I should be able to write such a symphony myself.*

This little window into Schumann's soul opens onto many of the qualities and constraints that make him a uniquely transparent representative of the genre of musical creator: his emotional desire for more expansive forms both in temporal and musical and tonal range, the underlying motif of song, the overcasting shadow of Beethoven, the literary metaphor. Also, 1839 seems to be the year of conception of a long project—to write the new large orchestral pieces that he hoped were within him. Like Wiles's long mathematical project, this one was also the work of many years. We will look in detail at one of its consequences, a remarkable work that he thought one of his best, though it remains remarkably understudied. As in theoretical physics, works of art, and ground-breaking literature, generalities can provide essential background, but cannot take us to the junctures where new ideas are conceived. For that we need to inspect an innovative musical work closely, as well as to appraise it from a distance. There is no alternative to some level of immersion at a technical level[32] if we are to perceive and evaluate whether the analytic and geometric structures of music and mathematics bear the relations that we suspect they do. This includes, as will a close reading of

[31] Anyone as literary-minded and educated as Schumann would not have written this as a careless synonym for rapture, in 1839. The 'seventh heaven,' in ancient and medieval cosmology, was the sphere of Saturn, at once the sphere of contemplation, and the link between the planetary and stellar spheres beyond which angels dwelt.

[32] Readers familiar with musical notation will find the excerpts add clarity, but their content is also described in the text. A complete score is available for download at https://imslp.org/wiki/File:TN-Schumann,_Robert_Werke_Breitkopf_Gregg_Serie_3_RS_15_Op_86_scan.jpg.

a mathematical result later, presentation of technical notation, ubiquitous in abstract forms of creation. In the light of what we have already heard about how such symbols perform as creative tools as well as pure representation, it seems appropriate to look at them, and their recombinations, on the page as well as to write about what they represent.

A musical close-reading: The Konzertstück for Four Horns and Orchestra (1849)

For a closely read case-study in musical creativity we look a decade onwards to the extraordinarily productive year of 1849. Perhaps an ear to the musical contemporary of Humboldt's *Ansicht an die Natur* of Chapter 3 will draw from that a literary context for a work that also seems to be taking in great vistas. Schumann was writing in Dresden as the violent revolutions of that year swept Europe—and at the very time they reached that city. It seems to have been Clara, while pregnant, who managed the removal of the family to safety in an outlying village, while Robert became distracted by an inner excitement of imagination. As he wrote in April to Hiller, 'For some time now I have been very busy—it's been my most fruitful year—only therein did I find a counterforce against the forces breaking in so frightfully from without.'

The counterforce itself needed to be a powerful one—and the romantic ideal in the form developed by Schumann in 1849 was perhaps uniquely tuned to express a combination of exuberance and inner strength. As the Edinburgh musicologist Hans Gál writes, Schumann possessed an 'urge to communicate but was by nature an introvert.'[33] Gál draws attention to one of Schumann's favourite musical markings, *innig*, a difficult German word to translate well. 'With intimate feeling' is one rendering, but it fails to capture the intense compressed interiority of feeling and

[33] Hans Gál (1979), *Schumann Orchestral Music*. London: BBC Publications.

life—perhaps *innig* is more an invitation to follow the composer into a state of 'dwelling inside with the music.' The question for Schumann is how one can live intimately and privately within, yet respond heroically (with a 'counterforce') to physical dangers without. One answer is to choose a voice that brings with it both the musical colours and narrative motifs that combine the centripetal force of *innig*'s intimacy, and the centrifugal impulse of *kräftig* (with might). The leading candidate in possession of those qualities in early 1849 was undoubtedly the horn. As the brass instrument with the mellowest and most inflected tonality, the one most closely and bodily entwined with the player, it also carried ancient tropes of the hunt—ichneutic energy and hopeful expectation, its signature calls invoking wide mountain landscapes. In his 1840 review of Schubert's C-major symphony, the only specific passage in the entire work to which Schumann drew attention was in the second movement, where the horn solo 'seems to come from a distant sphere.'[34] The horn provided a pathway to the unanimity of intimacy and romantic spaciousness that Schumann's project needed.

At just this point, the composer met with an unfamiliar but vital friend in the form of musical technology: until the early nineteenth century the horn, although able to paint with rustic arpeggio calls the despairs and adrenalin rushes of the hunt, had remained a primitive instrument, unable to access the full chromatic musical scale. It was limited instead to the (Pythagorean) harmonics—the whole number multiples of its lowest, or 'fundamental' note. Since the musical scale is a logarithmic one, these harmonics sound at successive intervals of the octave, then a fifth, a fourth, major third, flat minor third, then a sequence of near major seconds of dubious tuning. In this way, the accessible tones become far more closely spaced at the high end of the playable

[34] Robert Schumman, ed. Henry Pleasants (1965), *Schumann on Music*. New York: Dover, p. 166.

register, but span huge gaps in the mid and lower range. Early solutions were cumbersome: exchangeable inserts of tube ('crooks') differing in length could tune the instrument to different fundamental notes, providing that rapid changes of key were not required. Virtuosi of the eighteenth century had also developed the technique of 'hand-stopping' (the reason that horn players to this day maintain their right hand within the bell of the instrument). A rapid closing-off of the bell by the hand can lower the sounding note by as much as a semitone, but at the cost of a much more muffled sound. By the early decades of the nineteenth century at least one horn manufacturer had at last created a version of the valve-horn, by which extra sections of tubing could be inserted into or removed from the airflow within a fraction of a second by the press of a finger. By the 1840s, Schumann had the voice he needed. An experimental piece in early 1849, the *Adagio and Allegro* for horn and piano—an endearing favourite—proved his hopes well-founded.

Later in February 1849, the new instrument and its capabilities, and the desire for a statement of heroic romanticism against a fully orchestral landscape, combined to inspire an unprecedented piece, the *Konzertstück* (concert piece) for a quartet of solo horns and orchestra. The name does not really do it justice, for it is a fully developed concerto in three movements, drawing on the tonal range and power of a full symphony orchestra to accompany its formidable solo quartet. The innovative leap that opens the pathway to the detailed thematic and formal structure of the work is the conception of the new genre itself, arguably a creative act of higher order than the writing of even great exemplars within a genre. More usually, new forms appear by gradual evolution—it is not really possible to identify a single originator of the forms of symphony or concerto, for example. In the case of the string quartet, however, the identification of the extraordinary emotional range capable of expression from this unprepossessing little group, has to be ascribed to the genius of Joseph Haydn. Schumann was also an originator of genre:

there is no example of chamber writing for the now-standard piano quintet (string quartet of two violins, viola, and cello with a fortepiano) before his ground-breaking work in E-flat major of 1842. Schumann scholar and biographer John Daverio has suggested that an inspirational source for the quintet might have been a piano trio (also in E-flat) by Schubert.[35] In this case, the imaginative step that Schumann might have glimpsed was that the augmentation of the string duet of the trio into a full quartet would give a more adequate musical counterweight to the piano. In turn that would permit a work that continually transformed from dialogue to accompanied solo—not just bringing the piano into the foreground, but alternating with passages where the quartet constitutes a 'soloist,' accompanied in turn by the piano.

What solo quartet might produce sufficient strength to extend this formal experiment by replacing the piano with a full orchestra? Schumann was already enamoured by the romantic and heroic narrative history of the horn, by its tonality and newfound technicality. He does not record, even if he was aware of them, any other sources for his final choice of a horn quartet, but that ever-present shadow on the backdrop of his compositional life might play a part here. The ghost of Beethoven never left the stage of symphonic composition until arguably the Viennese school of the very last years of the century. A musician minded to write in heroic vein in the 1840s could not resist a conscious or unconscious visitation of themes from Beethoven's 3rd symphony (the 'Eroica,' originally dedicated to Napoleon). That work's scherzo takes an unexpected turn at its central point, introducing a trio of horns that transform the symphony's sinuous thematic material into a call to the hunt. The addition of the third horn to the classical orchestra's usual pair enabled

[35] John Daverio (2002), *Crossing Paths: Schubert, Schumann, and Brahms*. Oxford: Oxford University Press.

more chromatic playing (the third horn is able to use a different crook to the first two at times), but more significantly opens up the sound of a major triad on three horns, in canon or simultaneously. Suppose the addition of a fourth horn to make a full quartet—such a move would unleash the possibility of two, three, and four-voiced canons, would be able to play with quadruple as well as triple rhythms, or divide into two duetting pairs. Even the mathematics of combinatorics urges the idea. It works marvellously, which is why it is so hard to account for the almost solitary place that the *Konzertstück*, holds in this, the most sparsely populated of Schumann's invented genres. In a letter to Ferdinand Hiller on 10th April 1849, the incisive musical critic and self-critic wrote, 'It seems to be one of my best pieces.'

Let the music begin. The formative energies of Schumann's own complex dualities, the shaping of romanticism both literary and musical with which he charged himself, the constant mentoring influence of his continual disciplinary study of Bach, the

Example 1

exuberance of his Florestan and the sobriety of Eusebius—all combine in the work's dense structure. The *Konzertstück* is announced by a bold and accented falling minor third from the orchestra. From the beginning there is a statement of ambivalence between this minor third in its own key, and the other minor third which belongs in the triad of a major key. The horns, making a first attempt at their heroic role, answer (for now) in a fanfare to wake the dead (Ex. 1), and lead the orchestra chasing after them into a landscape contoured in F-major, finding the first strand of thematic material in Ex. 2).

Example 2

A full sonata form then unfolds as one of the most elaborate Schumann ever wrote. It also proves unconventional in many ways—the closest sonata type (technically type V) is the 'ritornello sonata,' in which soloist and orchestra (the ritornello) alternate in giving voice to the thematic and developmental material.[36] But right from the beginning there are signals aplenty that this concerto-sonata is moving into very new territory, for those highly compressed opening chords and fanfare replace an expected orchestral ritornello exposition. The soloists then return, but only after an unsettling harmonic reversal[37] in the orchestra. They introduce a characteristically minimal fragment of song (Ex. 3) that will prove germinal throughout the work. At first impression, this is a new idea, but developing the ambiguous note of the opening form it also echoes, by inversion, the statement in Ex. 2).

[36] For details of the technical analysis of the *Konzertstück*, I am indebted to Prof. Julian Horton of Durham University; there is until now no detailed published analysis of the work.

[37] Bar 15; technically a 're-opened half-cadence' from G7 to C.

Example 3

The horns then introduce a second theme in short fanfares, but this time using the first and third of the trio in canon. As soon as the upper register is struck, the rising horn-calls reverse into falling motifs and, not for the first time, the piece undergoes a rapid mood-swing[38] from the brave towards the reflective. The second subject (Ex. 4), seems to achieve this by patterning itself on an inversion of (Ex. 2), but deforming the rhythm, speeding up the dotted motif, and delaying the arrival at the new tonic.

Example 4

The harmony colours the mood with tonal ambiguity: A-minor vies with C-major but neither gains the upper hand. The expected cadence at the end of the second subject is removed altogether, maintaining the ambiguity of key, but also breaking the classical lines of the sonata form in favour of a much closer integration of soloists and orchestra. More triplet calls in canon recover the mood and the key twists in another reversal to a bold statement of F-major, which sets the full orchestra off on an exploration of musical peaks in a recapitulation of the opening material. Again, the lofty emotional ground cannot last, and the tonality, pitch, and harmony begin to be sombre again into the lengthy development. On the way down to the darker valley floor, the orchestra meets the first horn, again sounding Ex. 2, but now at a lower altitude. Joined by a second horn, the pair fail in an attempt to

[38] Effected by Schumann's hint at the usual 'medial caesura' before the new subject at bar 36.

call the ensemble back to high country, and a shadow falls suddenly across orchestra and quartet alike. Questions seem to be exchanged about where the path ahead lies, in E-minor, against shimmering strings, alert and fearful (Ex. 5).

Example 5 *showing upper strings (top) and solo horns I and II (bottom)*

Minor and major alternate with increasing frequency until, still in the gloom cast by the tremulous strings, a new and sweet melody of hope, paradoxically in the 'falling' mode of the F-major arpeggio, is sounded by the first horn (Ex. 6). It sounds far away and solitary, but in a place of light once more.

Example 6

This is new musical material but appears in a development section which normally would have worked only with motifs already introduced. Schumann's repeated breaking of the norms of concerto-sonata form serves to weave the texture tighter. The other horns and orchestra seem now able to follow. Reprises of the first horn fanfare (Ex. 1) now in the orchestra, then again in the horns, lead into a fortissimo *tutti* recapitulation of the original theme (Ex. 2).

This is no canonical recapitulation and simple return to the first statement, however, any more than the opening section assumed standard ritornello form, for in retrospect, the recapitulation had started before the explicit F-major return of Ex. 2. Anticipatory clues to this clever overlap begin to emerge in the previous triplet horn-calls developed from Ex. 1. Not only do these echo the soloists' entry at the very beginning of the piece, thus making it possible to hear them as part of a recapitulation, but the harmonic progression in the orchestra hints prophetically at what is about to happen.[39] Schumann creates a coded structure in which the same music may be read in two different ways:[40] in this case either as the close of a development section or as the opening of a recapitulation. This is, in coding terms, a direct musical analogy of the 'overlapping genes' of some viral DNA.[41] For the parsimonious viruses, the trick allows up to a threefold compression of information. The effect is similar at this critical point in the *Konzertstück*, where the theme of ambiguity combines with a sense of overwhelming temporal density of musical ideas. The music is the more explosive because its logic is so highly compressed.

The theme of Ex. 3 returns in a thicker interweaving with all the material so far: horn-calling triplets, inversions, canons, and the rising question and answer motif affords now a range of different rhythmic structures. Harmonic anticipation is used effectively once more in the very dense recapitulation—for example, the second subject (Ex. 4) now enters at an interval of a fifth lower. This directs the following pattern of modulations to land, not (albeit ambiguously) on C, as it did on its first appearance, but points finally towards F, the home key. All this material is developed through a very rapid and complex sequence into a closing orchestral *tutti* that does finally reach the original F-major,

[39] The harmony pivots on A from D-major to D-minor (bar 158—61), preparing the F (third of D-minor) that needs to burst out as the tonic at the recapitulation of bar 164.

[40] Formally a 'non-congruence'; examples begin to occur in Beethoven's piano sonatas.

[41] N. Chirico et al. (2010), *Proceedings of the Royal Society B: Biological Sciences*, **277** (1701), 3809–17.

but remembers the darker episodes that the movement has experienced. Significantly, the horns are not involved in the final resolution, another departure from concerto form as Schumann received it. We might have come home, temporarily, at the close of the first movement but the soloists have not—they have more travelling to do.

Such musical memory is significant in the work, and must partly motivate Schumann's unusual choice to elide the movements—the *Romanze* follows without a break. The first idea in its ABA structure is a contrapuntal canon in D-minor between the first and second horns (Ex. 7).

Example 7 Romanze

The thematic material, like the structure of the work as a whole, is linked to the first movement—the theme modifies the material of Ex. 3 and Ex. 4, but the musical metaphors are different. If the landscape of the first movement was more mountainous, here we are water-borne. As the *Romanza* develops so the strings ripple beneath the full horn quartet in quadruple, then triplet semiquavers. Fluidity washes rhythm across bar-lines as the formal $\frac{3}{4}$ accommodates moments of four and two, as Schumann had frequently done in his earlier pianistic writing. The bass-line drives subtle shifts of key that play games with the listener's memory. So, for example, the B-flat in the bass-line at the opening of the movement is the sixth of the D-minor key, but when it returns at the opening of the final A-section it is the tonic.

The second elided continuation between the second and third movements is part of Schumann's long journey through different keys in this way—we know that finding our way home means landfall onto F-major—but we are still at sea. As Julian Horton commented as we worked through the piece,[42] 'One of the marks of composers of first rank is their strategic genius of setting up problems and resolving them within connections of tonality.' It would be hard to formulate a better description of this shared skill between the master-navigators of mathematical and musical landscapes. The next subproblem is to decide which of the conceptual mountain passes that present themselves is most likely to lead to the goal, while low enough to scale. The choice is as much aesthetic as logical, demanding a continual double-vision of local detail and global destination, as well as an intuitive grasp of the high-dimensional connectivity of key, or of mathematical logic. It is as close to an explicit methodology of creativity within constraint as we have come.

The call to rejoin the musical journey is sounded by the trumpet, and introduces the *Sehr Lebhaft* of the finale, another sonata form to balance the first, but now increased twofold in complexity, range of key, and density of musical ideas. The first fragment (Ex. 8) is deceptively rich in possibilities. The tonality seems at first to have found the sought-after F-major, but it is an unstable vision in the mist, and A-minor becomes the attractor.

Example 8

[42] Julian Horton, Durham University, personal communication.

The theme is inverted, canoned, fragmented, and reconstituted by quartet and orchestra. Its answering phrase, when the theme enters again a fourth lower (Ex. 9), receives the same treatment, before the two are combined and developed in a bewildering series of modulations.

Example 9

After the breathless rising semiquavers on the first horn eventually engulf two, then three of the quartet, they bring the development to a close in an immense climax which sets the orchestra running in the dactyl rhythm (da-di-di, da-di-di, . . .), beloved of Schumann, in insistent mood. Without a chance to take breath, the vision broadens, and the 'fluid' ideas first used in the *Romanze* begin to make an appearance. This time, however, the waves are not gentle ripples supporting a gliding quartet of horns, but giant waves that carry them over crests and into troughs (Ex. 10)—note the visual appearance of the wave created by the system of notation evolved within Western music. The four-bar 'wavelength' is strikingly apparent.

Example 10

The mountainous landscape of the first movement and the watery ideas of the second have found a rather terrifying way of combining forces. In confirmation that the force of the first movement has met the fluidity of the second, reassurance is supplied in the eye of the storm by a most unexpected reprise of the *Romanza*'s central floating theme, by which the horns manage to still the continuing turbulence of the orchestra. The strategic tone-planning of the entire work inflects the reprise, for the theme is now in E-major rather than the original B-flat major. In another example of the genius of setting problems that encode their own solutions, Schumann seems to be exploring a tonal valley an entire tritone from his goal, normally a very distant and difficult journey away. But the orchestra soon finds it possible to insist on C-major.

Example 11 Final movement

Emerging from the ocean, the horns find the energy to call in unison for one final time, tossing about another virtuosic version of the opening semiquaver fragment, itself developed from Ex. 2, but this time inverted and cascaded from horn I to III (Ex. 11).

The upper three horns summon all the players to a climax in the recapitulation. At first hearing, that material leads the music into waves that threatened to drown the quartet out of control of their destiny, but this time it is they that make the waves. As if to make sure that the listeners' musical memory loses none of the work's syntax, nor forgets its experiential voyage of hope, the lostness, danger, and all the rhythmic and contrapuntal ideas are woven together again in a breathless final section that brings quartet and orchestra together in a transposition of material by a major fourth, and so to a rediscovery of the home key, but somehow, as T. S. Eliot, 'knowing it for the first time.'

Returning to Barthes' *Loving Schumann* (and how could one not, after such a gift?), we find his analytical, but general, comments on the composer's music illustrated almost perfectly by the *Konzertstück*. It is not programmatic, but so inspired by the real world that it is natural to invoke metaphors (I chose terrestrial and maritime landscapes, but others possessing the same contrasts and inner structures would have done). The complexity and multitude of ideas are evident throughout—yet it is not fragmentary. Rather, the complex network of connections identifiable in the process of its creation are reflected in the way that its elemental ideas relate to each other.

The *Konzertstück* is, as Barthes also claimed, representative of a certain nostalgia—Daverio and Gál both notice that it can be viewed as a romantic development, or homage, to the *Concerto Grosso* of Bach, Corelli, and their contemporaries. This late baroque form took the leading thematic lines, formerly the prerogative of a single solo instrument, and distributed them to several voices from the ensemble, as well as in answering canon to the orchestra itself. If he was holding this idea consciously in mind, and the regular study of Bach gives every reason to suppose that he was, then the formal constraints within which Schumann channels his imagination operate at the level of the entire work, as well as the sonata form of its outer movements, and within the inner structures of its harmony and metre. In this light, the *Konzertstück* would constitute a 'neoclassical' move within the heart of the Romantic period. Schumann finds that a path from the legacy of historical form can be transformed into the very road forwards into the new symphonic work of romanticism that he sought.

To carry the immense themes of heroism, darkness, and light, faithful to these forms that structure at all scales, from single bars to the architecture of the entire work, and all within a quarter of an hour of wonderful music, is a poetic as much as a mathematical achievement.

From music to mathematics

In addition to the possible numerical routes from music to mathematics, the measure of space and time, the ubiquity of number itself, and the resolution of relationships, more have been suggested by our close-reading of Schumann's innovative work: the creative imperative to construct a whole from its many parts, the creative potential of a singular focus, and the creative strategy of constructing intermediate problems and solutions. Notes, symbols, and numbers may combine in unimaginably numerous ways. From this perspective, the creative process breaks down into two tasks of first selecting the right components from the myriad possibilities, then secondly assembling them in a resolved order and with their proper valency. Schumann once wrote of the experience, an overwhelming and ultimately debilitating one, when he becomes conscious of the vast expanse of the musical cosmos:

> *Sometimes I am so full of music, and so overflowing with melody, that I find it simply impossible to write anything*

Mathematician, Henri Poincaré, in his celebrated work on mathematical creation already drawn on in this chapter,[43] also alights on the challenge posed by the sheer profusion of mathematical possibilities. His metaphor for the vast space of candidate-ideas is not the watery torrent of Schumann but the particulate, or rather atomistic, theory of matter. He writes conjecturally on the transition of conscious to non-conscious ideation in terms of what befalls the 'atoms' of mathematical ideas:

> *[they must be] something like the hooked atoms of Epicurus. During the complete repose of the mind, these atoms are motionless; they are, so to speak, hooked to the wall; so this complete rest may be indefinitely prolonged without the atoms meeting, and consequently without any combination between them.*
>
> *We think we have done no good, because we have moved these elements a thousand different ways in seeking to assemble them and have found no satisfactory*

[43] Henri Poincaré, *Mathematical Creation.*

> *aggregate. But, as a matter of fact, it seems as though these atoms are thus launched, so to speak, like so many projectiles and flash in various directions through space. After this shaking-up imposed upon them by our will, these atoms do not return to their primitive rest. They freely continue their dance.*

The image of a sort of mental Brownian motion of atomic ideas is compelling, but the creative processes of both music and mathematics must deal not only with the combinatorial magnitude of choice, but also with the problem of multiple scales. An extended development of mathematical reasoning needs to respect the 'atoms' represented by its symbolic statements, line by line, as much as a symphony needs to respect the part and place of every note. But each has an overall story to tell, and must cohere at other levels of theme, structure, and ultimate goal. Short two- or four-bar phrases need to make sense on their own. The same is true of pairs of equations, local subsections of mathematical development. The parallels hold at all intermediate levels of structure.

Mathematics, like music, explores vast and intricate structures as well as elements of detail. Mathematicians 'construct' proofs, but debate over how much of their work is discovery, how much invention. As with music there are true but dull results, trivial as well as profound. Some mathematics is pedestrian, some breathtakingly elegant. Both require a germ, a starting point that breaks the sterility of the blank paper or empty stave, and the choice of project is bewildering—possible starting points are too numerous to contemplate, but the potentially fruitful are rare.

The point of departure of a musical or mathematical composition may be at the 'atomistic' level. Schumann's core idea represented in Ex. 2 or its reflection in Ex. 3 of the *Konzertstück* constitutes a musical example. Perhaps an even starker case would be the use of musical versions of personal initials (such as B-A-C-H or D-S-C-H in Shostakovich's intensely personal eighth string quartet). The starting point for Andrew Wiles's vast project to prove Fermat's last theorem is the simple Pythagorean particle

$a^2 + b^2 = c^2$; Emmy Noether's explosive generalization of symmetry to physics began with a troublesome issue with the conservation of energy in the new mathematical forms of general relativity. But other scales also serve as hosts for starting points. Schumann's conception began as much with the desire to realize a symphonic *Concerto Grosso* of romantic ideas painted across a wide and pastoral landscape, as with rising and falling versions of a single thematic fragment. Wiles knew from the outset that his work would involve a rich synthesis of algebraic geometry and number theory, itself exploring wide swathes of the modern mathematical landscape, but at the same time incorporating vital fragments of single concepts.

It has been illuminating to talk with creative artists and scientists working today on their own awareness of the mental tools of invention. The question of scale is especially delicate in the case of wordless forms of art. Janet Graham is a composer and music therapist working in the north-east of England. Her music is contemporary, not 'tonal' in a classical sense, but (in her own words) 'possessing a tonal centre.' Her commitment to music as both art and therapy springs from her experience of its effectiveness, and also from her conviction that 'we are musical beings.' In her experience, music can 'dig out' deep-lying disharmonies in people who are damaged or hurt, in an echo of the medieval connection of music and medicine, and can express the creative energies that lie at those same levels. Her approach to musical creation at multiple scales drives a fascinating notational process—she often starts with sketching out a 'musical score' for a whole work, but at highly coarse-grained level. Soft pencil marks on a stave contain no detail of individual notes, or even fragmentary themes, but give the entire work a shape that can be visualized at a glance. Scribbles indicate climaxes and softer sections, hints of orchestration, the appearance of new material, and the development of old—these framework structures are typically the first she lays down. Only afterwards come the details of rhythm, pitch, and voice, then finally individually written notes. Those 'atomistic' elements at the fine-grained level of a work may spring from a different source from that which had inspired the work as a whole; Graham gives an interesting example of a word-fragment—'snow-sky'—that became translated into a wordless falling fifth as she worked inwards from the body of the whole piece to its molecular makeup of notes and their relations.

Compare that account of musical creation 'zooming in' through the scales of entire oeuvre to note, with another personal reflection:

> *I had to consider a sum of an infinite number of terms, intending to evaluate its order of magnitude. In that case, there is a group of terms which chances to be predominant, all others having a negligible influence. Now, when I think of that question, I see not the formula itself, but the place it would take if written: a kind of ribbon, which is thicker or darker at the place corresponding to the possibly important terms;*

This is an almost exactly similar account of a coarse-grained, personal, and impressionistic notation to that of Janet Graham, but now transposed into the key of mathematical argument. The writer is the French, and later American, mathematician Jacques Hadamard, whose classic account of mathematical creativity bears comparison to James and Beveridge in literature and experimental science, respectively.[44]

A mathematician's mind

A Mathematician's Mind is the title of the last edition of Hadamard's regularly reprinted book. Like James and Beveridge, he takes a personal approach, borrowing from the testimonies of other mathematicians, which he invited by means of a questionnaire, as well as drawing on introspection of his own creative process.

[44] Jacques Hadamard (1945), *The Psychology of Invention in the Mathematical Field* Princeton: Princeton University Press.

He does not address the musical-mathematical mapping explicitly, but his 150 pages are replete with musical analogy. He quotes in full, as we clearly must in this chapter, and for similar reasons, a well-known letter of Mozart that captures the 'landscape' view of a creating mind:

> *When I feel well and in a good humour, or when I am taking a drive or walking after a good meal, or in the night when I cannot sleep, thoughts crowd into my mind as easily as you could wish. Whence and how do they come? I do not know and I have nothing to do with it. Those which please me, I keep in my head and hum them; at least others have told me that I do so. Once I have my theme, another melody comes, linking itself to the first one, in accordance with the needs of the composition as a whole: the counterpoint, the part of each instrument, and all these melodic fragments at last produce the entire work. Then my soul is on fire with inspiration, if however, nothing occurs to distract my attention. The work grows; I keep expanding it, conceiving it more and more clearly until I have the entire composition finished in my head though it may be long. Then my mind seizes it as a glance of my eye a beautiful picture or a handsome youth. It does not come to me successively, with its various parts worked out in detail, as they will be later on, but it is in its entirety that my imagination lets me hear it.*

This is Hadamard's archetype for 'illumination'—the emergence from the non-conscious—of the shape of a desired work. Building on the earlier reflections of Henri Poincaré, he finds in Mozart's description the resonance of his own eagle's eye view of a mathematical proof, the details blurred by distance, but all the contours intact. He is familiar with the sudden 'gift' of a solution, effortlessly appearing (in his case, tellingly, at moments of awakening from sleep). Recalling the discovery of a method for evaluating the determinant of a matrix, he describes his impression:[45]

> *One phenomenon is certain and I can vouch for its absolute certainty: the sudden and immediate appearance of a solution at the very moment of sudden awakening. On being very abruptly awakened by an external noise a solution, long searched for appeared to me at once without the slightest instant of reflection on my part—the fact was remarkable enough to have struck me unforgettably—and in a quite*

[45] Hadamard, *Psychology of Invention.*

> *different direction from any of those which I had previously tried to follow. Of course, such a phenomenon, which is fully certain in my own case, could be easily confused with a 'mathematical dream,' from which it differs.*

As in James's and Beveridge's accounts, this is one of the stages along the journey that starts with a vision and germination, requires extensive—and often apparently fruitless—preparation, then a period of 'incubation,' when the conscious mind seems disengaged from the exercise. He is sympathetic to Poincaré's hypothesis that a form of non-conscious aesthetic is responsible for the filtering and construction of mathematical ideas below the surface of explicit thought and logic, for he knows of the role played by the emotions in the conscious aspects of doing mathematics. He even quotes with agreement Pierre Daunou, nineteenth-century French historian, who seemed to have had uncommon insight into the part played by emotions in the sciences as well as the arts:

> *In Sciences, even the most rigid ones, no truth is born of the genius of an Archimedes or a Newton without a poetical emotion and some quivering of intelligent nature.*

Hadamard also knows about the long and fully conscious final labour of 'verification,' borrowing and anglicizing the marvellous French verb *préciser* for such detailed working out of the final conception. This is the 'writing out' of the proof that, in Poincaré's case, took 'a few hours.' By this point the mathematician has apparent sight of the entire argument—there seem to be no more mountain ranges of difficulty to cross, but the final road must be marched nevertheless. A paradoxical quirk in mathematical accounts of this final stage in the creative process is the occasional complete reversal of the original insight. Poincaré found, for example, that his intuited proof of the non-existence of a class of mathematical function actually showed that they do exist. The great set theorist, Cantor, arrived once at a result so paradoxical that he confessed, in a letter to colleague Richard Dedekind, 'I see it, but I do not believe it.' Cantor, the pioneer of the rigorous mathematics of infinite numbers, had already shown (by his famous 'diagonal' argument) that the infinity of real numbers was demonstrably of higher order than that of rational fractions. This was not shocking to him—the surprise was his proof that the infinity of points in the plane (a two-dimensional space) was of the same order as that of real numbers on a line (a one-dimensional space).

Hadamard had other motivations for his essay than the desire to give public account of creative processes in mathematical thinking. He was also keen to take issue with a number of writers who have claimed authoritative knowledge of mental invention in general, yet whose views were contradicted by his own experience. His first adversary is nineteenth-century French philosopher of aesthetics, Paul Souriau, whose 1881 *Theory of Invention*[46] advocated a central role for pure chance. Souriau's 'atoms' of thought are in as chaotic motion as the molecules of the air we breathe. Their random encounters contrive the creation of the new by recombination. Everything is mechanistic, rule-driven. So even at the level of algebraic notation, Souriau asks whether mathematicians consciously follow the course of their ideas through the lines of mathematical manipulation—it is his rhetorical response, 'of course not!,' that draws Hadamard's disdain. Souriau 'does not seem to have consulted among professionals,' for if he had he would have been rewarded with rich descriptions of how mathematicians work, like painters, with one eye on the canvas and the other on the real or imagined scenes. Any single detail is prone to error, but calls attention to itself if it stands out incongruously from the whole picture. Such double 'impressionistic' vision is essential to maintain in any imaginative shaping of a whole from constitutive parts. A mistake of a non-practitioner is to assume that mathematical notation is constitutive of their author's thinking, rather than communicative of it. Hadamard would have smiled in recognition of Andrew

[46] Paul Souriau (1881) *Theéorie de l'invention*. Paris: Hachette.

Wiles's experience of wandering inadvertently into a vast 'mathematical landscape' in the course of a cognitive trail of individual steps. He has another thought-provoking metaphor for his craft's holistic conception:

> *The true process of thought in building up a mathematical argument is certainly rather to be compared with the act of recognizing a person.*

The act of recognition is, as we know, by no means a passive one—we 'see' actively in our projection of patterns onto the incoming impressions of the outside world. The implicit reference to structures on multiple scales once more is unmissable, and the relation of local to global structures, their coherence and harmony, the work at both conscious and non-conscious levels. It also echoes the eternal question that mathematics and music both face of where they reside on the spectrum from discovery to invention.

Hadamard enters the ring of another psychological dispute—specifically with those who contest the very existence of wordless thought. The philosophical and psychological viewpoints that promote language as prior to, and essentially constitutive of, all thinking is known as 'lingualism.' The opposing view considers thought prior to language; so that words become communicative rather than constitutive of thoughts. The debate is ancient in origin—Plato takes sophists (the lingualists of the ancient world) to task in the *Gorgias* over their insistence that thought without language is inconceivable. In modern times, Alexander von Humboldt's elder brother, Willhelm, reignited the issue in the 1820s by advancing an anthropological perspective that, outside discussion of mathematical thought, has dominated the debate since. In this view, differences in conceptual structures between nationalities are explained by parallel variations in the semantic structures of their respective languages. By extension language structures mind, not the other way around. The German orientalist and philologist of the following generation, Max Müller is, following the elder Humboldt, the opponent of Hadamard's Platonic position of the primacy of thought. Hadamard quotes from a lecture of Müller's:[47]

> *The idea cannot be conceived otherwise than through the word, and only exists by the word.*

This was not to assign any weight to the opinion, which to the mathematician is obviously in error, but to illustrate his incredulity that a professor of psychology could attempt such a universal and prescriptive statement.

The *Imagination Institute* workshop that we have drawn from at several points illustrated amply how mathematical scientists think in varied ways—some algebraically, some visually, others dynamically. Intriguingly none of that group named verbal reasoning as their main mode of conceptual thinking, though Hadamard finds Hungarian mathematician, George Polyá a rare exception in advocating appropriate language in the generation of fruitful ideas:

> *I believe that the decisive idea which brings the solution of a problem is rather often connected with . . . a well-turned word or sentence.*

Even this is not an expression of 'thinking with words,' but using the spring and elasticity of language to launch new leaps of thought, or to make connections between ideas. It is the same notion of verbal suggestion of the non-verbal that Schumann wrote about Berlioz, correcting a simplistic notion that programmatic music carried any directly representational sense. Here he is again on Beethoven's *Pastoral Symphony*:

> *[I]t was not one single, fleeting spring day that inspired him to his joyous outcry, but rather the dark converging mixture of noble songs above us (as Heine, I think, has somewhere put it), creation itself, with its infinite multiplicity of voices.*

Words are not translated into music any more than a single glance at a meadow and riverbank are, but they may provide one

[47] Max Müller (1888), *Introductory Lecture on the Science of Thought.* London: Longmans

possible source for the conception of the wordless. Hadamard's chief witness against Müller's monochromatic vision of language-based thought is Francis Galton. The Victorian polymath, known principally for his innovative work on statistics, especially in application to heredity, had offered himself as a counter-example to Müller's thesis, expressing the extraordinary difficulty that he has in putting his ideas into words at all:

> *It is a serious drawback to me in writing, and still more in explaining myself, that I do not so easily think in words as otherwise. It often happens that after being hard at work, and having arrived at results that are perfectly clear and satisfactory to myself, when I try to express them in language, I feel that I must begin by putting myself upon quite another intellectual plane. I have to translate my thoughts into a language that does not run very evenly with them. I therefore waste a vast deal of time in seeking for appropriate words and phrases, and am conscious, when required to speak on a sudden, of being often very obscure through mere verbal maladroitness, and not through want of clearness of perception. That is one of the small annoyances of my life.*

This is not an experience consistent with a theory that all thought is already in the form of language, yet it is the common experience of both mathematical and musical thinking. Galton's experience of operating on more than one 'intellectual plane' offers a geometrical way of systematizing our findings that visual, narrative, and wordless forms of thinking all offer distinct arenas for creative potential. It also seems to capture authentically the relatively smooth transport of thought within the respective planes, as well as the awkward translations between them. So, writing or painting about music is difficult, but not impossible; writing, or even algebraic description, of mathematics is a communicative and representational imperative, but it does not constitute mathematical thought itself.

The 'precising' of mathematical ideas, however—their necessary linking to other planes of representation by the labour that Galton found so taxing, does have a vital connecting role in the imaginative construction of new ideas, the process that Hadamard terms 'relay results.' He fears that this statement is so trivial as to be hardly worth stating, but it is the mathematical version of a truth about the creative process so general and so important that he finds himself mulling over example after example. The great nineteenth-century Scottish physicist, James Clerk Maxwell, would urge mathematicians to formulate their thinking in 'words without the aid of symbols,' not because he would sympathize with the lingualists, but because he knew the creative force of communicating ideas. Through his long and fruitful connection with Michael Faraday, a very different thinker, he had experienced relay results at first hand. Once a new idea is worked through into a precise formulation, only then can it become the seed from which further results can germinate. The relay-result is the connecting rod that revolves the history of culture. It is the instantiations of baroque or classical musical ideas by Bach and Beethoven that allow Schumann to conceive of new forms of Romantic music. It is the driving idea of unprecedented precision of measurements in the motions of planets by Tycho Brahe that constituted the key for Johannes Kepler's systematic laws of planetary motion, and those in turn that made possible the great generalization of Isaac Newton's law of gravitation.

A mathematical close-reading and a beautiful connection—the fluctuation–dissipation theorem

To earth these generalities, we need to attempt in mathematics, as we did in music, the paradoxical task of contemplation in words of this intrinsically wordless art. At several points Andrew Wiles has provided the deep insights that come from his protracted and profound mathematical journey into the wonderful resolution of Fermat's last theorem. For readers who wish to explore that story more fully, the excellent written lay account by Simon Singh[48]

[48] Simon Singh (1997), *Fermat's Last Theorem*. London: Harper-Collins.

provides a rich and beautiful entry into the years of illumination, struggle, incubation, verification, and eventual clarification of a result that draws on much of modern mathematics in unforeseen ways to reach its final destination.

With the goal of understanding scientific creativity more fully, contrasting and fruitful examples can be taken from mathematics that springs directly from encounters with the natural world. Emmy Noether's beautiful and poetic link between symmetries and conservation laws is an example. I hesitate to differentiate using the terms 'applied' and 'pure,' for many reasons; 'pure' mathematics that seemed at first to have no possible correspondence with nature, has often later found an essential place in physics. Furthermore, it also seems that geometric, algebraic, and other quite abstract planes of imagination are operative in all types of mathematical thinking, whether originally sparked 'relay'-fashion from other abstract results, or from natural observation and question.

There is another beautiful result that hovers over much of physics and even engineering, as well as within the fundamental structure of dynamical systems and statistical mechanics. It illustrates Hadamard's careful collection and systematization of creative pathways in mathematics, and illustrates the 'relay' form of narrative, as well as structure and coherence on many scales, and the 'focal point' that express commonalities between mathematics and music. It even captures the apophatic theme of letting the unknown and unknowable speak. It is amenable to geometric-visual conception as well as abstract-algebraic and intuitive-dynamic. It is aesthetically beautiful, and goes generally by the name of the *Fluctuation–Dissipation Theorem.*

The title may appear technical and off-putting but the core concept, one of the twentieth century's great insights of mathematical physics, is actually very simple to grasp. Like many aesthetically appealing creations of art or science, it embodies a deep connection between two apparently distant ideas. Just as in the case of our close musical analysis, there will be some examples of notation displayed as well as the narrative text. This should not be off-putting: for those able to read it, it will help, and for those unfamiliar with mathematical notation, as with the musical, it can be read for its visual structure and pattern, without a detailed mastery. The testimony of creative mathematicians that symbols participate in their imagining is enough to arouse our interest in their arrangements and rearrangements on the page.

A concrete and simple example will serve to illustrate the idea, beginning with the 'Brownian motion' we have already encountered in the dance of polymer molecules in Chapter 2. A microscopic bead suspended in a fluid can be caused to move by applying a force to it. This is conveniently done by including magnetic materials within its composition and arranging that a magnetic field be applied to the bead's container. Under a microscope the bead can be observed to drift at a steady speed v in the direction of the magnetic force. The stronger the force f, the faster is the drift. But the fluid may be more or less viscous—in this case a higher viscosity will proportionally slow down the bead, creating more drag on it (the rate that drag force grows with velocity is usually notated by the Greek letter zeta ζ). So, we write a simple equation that formalizes the statement, 'the velocity of the dragged bead increases in proportion to the force imposed on it, and reduces in proportion to the drag of the fluid surrounding it':

$$v=\frac{f}{\zeta}$$

Some readers of the equation will immediately visualize it, perhaps in the form of the adjacent picture. Now suppose that the magnetic field, or in general the source of the force dragging the bead along, is turned off. The first result apparent to the observer peering down the microscope is that the bead abruptly comes to a stop. But then, providing that the bead is small enough and the microscope sufficiently powerful, the jiggling Brownian motion will become apparent. The bead will move of its own accord (actually through the constant buffeting by the much smaller molecules of its suspending fluid, themselves in thermal agitation and random, though unseen, movement). This motion is of a completely different character to the forced motion of drag. It has no privileged direction of transport, but makes tiny random steps in arbitrary directions, each step independent of the last. This is the 'random walk' of *diffusion.* On average a collection of such beads will not have moved more to the right or to the left of their starting points after a time interval, t, but each will in general have moved some distance away r. The formal, and equally simple, mathematical formulation of this statement considers the average of the *squared-distance* moved for all such beads. This is always a positive number; we might write it as $\langle r^2 \rangle$. The angular brackets are a normal notation for the average of the quantity inside them, taken over a population of many examples. Now the statement of diffusive behaviour is just that 'this average square distance moved grows proportionally with time,' again with a possible visualization:

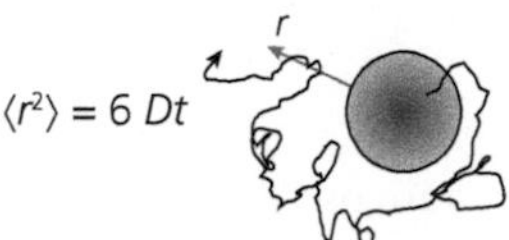

The rate at which the bead will randomly explore the fluid around it is described by the 'diffusion constant,' given the symbol D (the numerical coefficient of 6 is there by convention).

In one of Einstein's revolutionary papers of 1905, he solves the riddle of Brownian motion for the first time since Brown's elegant and exhaustive experimental account of it in the 1830s. We saw in Chapter 2 that Einstein realized that the molecules of the fluid in which a bead is suspended are in this constant random motion, as required by their possessing a temperature—the higher the temperature, the faster the average motion. In consequence there would be a slight imbalance of molecular collisions on any two opposite sides of the bead, and a small net propulsion away from the side of greater intensity of collision. Although a moment later, the imbalance might well be in the other direction, this is enough to secure the sequence of tiny steps that drives diffusion. Brownian motion became a window into the molecular world, a visible sign of its (as yet) invisible presence.

Einstein's remarkable *Annalen der Physik* paper[49] contained an extraordinary element of mathematical creativity that speaks strongly to questions of innovation. He wrote down a customary 'equation of motion' for the bead. This is a mathematical recipe for calculating the position of the object at a small increment of time in the future, given knowledge of its position and velocity at the present. In simple form, it gives a small change in position, written Δx, that takes place over a small time-interval, Δt. The incremental change Δx is typically calculated from the positions, velocities, and forces between the objects captured within the mathematical model for them. For example, Newton had written down such an equation of motion for planets orbiting the sun in terms of their gravitational interaction—summing all the increments Δx around an orbit led to the prediction that their orbits would assume the geometrical shapes of ellipses. The radical element to Einstein' equation of motion was that an element of his Δx values was *random* and unpredictable. Arising from the unknown myriad microscopic trajectories of the molecules, themselves not explicitly calculated, there was no route to calculate what Δx might be at any stage. There was no law of gravity, or any equivalent, that could be used to calculate a trajectory subject to Brownian motion. However, Einstein understood that there is a great difference between knowing nothing, and not knowing everything, about a quantity. Furthermore, his love of the science of thermodynamics contained the insight we have already encountered in the Brownian motion of polymer molecule, that

[49] A. Einstein (1905), Über die von molekularkinetischen Theorie der Warme geforderte Bewegung von in ruhenden Flüssigkeiten suspendierten Teilchen. *Annalen der Physik*, 17, 549–60.

'well-behaved' ignorance in physics can contain nascent advantages. Entire phenomena such as the pressure of gases, or the elastic force from a stretched polymer, emerge from the sum of many random events. They can then, somewhat miraculously, become predictable, not because every contributory event is predictable, but because their average effect is. The very notion of 'entropy'—a thermodynamic framing of ignorance—is the nexus of physics and the apophatic, yet it explains and even predicts.

By these means, Einstein invented the 'stochastic differential equation'[50]—an equation of motion that contained a term corresponding to a random force, whose specific history or instantiation is unknown, but of which some set of average properties is known. His well-attested claim that his dominant imaginative powers were visual is supported by this classic paper. He begins on familiar territory, with the perfect gas law and the measurably constant and predictable pressure that it describes. The next move is critical—he invites a reader to imagine the underlying cause of this pressure as the random collisions of molecules of gas with the walls of their container. This reminder of how an apparently smooth, static, and predictable property can emerge from an underlying discontinuous, dynamic, and unpredictable submerged world of molecular motion, sets up the imaginative step into an intermediate world. For the 'microscopic' scale of the diffusing beads is neither completely static and smooth, like the fully macroscopic vessel, nor is it as utterly unknown as the trillion molecular trajectories themselves. The Brownian motion of the bead shares some smooth predictability with the macroscopic, in terms of its average, yet participates in the stochastic nature of its motion with the fully molecular world. The new type of mathematical equation of motion built its hybrid nature from both classical and statistical elements—the impressive

[50] Others, including Smoluchowski and Bachelier, conceived of the same idea in the period 1900–5, but independently, though the 'coincidence' speaks of the ripeness or readiness within a field's discourse for new ideas when the appropriate questions have surfaced.

achievement was Einstein's orchestration of them into a consistent whole through a courageous symbolism that captured and tamed random and unknown forces.

One more surprising and beautiful result appeared, as if from nowhere, from Einstein's creation of conceptual bridges from the microscopic world downward into the molecular and upward into the macroscopic. For it was possible to study mathematically the behaviour of his buffeted beads, not only in the case of free diffusion, but also under the direction of an external force. Doing this opened a window onto the origin of the fluid drag that opposed the force (as we notated by ζ before), in the same way that the same theory had revealed the underlying causes of diffusion (characterized by the diffusion constant D). For the first time, it was possible to see that the sources of drag and diffusion are the *same*—the multitude of molecular thermal collisions with the bead give rise to both. Their close relationship takes a quantitative form, known as the Einstein relation:

$$\zeta = \frac{k_B T}{D}$$

The faster the diffusion D, the smaller will be the drag ζ. The number that translates between the two is a familiar one in the physics of heat: the typical thermal energy of a single molecular degree of motion, $k_B T$, formed from the absolute temperature (T) and Boltzmann's constant k_B, which simply sets the scale for temperature. The mathematics could hardly be simpler at this level of result (the derivation is a little more sophisticated, but well within an elementary calculus course), but the consequence is a deep connection between two apparently distinct physical behaviours. At first blush, the free and random diffusion of a particle observed over time (characterized by D) would seem quite independent of its response to a force propelling it in a particular direction (characterized by ζ). But Einstein saw that they sprang from the same source, and had to be deeply related. In other words, if a fluid can be said to possess a temperature (and all the properties of thermodynamic equilibrium that implies) then there is no freedom to choose *both* the diffusive *and* the drag behaviour of a suspended particle independently. Its thermal *fluctuation* is geared to its *response* to an external force by a hidden common cause.

An electrical analogy—Johnson–Nyquist noise

Harry Nyquist was a Swedish-born engineer, who as well as holding 138 patents as an enormously productive inventor, was also responsible for some of the deepest theoretical results of the last century in signal processing, stability, and noise. After emigrating in 1907 at the age of eighteen to the USA, and after a degree in engineering at Yale, he worked for AT&T research laboratories in New Jersey from 1917 to 1934. He is responsible for the next stage in the story of the fluctuation–dissipation theorem, as it takes on the form of Hadamard's class of 'relay results.' His case also illustrates once more the fruitful complexity of relationship between theoretical physics and industry. A common and glib impression is that industrial research is conducted in too messy and profit-oriented an arena to benefit fundamental science in any way. New academic research may indeed be 'applied' in industry, but much governmental research strategy the world over fails to recognize the history of intellectual flow in the other direction. This stream is extremely rich, especially in the twentieth century, although the leading example of thermodynamics emerging from the nascent technology of steam power in the nineteenth ought to have prepared us to expect as much. Industry has turned out to be replete with interesting and extremely challenging problems. It takes a good deal of insight to translate them into something that can be tackled scientifically; this act of translation is itself a creative and demanding process. A strong example that we have already met is polymer science, whose primary data arose from the post-war plastics industry.

The communications industry, when Nyquist was working in it,[51] developed a keen interest in 'noise'—because of its interference with 'signal.' The noisier an environment or system, the less information can be transmitted. This is a common experience as well: at a loud party one has to shout to make oneself understood.

An irreducible source of noise in electrical transmission circuits had been realized, as early as 1918, as an electrical analogue of Brownian motion.[52] Another Swedish émigré, John Johnson, had studied at Yale with Nyquist, producing a thesis on ionization, using the new cathode ray vacuum tubes. A decade later, after they had both joined the same company, Johnson was using his vacuum tubes to measure the thermal noise in electrical circuits. Just as particles diffuse freely in a thermal medium, small electrical currents flow back and forth in circuits even in the absence of batteries or generators, arising from the thermally agitated motion of electrons in the wires. His measurements adopted a critical degree of refinement over any that had been applied to the Brownian motion of particles, however, for he was able to measure, using tuning circuits analogous to those used in radio receivers, the noisy fluctuating currents at each frequency independently. Rather than just one number characterizing the random currents, analogous to the diffusion constant D, there were as many 'current-diffusion constants' D_v as there were different frequencies of signal (which he notated by the subscript Greek letter v, 'nu').

The same circuits that sustained thermal noise when left to themselves, could also be driven by an applied voltage. For each frequency, the current within the circuit depends on the applied voltage through a resistance R_v, in much the same way that the velocity of a Brownian particle depends on the force applied to it through its fluid drag (ζ). Johnson found his famous result (rewritten here to emphasize its connection by analogy with

[51] His laboratory was soon to become AT&T's Bell Laboratories.

[52] W. Schottky (1918), 'Uber spontane stromschwankungen in verschiedenen electrizitatsleitem,' *Annals of Physics*, **57**, 541–67.

Brownian motion) by comparing the thermal currents and the resistances at each frequency:

$$R_{\nu} = \frac{4k_B T}{D_{\nu}}$$

A visual comparison of this little equation with the 'Einstein relation' given earlier, for drag and diffusion of a particle, will identify the structural similarity (apart from a conventional factor of 4), supporting an exact mapping from current to diffusion (on the right-hand sides) and from resistance to drag (on the left).

Johnson presented this experimental result to Nyquist who, within a month, had written a succinct and beautiful explanation.[53] Although there are very few references in his paper (just one review article, the original experimental paper, and Johnson's back-to-back article), Nyquist is clearly thinking in the tradition of Einstein; his method is full of thought experiments. The engineer invokes electrical circuits, and even long signal lines that enable two circuits, connected at a distance, to exchange energy with each other. Nyquist's familiarity with patenting inventions comes to mind; anyone familiar with patent documents will not fail to recognize the style of chosen examples to make specific claims, then a move to generalize to systems having a wide set of properties. The power of this genre of document to structure thought, and its natural discipline of reasoned movement from the particular to the general, touches on the connection between literary form and science we explored in Chapter 4. One wonders if Einstein's own experience in the Bern patent office might not have been entirely lost time.

Nyquist's paper is a beautiful translation and generalization of the Einstein analysis of Brownian motion. Its core idea comes from the statistical thermodynamics of Boltzmann and Gibbs of the nineteenth century, who showed that each dynamical

[53] H. Nyquist (1928), 'Thermal agitation of electric charge in conductors,' *Physical Review*, **32**, 110–13.

mode of a system is on average excited by heat to carry a fluctuating thermal energy, whose average is of the order of the characteristic value $k_B T$. By requiring this to be true of the electrical and magnetic fields within any circuit, the general form emerges connecting the fluctuation and the resistance at each and every frequency.

A universal truth

Mathematics, whether originally stimulated by the properties of abstractions such as pure numbers, or from material observations from nature, often proceeds from specific cases to the general. One seems at first to be in a darkened room, stubbing toes and striking shins against scattered items of furniture. In the case of Fermat's last theorem, the failure to find, over the course of three centuries, a single example of four whole numbers a, b, c, and n that satisfied $a^n + b^n = c^n$ for n greater than the value 2 of Pythagoras' theorem, suggested very strongly that something was going on. Theorems are often 'conjectures' for a long time before they are proved, either because of multiple instances that point towards them, or because of the mysterious intuitive conviction of trained minds. A current example is the 'Riemann hypothesis,' referred to in Chapter 2 by theoretical physicist Michael Berry. It is still an unsolved and important mathematical problem—its eventual truth or falsity drives strongly opposed convictions. Like the Fermat result, it makes predictions in an infinite number of cases, and as with the situation before Wiles's work, numerical calculations have now checked vast, but of course still finite, numbers of cases. If a deep and general solution is found to a puzzle like those of Fermat or Riemann in a way that delivers insight as well as a formal demonstration, it is like turning on the light in the darkened room. The first cases that caused mathematical bruises during the initial stumbling about are still there, but now perceived in clear relation to each other. New examples appear immediately, and the connections between them spring into view.

The two cases of Brownian motion and electrical circuit noise are profound in themselves, but hint at something more universal still a structure at a deeper level, of which these two cases are examples. The imaginative task of 'turning on the light' is the great challenge, for it calls for the creation of structure as yet unseen. We know what a particle and a fluid are; we know what a current and a circuit are; we might have a dawning notion that these are examples of objects much broader and more general in conception. Those are just the objects that a mathematician needs to call up essentially *ex nihilo*, endow them with function and relation, and show that all previously discovered examples are derivable special cases. Two physicists at Pennsylvania State University, Herbert Callen and Theodore Welton, told this story, and provided the first articulation of the general result in 1951.[54]

The separation in time between Nyquist's work and their own is testament to a long period of 'incubation,' in which a community of scientists were thinking and conversing about a larger vision for response and fluctuation. They summarize it well in the introduction to their paper:

> *It has frequently been conjectured that the Nyquist relation can be extended to a general class of dissipative systems other than merely electrical systems. Yet, to our knowledge, no proof has been given of such a generalization, nor have any criteria been developed for the type of system or the character of the 'forces' to which the generalized Nyquist relation may be applied.*

The task of generalization was assisted by framing their development within a quantum mechanical, rather than the simpler classical, system. Although an ostensibly more difficult task,

[54] H. Callen and T. Welton (1951), 'Irreversibility and generalised noise,' *Physical Review*, **83**, 34–40.

the strategy offered the advantage of a very developed formal mathematical language for *general states*, their energies and time-dependence, from the intensive work initiated by Schrödinger, Heisenberg, and particularly British physicist Paul Dirac from the mid-1920s onwards. Notations for 'generalized' fluctuation and dissipation were available on the journal shelves as a result. The Callen and Welton paper is an example of how formal structure can assist, rather than restrain, creativity by suggesting avenues and objects to manipulate. What is essential is to work with the tools, 'setting up formal problems and solving them,' while maintaining the clear-sighted vision of the destination and 'focal point,' to borrow the musical language of Horton and Schumann.

The Callen and Welton paper is a work in three movements that plays with different perspectives on a theme. In the first, they remind their readers how gentle disturbances of a physical system cause occasional jumps from one of its stable energy levels to another, then show how this effect can be reinterpreted as the familiar 'dissipation,' like a dragged bead or a driven current. Starting over again in a 'second movement,' they introduce apparently new material, showing how the same quantity must also fluctuate within a system at finite temperature. This involves a very adroit conceptual and notational move, which is worth recording here even for those unfamiliar with the notation itself, for it will be immediately apparent how different the generalized quantum mechanical tools appear from Einstein's statistical approach to fluctuations.

The authors write Q for a general measurable quantity of a dynamical system (this could be the position of a bead, or the current in a wire). Q might be forced to change by an external driving force, or to fluctuate. Motivated by the simple case of diffusion, they choose to calculate an expression for the average square (so always a positive number) of its rate of change, written Q, when the system is in the state $|n>$ of energy E_{n}. In a stroke of insight, the squaring of Q generates a sum over all other possible states $|m>$, which in their notation looks like:

$$\langle n|\dot{Q}^2|\,n\rangle = \sum\nolimits_m \langle n|\dot{Q}|\,m\rangle\,\langle m|\dot{Q}\,|n\rangle$$
$$= -\frac{1}{\hbar^2}\sum\nolimits_m \langle n|H_0Q - QH_0|\,m\rangle\,\langle m|H_0\,Q - QH_0|\,n\rangle$$
$$= \frac{1}{\hbar^2}\sum\nolimits_m (E_n - E_m)^2\,|\langle m\,|\,Q|\,n\rangle|^2$$

The stages in deduction here are all from standard undergraduate quantum mechanics, using Paul Dirac's elegant notation for states $|n>$ and mathematical operations on them. The mathematical structure is called a 'vector space'—a generalization of the ordinary two- and three-dimensional spaces we are familiar with in geometry, but to higher dimensions. The choice to work in terms of the rate of change of a fluctuating quantity, rather than the quantity itself, is also a wonderful idea. It calls on another general ingredient from quantum theory, which gives the rate of change of any measurable quantity in terms of a sort of mathematical two-step dance with the energy, written H_0 (so look for the appearance of the rhythmic and symmetric $H_0Q - QH_0$ in the equation). The approach seems perverse at first, for the habitual way to write the average of a quantity Q is to find a set of probabilities $P(Q_i)$ for all its possible values and sum over them. The calculation would then look something like this:

$$\langle n|\dot{Q}^2|n\rangle = \sum\nolimits_i \dot{Q}_i^2 P\left(\dot{Q}_i\right)$$

But the development that Callen and Welton achieve is an elegant example of a creative journey within mathematical physics that, while keeping this distant end point in view, descends into the jungle of possibilities in front of it, not by such a direct route, but by an imaginative and informed detour. The expression that they derive bears an immediate physical insight, for the sum (denoted by the large Σ) is over all the possible states of the system—the dynamic image it calls to mind is that of fluctuations by jumps from the initial state $|n>$ into other states $|m>$ and back again.

In a 'third movement,' the paper transforms the expression for the fluctuations into the 'first movement' formula for the power dissipated by the same general system when the jumps from one state to another are excited by an external voltage. The differences in the energy levels $(E_n - E_m)$ enters there in just the same way as for the fluctuation calculation, as the authors must have intuited as they worked through their approach. In an illuminating illustration of how notational manipulation can engage creative mathematical thought, the connection is made by writing this difference in energy of the 'home' state to other states in terms of the energy of a photon of light with the same energy. The universally accepted and familiar notation for this is the product of Planck's constant and the photon frequency: $\hbar\omega$. The sum is also now written in a continuous way, since light of all frequencies is available, working in a distribution of energy states, $\rho(E_n)$. The central expression now looks like:

$$\langle n|\dot{Q}^2|n\rangle = \int_{-\infty}^{\infty} d\omega\, \hbar\omega^2 \, |\langle n + \hbar\omega\, |Q\,|n\,\rangle\,^2 \rho(E_n + \hbar\omega)$$

The new notation is really just a semantic, not a structural shift, from the previous expression. The material is the same, although its interpretation is different. Akin to the harmonic ambiguity of Schumann's writing, where the route to resolution was achieved by constructing a passage that could be heard as both a recapitulation and a close, the double-meaning here also brings this work to its closure. Now the 'jump' can naturally be interpreted as either a self-induced fluctuation or the response to incoming radiation (dissipation). The final move, hoped-for all along and now falling out easily, is to demonstrate that the expressions for the dissipation and for the fluctuation are essentially the same, and at all frequencies, connected via the fundamental thermal energy $k_B T$ in the now-familiar way.

Everything in nature has some degree of compliance; nothing is infinitely rigid. This deep result tells us that, in consequence, everything is also in motion, at our common length-scales usually imperceptibly, but with effects that endow materials with properties we are familiar with (the softness of the flower petal becomes a brittle rigidity on freezing). Such compliances link in every case to corresponding fluctuations. Since the first general proof of the fluctuation–dissipation theorem, other ways of deriving it have been offered. It is another quality of deep results that they spawn new insights and questions—they are never ends in themselves. At the very simplest level, the theorem is a gift: one never has to calculate both of the quantities that it links together for every new case. Sometimes the fluctuations, sometimes the dissipation, is more straightforward—the other now comes for free. But at its heart is an elegant truth, that there lies a deep connection between the way something possessing constant internal motions fluctuates, and how it responds to a push from outside. An elegant modern proof of the theorem captures this idea directly by recasting a local fluctuation in one part of a system as if it were an external force on the remainder. In this way, the connection between the internal fluctuations and the response to a force becomes explicit—for they are visibly the same thing. Such growing insights suggest that an extension to the theorem might apply to any dynamical system, even if not in thermal equilibrium. Turbulent fluids, even living systems themselves, may entertain the analogies of response, and inherent fluctuation. The signs are that underlying patterns link those as well.

The theme of creativity through discovering deep connections between previously disjoint ideas or quantities has grown in strength over the last two chapters. Noether's theorem identified a conservation law for every symmetry; the fluctuation–dissipation theorem a sort of free diffusion for every forced motion with drag. There are 'homeomorphisms' with connections in musical creation as well.

Music, mathematics, and wordless creation

Thinking though, and writing down in close sequence, the stories of our 'deep dives' into music and mathematics has been, itself, an experiment. I chose Schumann and the *Konzertstück*, and Einstein, Nyquist, Callen, and Welton's fluctuation–dissipation theorem because I find them both to a high degree aesthetically pleasing, and perhaps because they have not had as much general exposure as they deserve. The 'experimental' question turned on how similar would be the impressions of analysing their structure, achievements, aesthetics, and what might be deduced of their creation process.

Conceivably the two stories might have failed to resonate at all, but when heard within a universal narrative of creativity, and even at the level of their semantic strategies, they clearly do. The objection that there is no creativity in mathematics, for true results are there to be discovered not created, evaporates when the process of groping for a proof is substituted for the simple result. Proofs and deductions are many, and infinitely variable. They have initial statements, second-subjects, development sections, recapitulations, and climaxes. They draw on existing material but represent it in new ways, or combine apparently incommensurate notions in new and powerful harmonic relation. They engage their readers in a temporal sequence of expectation and surprise. The musical metaphors are never forced; they are the ones that work. In both cases, we have also found that the dynamic and creative interaction between ideas and their notation is necessary in the process of thinking about mathematics and music, as well as in creating and communicating them.

Oxford mathematician, Marcus du Sautoy, has made a similar conjecture about the connections between mathematics and music at the level of proof and composition. His collaboration with the Manchester-based composer, Emily Howard, has involved a very different deep-dive into this idea, including the creation of a series of short compositions each inspired by a specific mathematical proof (such as Pythagoras's theorem or the irrationality of $\sqrt{2}$). Whether such extremely close correspondence stands up to close scrutiny remains to be seen. Howard's own testament to the inspiration she receives from mathematical ideas is much more in line with Schumann's more distanced description of the cognitive lineaments between music and its inspiration:[55]

> *It's the resulting collision and union of disparate ideas from diverse sources that excites me, and the subsequent translation of these hybrid ideas into sound is essentially the crux of my creative process.*

Creating music begins with number and relation in rhythm, pitch, and volume. But it draws more deeply on an ancient conviction with mathematics that the universe itself is both numerate and musical in some sense. Both music and mathematics illuminate wordless spaces within the human mind. Both tend to draw creative energy from their 'singularities,' whether Schumann's 'musical centre,' or the drive to unify material structure and dynamics within Einstein and Nyquist.

Ultimately, although we have tried to say a great deal about them, we are left, perhaps, with the wordlessness in which they dwell as a necessarily apophatic road into their contemplation and enjoyment. Although both draw frequently on the other visual and textual planes of imagination, they both need ultimately to take leave of them. At no point, however, does either let go of the impulse and response of emotion and affect. On the contrary, both mathematicians and musicians have given witness to

[55] http://music.britishcouncil.org/news-and-features/2017-06-08/profile-emily-howard.

the deep role of aesthetics at every point of their creative journeys. Sometimes they are emotionally overwhelmed by it. The entanglement of emotion and cognition in science and art must be the next stage of our journey.

7 Emotion and Reason in Scientific Creation

The works within our capacity consist either in the mind's sight, or in the desire of the same, or in bodily motions, or in the dispositions of these same motions. Sight first looks; then it verifies what has been looked at or cognized, and when the fitting or harmful have been verified within the mind or within sight, desire strains to embrace the fitting, or retreats into itself to shun the harmful.

Robert Grosseteste, *De Artis Liberalibus* (On the Liberal Arts)

The theme of duality has haunted our hunt for the sources of creativity in the sciences. Dualities appear in even stronger relief when we use the visual, literary, or musical arts as a comparative grid. At every turn we seem to have found pairs of activities or of abilities in creative tension. Furthermore, these 'connected dualities' seem deeply embedded within the process of creativity itself. A non-exhaustive list of examples encountered so far might include:

- external (Platonic 'Muse') and internal (Aristotelian *poiesis*) sources of ideas;
- holistic and reductionist approaches to perceiving and understanding the world;
- the Kantian duality of 'natural law' above us and the 'moral law' within us;
- active and passive modes of creation;
- imagination and reason in creative thought;

The Poetry and Music of Science. Tom McLeish, Oxford University Press.
 DOI: 10.1093/oso/9780192845375.003.0007

- conscious and non-conscious thinking;
- representational tension between 'canvas' and 'object';
- mathematical representation and the physical system itself;
- the neurological lateralization of left and right hemispheres (without suggesting that this structure necessarily maps onto any other dualities);
- the fields of Arts and Sciences themselves.

In the telling the story of science, we have found typically one pole of each of these dualities to be conventionally enhanced, the other muted. So, for example, reductionism captures more airtime than holistic thinking in science. The same is true of conscious work, the 'workshop,' or experimental phase, of developing a new idea, which enjoys much more open discussion than the partially non-conscious but vital work of ideation. Our finding of multiple dualities within creative thinking intriguingly mirrors a conjecture within cognitive neuroscience of distinct processes of 'System 1' and 'System 2.'[1] The first is fast, autonomous, intuitive, and integrative; the second serial, conscious, and rule-based.

There is yet another duality, long recognized as intrinsic to human action, that seems to have received another lopsided treatment in contemporary discussions of science—that between the emotional or affective and the rational or cognitive. The role of affect is by no means restricted to the response to scientific achievement, the context in which it is most frequently discussed. The 'awe and wonder' reaction to cosmology, in particular, is a commonplace strategy in its popularization. Much of the 'sci-art' movement falls into this category. There is much less discussion of the interplay of reason and emotion in the

[1] K. E. Stanovich and R. F. West (1999), 'Discrepancies between normative and descriptive models of decision making and the understanding acceptance principle'. *Cognitive Psychology*, **38**, 349–85; J. St. B. T. Evans and K. E. Stanovich (2013), 'Dual-process theories of higher cognition: advancing the debate'. *Perspectives on Psychological Science*, **8**(3),223–41.

scientific creative process itself, in spite of the recorded flashes of insight from Poincaré and others that point to its deep but obscure importance.

Mainstream scientific commentary tends to express caution about the emotional in the scientific process. Richard Dawkins, for example, in *The God Delusion*[2] wants to distinguish and distance truth-seeking from feeling:

> *I don't want to decry human feelings. But let's be clear, in any particular conversation, what we are talking about: feelings, or truth. Both may be important, but they are not the same thing.*

Total denial is not uncommon—a BBC television documentary in the *Horizon* series screened in 2016 featured a scientist involved in the search for extra-terrestrial intelligence who affirmed straight to camera, 'there is no emotion in science.' In spite of quieter voices insisting that a search for beauty, and so the inseparable engagement of the emotional response, remains an essential guide to science,[3] such covering-over of the role of the unconscious, affective, and integrative is assisted by reluctance to address the imaginative ingredients of science, or the twin roles of passive and active engagement of cognition.

A more nuanced contemporary view of how the emotions work in science has been expressed by Peter Medawar, whom we met in the introduction in Chapter 1, criticizing the notion that science works through nothing but tidy method. He repeatedly drew attention to the issue of hypothesis creation, and the role of the non-conscious mind. A key passage is worth quoting in full:[4]

> *The weakness of the hypothetico-deductive system, in so far as it might profess to offer a complete account of the scientific process, lies in its disclaiming any power to explain how hypotheses come into being. By 'inspiration' surely, by the 'spontaneous conjectures of instinctive reasoning,' said Peirce: but what then? It has often been*

[2] Richard Dawkins (2006), *The God Delusion*. London: Bantam.

[3] A famous example is Paul Dirac: 'It is more important to have beauty in one's equations than to have them fit experiment' in 'The evolution of the physicist's picture of nature', *Scientific American*, May 1963, **208**, p. 47.

[4] Peter Medawar (1984), *Pluto's Republic*. Oxford: OUP.

> *suggested that the act of creation is the same in the arts it is in science;*[5] *certainly 'having an idea'—the formulation of a hypothesis—resembles other forms of inspirational activity in the circumstances that favour it, the suddenness with which it comes about, the wholeness of the conception it embodies, and that fact that the mental events which lead up to it happen below the surface of the mind. But there, to my mind, the resemblance ends. No one questions the inspirational character of musical or poetic invention because the delight and exaltation that go with it somehow communicate themselves to others. Something* travels*—we're carried away. But science is not an art form in this sense—scientific discovery is a private event, and the delight that accompanies it, or the despair of finding it illusory, does not travel.*

Medawar does not dismiss the affective in scientific creation, but he does internalize it. He seems to deny the possibility of realizing a 'commons' of scientific metaphor that Wordsworth considered the prerequisite of a poetry of science. Medawar does not quite declare emotions to be consequent to, rather than generative of, ideas, and they are so closely entwined as to be suggestive. Perhaps the 'private' nature of the emotional he detects in science is not so much intrinsic to a categorical difference from artistic creative process, as to the conventional privacy attached to all talk of inspiration in science that we noted in Chapter 1. He does articulate another dualism—of delight and despair—that we have met repeatedly. But surely these are also shared emotions. An experience I once had, of sharing a visual and mathematical dream with a collaborator, was accompanied by shared wonder and delight. We need to ask whether the failure of science to create pathways by which this delight can 'travel,' in Medawar's language, is intrinsic or conventional, or simply occasional. But his language is suggestive of a dimly perceived connection.

If affect is participative in the process of originating ideas, then we might expect it to take the form of 'leading' emotions, such as desire, rather than 'following' ones, like delight. Expressions of desire are even harder to find in the open literature of scientific discovery, though they do emerge when personal accounts are

[5] For example, Jacob Bronowski (1965), *Science and Human Values (revised edition)*. New York: Harper.

invited, especially within very long term and arduous research programmes. In a rare moment of introspection, for example, Einstein recounted in 1934 the experience of his ten-year search for a theory of gravity, which eventually became the general theory of relativity in 1915, in turn inspiring Emmy Noether's work[6] on its energy conservation:[7]

> *The years of anxious searching in the dark, with their intense longing, their alternations of confidence and exhaustion and the final emergence into light—only those who have experienced it can understand.*

Here in a powerfully condensed single sentence is a window onto the driving emotions of longing and anxiety that accompany the creative search. Yet this glimpse into the affective experience of surely one of the most significant intellectual acts of scientific theory-creation is surprising in the light of Einstein's de-emphasis of emotion in relation to the intellectual at almost every other turn. In his later *Autobiographical Notes*, he downplays the significance of anything but thought itself in his life-course:[8]

> *The essential being of a man of my type lies precisely in what he thinks and how he thinks, not in what he suffers.*

It took just a decade for his first lucid perspective onto the emotional drive to become veiled once more. Such silence on the role of the specific emotion of desire in the pursuit of science, in our own time, contrasts with a much greater openness in earlier ages to a tight relationship between desire and rationality, between judgement and rightly disposed love. Not, therefore, for the first time, will we have to listen to voices within the history—including the long-history—of science in order to find

[6] See Chapter 5

[7] Albert Einstein (1934), *Notes on the Origin of the General Theory of Relativity*, published in *Mein Weitbold*. Amsterdam: Querido Verlag and reproduced in Albert Einstein (1954), *Ideas and Opinions*. New York: Bonanza Books.

[8] Albert Einstein (1949), Autobiographical notes, in P. A. Schipp, ed. *Albert Einstein: Philosopher Scientist*. Evanston, Illinois: Library of Living Philosophers, pp. 3–95.

perceptive thinkers who are willing to give voice to the creative energies of emotion as they support and drive cognitive processes of rational discovery. The covert partnership between emotion and cognition closely parallels that other hidden partnership between imagination and reason that surfaced most clearly in the relationship of *poiesis* and *theoria* in Chapter 5.

We found in that story, and at several other points as well, the 'distant mirror' of medieval philosophy, not the antiquated irrelevance or historical curiosity that many expect it to be, but a helpful reflective tool. It sets the hidden assumptions of our own time into relief. The high middle ages also benefitted from a more developed introspection than is encouraged in the formation of a scientist today. In our exploration of the visual imagination, for example, its old theory of vision made us think hard about how the inward stream of impressions from the world is met with our outward reconstruction to form the total process of perception. From that vantage point we were then able to see more clearly how the visual metaphor, and even the visual cognitive apparatus, is recruited in the work of making new science as well as new art. Here is another 'connected duality'—of the incoming and outgoing, the intromissive and the extramissive, that invites us to wonder how it might connect with the others listed at the start of this chapter.

Christian thought throughout most of its history, and in particular during the medieval period, has been deeply reflective on the role of desire. Furthermore, and surprisingly to many today, this historical narrative is by no means all negative. It is the object of desire, and the manner of its expression, that determine its good forms from the bad. Desire plays a role in another great duality that occupied medieval thought continually, from Augustine to Anselm to Aquinas. Rushing too soon into a translation of the two Latin terms employed for its poles would be a mistake, for they seem to encompass many aspects of the dualities that underlie creation. They are the human attributes referred to as *aspectus* and *affectus*.

Returning to our thirteenth-century guide, Master Robert Grosseteste of Oxford and Lincoln, we meet the tussle between *aspectus* and *affectus* in human works—what we do—in his treatise on education *De artis liberalibus* (On the liberal arts). The purpose, he explains, of the curriculum of the seven liberal arts is to direct our desires, thoughts, and actions to appropriate ends. Without the exercise of rhetoric, grammar, and logic (the 'trivium' of first-level disciplines), of astronomy, music, geometry, and arithmetic (the 'quadrivium' of more advanced, mathematical arts), we are prone to error. We talk loosely today of acting from our 'head' or our 'heart,' as if these were exclusive modes of motivation. But for a medieval thinker, the emotions and rational thought were intrinsically intertwined. In many ways we are less subtle now.

Affectus comprises the will, desire, or divine speculation. It has a greater projection onto the emotional than does *aspectus*, and less onto the cognitive or rational. It is the immediate motivator of motion or action. It responds more to the internal than the external, but can be directed by *aspectus*.

Aspectus is intellectual apprehension. It has a greater projection onto the cognitive and rational, and less onto what we would term the 'emotional' today. It acts on *affectus*, rather than being acted on by it. It possesses a sense of inner perception, gently invoking the metaphorically visual 'aspect' of cognition.

Grosseteste writes within the intellectual and theological context of his age, of course. In the duality of *affectus* and *aspectus* he follows earlier thinking. Anselm of Canterbury[9] anticipates Grosseteste by a century in describing the priority of sight *before* the emotion of love can engage:

> *Of these, sight (aspectus) is like the exterior [part], because nothing arrives at desire (affectus) that does not first occur to sight. For nothing is loved unless it is first known.*

[9] Anselm, *Cur Deus Homo* 2.13.

Describing a striking example of a medieval 'aha' moment, Anselm's biographer Eadmer records a delightful personal experience of the connection between desire and understanding.[10] One night before Matins, Anselm lay awake in bed wondering about how the biblical prophets could know both past and future. He had greatly desired to understand this for a long time. As he stared at the wall, he had the impression that he could suddenly see right through it, to the church where monks were preparing for the office, one of them ringing the bell. Then Anselm understood,

> *that it was a very small thing for God to show to the prophets in the spirit the things which would come to pass, since God had allowed him to see with his bodily eyes through so many obstacles the things which then were happening.*

The memory resonates with many of the features that have surfaced repeatedly in experiences of creative vision: it occurred at a 'liminal' moment (in this case the passage from night to morning and from sleeping to waking), the apparent reception of insight as a passive 'gift' following a long period of deliberate puzzling. The moment of insight also took 'cross-modal' form (in this case of a visual metaphor for cognitive insight), and was accompanied by an affective shift: Anselm's desire turned to joy—a specific example of the core *intellectual* experience of joy he records in the prefaces to his philosophical works, the *Monologion* and *Proslogion.*

To comprehend these medieval thinkers today we need to understand the overarching narrative framework within which they worked. In particular, their theologically Christian worldview interpreted the biblical story in the light of experience to suggest that the human condition is 'fallen'—it fails to attain morally, and even mentally and physically, what it was created to be, or what it can be. Medieval scholars, drawing on Augustine and the earlier Greek Church Fathers had developed an extended

[10] R. W. Southern (1979), ed. and trans. *Life of St. Anselm, Archbishop of Canterbury, by Eadmer*. Oxford: Clarendon Press.

interpretation of the Genesis story in which the first humans disobediently eat of the fruit of the 'Tree of knowledge of good and evil.' They suggested that beyond the moral aspect to Adam's fall there was also an intellectual one: that an original, deep, and intuitive grasp of the workings of nature had been lost through human disobedience. This was not, happily, the end of the story, but rather the beginning of a new chapter. One of the tasks now entailed on humankind is the reappropriation of insight, imagination, and understanding of the natural world, as well as right action within it. Such reconstitution of powers was neither simply a divine gift, nor to be won through human effort alone. It had to work within the story of grace, but work it must. Our senses, rational thought, and desires, although dulled, are left with us for a reason. This is the background to the 'virtuous circle' of three powers tightly described in the quote at the head of this chapter. Sight (*aspectus*) rightly directs desire (*affectus*), which in turn incites proper actions, or good works. It became the background for the 'ladder' to recovery of natural understanding that we encountered with Bonaventure in Chapter 5, and eventually the justification of experimental method in Francis Bacon in Chapter 4. Elsewhere Grosseteste makes more explicit the completion of the circle—that good works can themselves add to the illumination of the world that is necessary for the (in)sight of *aspectus* to operate.[11]

A clear account of this calling to reclaim the mind is found in Grosseteste's commentary on Aristotle's *Posterior Analytics*.

> *Since sense perception, the weakest of all human powers,*[12] *apprehending only corruptible individual things, survives, imagination stands, memory stands, and finally understanding,*[13] *which is the noblest of human powers capable of apprehending the incorruptible, universal, first essences, stands!*

[11] In his extended commentary on Genesis, chapter one, the *Hexaemeron*.

[12] Recalling St Paul's categories in 1 Cor 1v7.

[13] This may be an abbreviation of a five-step 'ladder of intelligence', met with in Chapter 5, and detailed by Isaac of Stella in his Sermon 4 on the Feast of All Saints (trans. B. McGinn, 1977. Cistercian Press): 'For the soul too, while

Human understanding (*aspectus*) is now inseparable from human emotion and loves (*affectus*); the inward turning of the latter at present dulls the former. However, there is an avenue of hope that the once-fallen higher faculties might be re-awakened: engaging the *affectus*, through the still-operable lower senses, in the created external things of nature allows it to be met by a remainder (*vestigium*) of other, outer *light*. So, a process of re-illumination can begin once more with the lowest faculties and successively re-enlighten the higher. In other words, science becomes a moral and spiritual exercise in personal and corporate healing and flourishing.

There are two remarkable contrasts between this frank discussion of learning and the work of the mind in the early thirteenth century and the majority of similar introspections of our own time. The first is the equality and the frank interdependency of thought and feeling. Grosseteste even attributes scientific shortcomings, not to faulty *reason*, but to insufficiently excited *desires*. So for him for example, Aristotle's failure to conceive of a temporal beginning to the cosmos is not a falling short in *aspectus* but from a disorder in his *affectus*. The second is the equally striking linking of moral and intellectual action. Philosopher Steve Fuller has suggested that the rational faculties eclipsed the emotional and moral in discussions of human reason only after Kant.[14] But the earlier theological framing creates the space to endow the intellectual curriculum with moral purpose:[15]

> *Now, there are seven arts that purge human works of error and lead them to perfection. These are the only parts of philosophy that are given the name 'art', because it is their effect alone to lead human operations towards perfection through correction.*

on pilgrimage in the world of its body, there are five steps towards wisdom: sense-perception, imagination, reason, intelligence and understanding.'

[14] Steve Fuller, Mark de May, and Steve Woolgar, eds. (1989), *The Cognitive Turn: Social and Psychological Perspectives on Science*. Dordrecht: Springer.

[15] Gasper, Giles E. M. et al. eds. *The Scientific Works of Robert Grosseteste Vol. 1.: De artis liberalibus.*

It is this holistic notion of the mind that saves duality from becoming dualism. This reflection on the medieval curriculum in Grosseteste's *De artis liberalibus* is one of the most developed of its kind. The text contains remarkable insights into the operation of the voice and the perception of sound through vibrations in the air (as the underpinning natural science of the vital art of rhetoric). The treatise acknowledges the importance of number in descriptions of nature. Motion is given mathematical treatment, including an explanation of why rational numbers are needed to describe it. The phenomenon of movement carried, for medieval writers, a strong metaphysical as well as physical sense, for it indicated both intention and change within time. Grosseteste deduces that perception by the mind of the special form of motion that gives rise to musical tones is a translation of their pitches into numbers, then recognition of those numbers by the memory. The understanding of number, through proportion, thence into music and harmony, becomes a chain of connectivity that starts with the desire for the good, and ends with an exploration of the laws of motion in the heavens and on Earth as an imperative that is both mental and moral:

> *And since measured inflection of the human voice and of the gestures of the human body is regulated by the same proportions as those by which the sounds and movements of other bodies are [regulated], the art of music considers not only the harmony of human voice and gestures, but also [the harmony] of instruments and of those the pleasure of which consists in motion or sound and with these the harmony of the heavenly bodies.*[16]

All this is to be accomplished though the mind, and the rectification by the disciplines of both its sight and its desires. It seems strange to talk of the emotional and the cogitative in such tightly woven textures that it becomes impossible to tease them apart in a clean way. But perhaps the distant thought-world of the twelfth- and thirteenth-century renaissance had a clearer holistic introspection into cognition than we possess today.

[16] Robert Grosseteste, in Gasper *et al.*

Any discussion of duality in cognition and action, especially contrasting affect and cognition, immediately triggers thoughts in our own time of cortical lateralization—the celebrated dichotomy of the human left and right hemispheres of the brain, and of their differentiated function. Literary scholar and neurologist Iain McGilchrist has surely written the *summa* of lateralization in his extensive survey of the neurological and cognitive duality in his book, *The Master and His Emissary*.[17] Insistent on taking leave of the simplistic and discredited identification of the left brain with cool scientific logic and the right with artistic creativity, McGilchrist provides a neurologically supported and nuanced account of the dualities that are notwithstanding present in the asymmetric human cortex. The emotional and cogitative, the holistic and reductionist, and other dualities become viewpoints within a much more complex story.

It is tempting to conjecture that Robert Grosseteste's eight-centuries-old discussion of *aspectus* and *affectus* might represent the recognition in a former age of the delicate interplay of left and right brain hemispheres. What is true neurologically today will have been the case in the twelfth and thirteenth centuries, since cortical evolutionary timescales are very much longer than the intervening centuries. We do, however, need to translate between metaphors, from a theological to a secular framing, and from one set of philosophical structures to another (and, in addition, from Medieval Latin to English, itself a far from straightforward operation). Listen, with that caveat in mind, to McGilchrist warning against an over-reliance on pure analytical over integrative thinking:

> *Our talent for division, for seeing the parts, is of staggering importance—second only to our capacity to transcend it, in order to see the whole. These gifts of the left hemisphere have helped us achieve nothing less than civilisation itself, with all that that means. Even if we could abandon them, which of course we can't, we would be fools to do so, and would come off infinitely the poorer. There are siren voices that*

[17] Iain McGilchrist (2009), *The Master and His Emissary*. New Haven: Yale University Press.

> *call us to do exactly that, certainly to abandon clarity and precision (which, in any case, importantly depend on both hemispheres), and I want to emphasize that I am passionately opposed to them. We need the ability to make fine discriminations, and to use reason appropriately. But these contributions need to be made in the service of something else, that only the right hemisphere can bring. Alone they are destructive. And right now they may be bringing us close to forfeiting the civilisation they helped to create.*

McGilchrist might, in former ages, have expressed his warnings against the destructiveness of deploying left-brain analysis alone, as a failure to follow up 'seeing' with 'loving' or 'right desiring,' or perhaps invoking the full subtlety of the medieval argument, as a failure to set in motion the virtuous circle of right seeing, judgement, desire, and action. The *aspectus* seems in most respects to arise from left-brain properties. It analyses, judges—invokes the detailed work of language, and abstracts from the particular to the general. The *affectus* deploys more holistic and immersive powers. It covers the inarticulate yet vital work of desire and motivation. But it also integrates, perceiving wholes rather than parts. Vitally, this integrative role of the right hemisphere applies not only to object but also to subject. Our very sense of self, including the integrated structures that bind together the linguistic and analytical pieces of left-hemisphere cognition, arises from the right hemisphere. This weaving together of the psychological self can occur in the context of emotion, as becomes all too obvious in careful studies in patients with dementia. As Douglas Watt[18] puts it, 'emotion binds together virtually every type of information the brain can encode . . . [it is] part of the glue that holds the system together.' Here we seem to be calling on the multiple valency of *affectus*, which is not only a motivating force for individual acts, but a cohesive one for the whole human being.

McGilchrist acknowledges that measures, even definitions, of creativity are highly problematic, in agreement with the other

[18] Douglas Watt (1999), 'At the intersection of emotion and consciousness', *Journal of Consciousness Studies*, **6**, *6–7*.

literature we have surveyed. Cognitive neuroscience is itself undecided on the role of affect in creativity. One problem seems to have been a confusion between the specific mood of desire to solve a problem in a creative way, which has a positive correlation on performance,[19] and a general mood of emotional positivity, which has been recorded to have a negative effect.[20] Laboratory-scale problem-solving approaches in psychology seem limited in how much they apply to long, laborious, reflective, and imaginative creative projects. Most of the attributes related to creativity (making connections between initially distant ideas, thinking flexibly) seem to be dominated by neurological activity in the right hemisphere. Yet that is not the entire story, for there is evidence that when communication between the two hemispheres of the brain is reduced or prevented (by lesion in the central connecting region, the *corpus callosum*), creative powers are actually reduced. Such neurological findings resonate strongly with the ancient conclusion that a mutually informing duality of *affectus* and *aspectus* is required to acquire the imaginative insight which is the final goal of the liberal arts.

A modern inheritor of the medieval love of systematizing and categorizing human cognitive dispositions in detail is the neurologist Antonio Demasio. He is also, intriguingly, a leading voice in research on the emotions, and on their essential role in all thought:[21]

> *The mechanism of primary emotions does not describe the full range of emotional behaviors. They are, to be sure, the basic mechanism. However, I believe that in terms of an individual's development they are followed by mechanisms of secondary*

[19] A. Isen, K. Daubman, and G. Nowicki (1987), 'Positive affect facilitates creative problem solving', *Journal of Personality and Social Psychology*, **11**, 47–52.

[20] G. Kauffman (2003), 'The effect of mood on creativity in the innovation process', in L. V. Shavina, ed. *The International Handbook on Innovation*. New York: Elsevier Science, pp. 191–203.

[21] Antonio Demasio (2006), *Descartes' Error: Emotion, Reason and the Human Brain*. New York: Random House, Ch. 7.

> *emotions, which occur once we begin experiencing feelings and forming systematic connections between categories of objects and situations, on the one hand, and primary emotions, on the other.*

Here are new echoes of the role of emotional faculties in supporting logical operations such as 'forming systematic connections between categories of objects.' Demasio also delivers a challenge to a neatly separable duality between emotion on the one hand, and cognition on the other. His 'secondary emotions' begin to look like bridges between 'primary emotions' and the cognitive work of apprehending 'categories of objects and situations.' The pattern bears uncanny similarity to Grosseteste's medieval account of the role of 'bodily actions' in negotiating the energies generated by the poles of *aspectus* and *affectus*. The deepest of medieval and modern reflections support a central place for the emotional in the creation of the new.

Early modern echoes

We do not need to go as far back as the thirteenth century to find examples of integration of cognition and emotion in judgement and creative, including philosophical, thought. Historians of science, Rosfort and Stanghellini, comment on the stark contrast between seventeenth- and eighteenth-century radical thought represented by Spinoza and Hume:[22]

> *If one reads Spinoza's geometrically and stringently structured Ethics (1677) together with Hume's eloquent Treatise (1739–40), one cannot but notice the pronounced difference in atmosphere. Spinoza's radical rationalism, with its logical denunciations of the passions, saturates the structure and expression of almost every sentence, and Hume's no less radical empiricist scepticism towards our rational capacities allows his passions, sentiments, and taste to animate the text.*

Here is an example, by contrast, of Hume's identification of emotion as a guide of judgement of the truth:

[22] R. Rosfort and G. Stanghellini (2012), 'In the mood for thought: feeling and thinking in philosophy', *New Literary History*, **43** *(3)*, 395–417.

> *The difference between fiction and belief lies in some sentiment or feeling, which is annexed to the latter, not to the former, and which depends not on the will, nor can be commanded at pleasure. It must be excited by nature, like all other sentiments; and must arise from the particular situation, in which the mind is placed at any particular juncture.*[23]

In his earlier treatise on human understanding, Hume is even stronger on one deception of enlightenment rationality—that the emotions play no part in the philosophical search for truth:

> *By reason we mean affections . . .; but such as operate more calmly, and cause no disorder in the temper: Which tranquillity leads us into a mistake concerning them, and causes us to regard them as conclusions only of our intellectual faculties.*

This is a remarkable claim of a centrality of passion in the acquisition of scientific knowledge itself, not purely in the aesthetic enjoyment of its fruits. Yet it is a truth, if a suppressed one, that all scientists know. Furthermore, it presents an important diversity within the experience of affect: as well as the violent passions of joy and grief, there are the quieter ones that may even, for their persistent nature, be mistaken for the rational (*aspectus*). A salient example would be the 'aesthetic emotion'—the appreciation of beauty, which threads its way through the testimonies of both artistic and scientific creation, both as motivation and response. Keeping alert to such quieter emotions is important in gathering personal evidence on the creative phase of science, but doing so is to swim against the tide of current scientific narrative. If Hume was able to report, and even to analyse, the emotional thread of natural philosophy without controversy, it was because science was more honest about its passions in the eighteenth century.[24]

A recognizable connection between the character of rational thought, and its reciprocal relationship with the emotional and

[23] D. Hume (2007), *An Enquiry Concerning Human Understanding*. P. Milligan, ed. Oxford: Oxford University Press, p. 35.

[24] Richard Holmes (2008), *The Age of Wonder: How the Romantic Generation Discovered the Beauty and Terror of Science*. HarperCollins.

the moral, can be found in the reflections of the late-eighteenth-century British chemist, Joseph Priestley, whose likening of novels to scientific instruments we noted in Chapter 4. Priestley was a remarkable polymath, a devout Unitarian minister who connected his endeavours by developing a theology of scientific exploration, combining thoroughgoing enlightenment rationalism with (heterodox) Christian theism. Within a lifelong interest in the physical chemistry of gases, he stands most celebrated as a co-discoverer of oxygen alongside his French counterpart, Antoine Lavoisier. In the preface of a 1769 volume of reports on a practical approach to electricity, he commented on the metaphysical motivation for science:[25]

> *A Philosopher ought to be something greater, and better than another man. The contemplation of the works of God should give a sublimity to his virtue, should expand his benevolence, extinguish everything mean, base, and selfish in his Nature, give a dignity to all his sentiments, and teach him to aspire to the moral perfections of the great author of all things.*

A rational exercise, 'the contemplation of the works of God' here gives direction, and also *vision* (*aspectus*) towards the *affect* of aspiration (to 'moral perfections'). The embedding of a scientific imagination within a much larger narrative of human advancement towards the divine constitutes a modern echo of the narrative described by Anselm and Grosseteste in their own times, but emerging within the new experimental programme of enlightenment science.

As testified by scientific histories of the long eighteenth century, such as *The Age of Wonder*, by Richard Holmes, and *The Invention of Nature*, Andrea Wulf's biography of Alexander von Humboldt, we find an age where the explicit role of the emotions in science was more readily acknowledged. Humboldt's eager longing to scale the highest Andean peak in order to map the flora

[25] Joseph Priestley (1769), *The History and Present State of Electricity, with Original Experiments*. London: J. Johnson and J. Payne

there against those at lower altitudes, but at higher latitudes, is expressed with passion in his letters.[26]

> *What we glean from travellers' vivid descriptions has a special charm; whatever is far off and suggestive excites our imagination; such pleasures tempt us far more than anything we may daily experience in the narrow circle of sedentary life.*

Indeed, his lifelong love of nature seemed to replace the need for any lifelong human relationship. Also crucial is his manifestly holistic approach to modelling nature. His is the original idea that environment is connected, and that unrestricted human agriculture can do huge unintended damage.

> *In considering the study of physical phenomena, not merely in its bearings on the material wants of life, but in its general influence on the intellectual advancement of mankind, we find its noblest and most important result to be a knowledge of the chain of connection, by which all natural forces are linked together, and made mutually dependent upon each other; and it is the perception of these relations that exalts our views and ennobles our enjoyments.*[27]

Michael Polanyi points out that all such scientific judgements ultimately rely on the personal biographies and priors of those who make them. He also suggests that the personal and emotional is generative of the all-important first phase of scientific method—the conception of new ideas in the first place:[28]

> *We cannot ultimately specify the grounds (either metaphysical or logical or empirical) upon which we hold that our knowledge is true. Being committed to such grounds, dwelling in them, we are projecting ourselves to what we believe to be true from or through these grounds. We cannot therefore see what they are. We cannot look at them because we are looking with them.*

Affect accompanies scientific discovery from conception to maturity. In accounts of mathematical discovery, metaphors from journeys of exploration abound. The anticipation, exhilaration,

[26] Andrea Wulf, *The Invention of Nature.*

[27] Alexander von Humboldt (1997), *Cosmos: A Sketch of the Physical Description of the Universe.* E. C. Otte, trans. Maryland: Johns Hopkins Press.

[28] Michael Polanyi, Harry Prosch (2008), Meaning. Chicago: University of Chicago Press. p. 61,

and struggle of a trek towards a distantly glimpsed mountain peak is how mathematician Marcus du Sautoy describes the experience of intuiting a theorem, then the long process of laying out a proof towards it.[29] His language is even reminiscent of Humboldt's expressions of desire as constitutive for the exploration of uncharted territory.

The long trek can sometimes comprise a mass expedition of the many, lasting years or decades. A recent example in mathematics is the discovery of a surprisingly simple proof of the Gaussian Correlation Inequality (GCI) by retired German statistician, Thomas Royer.[30] Originally formulated in the 1950s, the theorem describes how the likelihood that two attributes of a random distribution have specified values increases when they are correlated. A proof of this compelling notion eluded the many mathematicians who had sought it, some using very sophisticated tools indeed. Tellingly, Royer's work had been highly applied—he had years of experience behind him resolving implications of drug trials for the pharmaceutical industry. Correlations crop up here all the time—human height is correlated with weight, health with income, fitness with diet. In a glorious example of the phenomenon of the sudden and apparently effortless emergence into consciousness of a creative act performed in the hidden regions of our minds, he records the sudden and emotional realization of a proof while brushing his teeth one morning:

> *The 'feeling of deep joy and gratitude' that comes from finding an important proof has been reward enough. It is like a kind of grace. We can work for a long time on a problem and suddenly an angel—[which] stands here poetically for the mysteries of our neurons—brings a good idea.*

[29] https://www.theguardian.com/books/2015/jan/23/mathematicians-storytellers-numbers-characters-marcus-du-sautoy.

[30] https://www.quantamagazine.org/statistician-proves-gaussian-correlation-inequality-20170328.

The longer the search for such a result, the greater the joy. Hume anticipates in *A Treatise on Human Nature* the excitement and surprise of the hunt:[31]

> *there cannot be two passions more nearly resembling each other, than those of hunting and philosophy.*

However, in adopting the metaphor,[32] he extends the passion of 'philosophy' from enjoying its fruit back into the process itself. Hume knows that the emotional propels the cognitive. The sensitive introspection of the twelfth- and thirteenth-century scholars is still stirring to those in the enlightenment, like Hume, who are alert to it, but has become dulled in our own time. As Oxford lecturer in science and religion, Donovan Schaefer, has beautifully put it:[33]

> *Science is laced with emotion, not just as its fruits, as Dawkins claims, but in its everyday fabric. Everyone feels this. Everyone knows it. But we are held back by a structure of common sense—again typified by Dawkins—that says that knowledge is emotionally inert. In this, he is insensate to the internal limit of science: its total dependence on our irrevocably sensualized epistemology.*

A return to the honesty of Hume, Humboldt, Darwin, or of Peter Medawar would also constitute a re-embrace of *aspectus* and *affectus*—the essential interlacing of insight, judgement, desire, and love. For the integrated work of emotion and cognition is as needed today as when they wrote, both for scientific imagination in the first place, and for its celebration in constituting the wholly human.

[31] David Hume (1960), *A Treatise of Human Nature*. L. A. Selby-Bigge, ed. Oxford: Oxford University Press, p. 451.

[32] For a modern reflection, see Roger Scruton (1998), *On Hunting*. London: Random House.

[33] Donovan O. Schaefer (2017), *The Wild Experiment: Emotion, Reason, and the Limits of Science* in *Are There Limits to Science?* Gillian Straine, ed. Newcastle: Cambridge Scholars Publishing.

Scientific testimony to the emotion of ideas

Leo Esaki is a Japanese physicist, who shared the 1973 Nobel Prize for the discovery of electron tunnelling and the elucidation of its quantum mechanical properties. This phenomenon is one of the many counterintuitive ways in which matter behaves when governed by quantum, rather than classical, behaviour: when 'tunnelling,' particles reach places and states that classical physics would forbid, pathways on which a classical particle would possess a meaningless 'negative energy.' Yet in quantum mechanics, negative energy does have a meaning, and it translates into a physical leaking through the barrier and a finite possibility for occupying classically forbidden territory. Today, many vital electronic components rely on this effect for their function. Esaki wrote candidly about the role of emotions in driving science, in an essay on twentieth-century culture:[34]

> *science too has dual characteristics. It has a logical, objective, cool, and rational or rigorous face—the aspect of the finished product that appears in manuals and is presented to the public in conventions and conferences. . . The other face is fantastic, subjective, individualistic, intuitive, and lively, and reflects the process by which the new is created. It is a process based on perception and inspiration, obviously supported by an acute mind. Scientists can use their imagination to grope forward, in a desperate struggle of trial and error, seeking out the secrets of the universe. If by chance they find a solution and their efforts are rewarded, then they can be truly happy: that rarely happens. This is the creative process which is the essence of science.*

'Fantastic,' 'subjective,' 'lively', 'desperate', 'truly happy'—these words signify experiences that flow in the field of *affectus*—and here they are playing out in a scientist's reflection on the progress of twentieth-century science itself. Nor are they emerging as emotional by-products of scientific discovery, but in this account clearly run as warp to the methodological weft of the very process of discovery. Esaki elsewhere refers to the French molecular

[34] Leo Esaki (1956), Prepare menti et cuori al XX seculo, in *Dall'informazione all cultura, Dieci Nobel per il futoro*. Venice: Marsilio, pp. 70–1.

biophysicist Francois Jacob, another Nobel Laureate, who talked about 'day science' and 'night science' to describe the logical and imaginative sides of science respectively.

Few scientists, when pressed, deny the need for the play of imagination in the creation of hypotheses quite distinct from the process of their evaluation, but fewer admit as readily or clearly as Esaki to the role of the emotions during science's 'night-time hours.' Even the official biographical accounts and psychological studies of science have played down the part played by emotions. Historian of science, Thomas Soderqvist, writes in a collection of essays on the history of scientific biography, 'scientific knowledge is socially, linguistically, and rhetorically contextualized, but rarely seen as having anything to do with the passions of the scientist, an attitude that has also spilled over into social biogra phy.'[35] Yet on the rare occasions when scientists themselves speak about the affective arc of doing new science, the language could not be clearer on an active role of the emotional, especially in the first, creative, phase of research that we are trying to hunt down:

> *You go through this long, hard, period of filling yourself up with as much information as you can. You just sort of feel it all rumbling around inside of you. Then . . . you begin to feel a solution, a resolution, bubbling up into your consciousness. At the same time, you begin to get very excited, tremendously elated—pervaded by a fantastic sense of joy. But there's an aspect of terror too in these moments of creativity. Being shaken out from your normal experience enhances your awareness of mortality. It's like throwing up when you're sick.*[36]

The extended metaphor of digestion, even of the pathological kind, is strikingly visceral here. The quotation also implies more than it states: there are the 'uphill' emotions of desire, self-assertion, or longing, implied in the 'long, hard' period of 'filling up.' Without these driving emotions, the creative phase of science would not be sustainable. The 'information' that constitutes

[35] Thomas Soderqvist (1996), Essential projects and existential choice' in *Telling Lives in Science*. Michael Shortland and Richard Yeo, eds. Cambridge: Cambridge University Press.

[36] Anonymous source, quoted in Soderqvist, 'Essential projects'.

the phase of nourishment in such experiences takes the form of data, published experiments, conversations with others, calculations and theoretical ideas, a search through misty pathways of thought for the right articulation of a question, and the spark of possible answers. When these begin to arrive, the 'downhill' emotions of elation and joy begin to replace those of desire, but even these are admixed with fear—the witness even calls it 'terror.' As elation and hope mount under the guiding star of an idea that might just be right, there is an equal fear of the hurt, both internal as disappointment and external as manifest folly, which surely follows if the idea should turn out to be mistaken. Sometimes this fear can actually incapacitate and prevent a bold conjecture from ever seeing the light. Soderqvist expresses these conflicting emotional structures that mirror the tension at the cognitive level between successful and unsuccessful resolutions:

> *The passions embody the realisation of the tension between the conditions for self-assertion: fear, despair, vanity, pride, jealousy and envy are the results of a failure to achieve empowerment: hope, faith and love are expressions of our success in this respect.*

He is surely right in the identification of 'tension' throughout the unresolved period of gestation, but he says more than this, for 'self-assertion' is a psychological category—neither purely emotional nor cognitive, if notwithstanding highly personal and identity-making. When a scientist re-creates in imagination an image of nature by means of a theoretical model, the process of establishing its abstracted structure is also an assertion of self. More is true: for when the internally formed model begins to interact with those enhanced perceptions of the world opened up by experiment, the act creates new connections between self, nature, and community. A deep complex of relations is constructed from the very beginning of the scientific act, which demands considerable personal and emotional investment.

I have had personal conversations with very gifted people who have chosen career paths distant from science, not because of any inability to have succeeded, but seemingly because of a similar fear in the face of the physical and mathematical universe. There is something wonderful about any insight into the reality of this 'sublime' (in the possibly terrifying sense of Kant or Burke) experience, but tragic if its result is to bar the way to a personal engagement with nature through gifted human aptitude.

An aspect of scientific creation not captured by the idea of 'self-assertion' is the felt experience of 'gift' that accompanies insight. Having heard Einstein relate the emotional structure of his journey to general relativity, and the spectre of its possible violation of the conservation of energy exorcized by Noether, it behoves us to hear Werner Heisenberg in autobiographical mode. He is relating the sleepless night he spent checking that the same ghost would not be summoned by his radically new atomic theory (which would later come to be known as 'quantum mechanics') in its mathematical representation of matrix mechanics:[37]

> *At first, I was deeply alarmed. I had the feeling that, through the surface of atomic phenomena, I was looking at a strangely beautiful interior, and felt almost giddy at the thought that I now had to probe this wealth of mathematical structures nature had so generously spread out before me. I was far too excited to sleep, and so as a new day dawned, I made for the southern tip of the island, where I had been longing to climb a rock jutting out into the sea . . . I now did so without too much trouble, and waited for the sun to rise.*

Was Heisenberg aware, I wonder, of the enacted poetic metaphor of his climbing the rock into the sunrise? He draws no breath between the account of the end of his long search for a consistent theory of electrons in the atom, the sudden realization of the vast new landscape of physics now before him, and the act of answering another longing and its reward of an ocean vista. Even the relative ease of final execution once clarity of sight is achieved, commonplace in the final solution of mathematical

[37] Werner Heisenberg (1972), *Physics and Beyond: Encounters and Conversations*, Arnold J. Pomerans, trans. New York: Harper Torchbooks, p. 61.

challenges, slips naturally into his account. Sometimes actions really do speak louder than words, especially on how we climb the peaks of mental imagination.

If the witness of more sensitively introspective past ages, and the most thoughtful and honest reflections of more recent scientific thinkers have made it impossible to refute the contribution of emotions in the conception of scientific ideas, we are still left with more questions than answers. Most of them swarm around the issue of the non-conscious region of mind within which radically new ideas in science, as well as in art, seem to be assembled. How could emotions, by definition felt consciously, operate with this hidden realm 'below the surface,' universally implicated in the creation of new ideas and solutions? Or if it were, what would the idea of a 'non-conscious emotion' mean, and how would we know anything of its operation? Talk of the 'uphill' or 'driving' emotions of desire and longing surely accompanies only the conscious struggles of that extended and necessary period prior to the wonderful non-conscious work of insight that re-emerges days, weeks, or months later into the light. They might be part of the admixture of *aspectus* and *affectus* that constitutes an effective recipe for subterranean mental work, but how could we ever tease apart such shadow concepts as the 'crypto-affective' and the 'crypto-cognitive' once they are hidden from mental sight?

Such a pathway of inquiry constitutes another route to the hypothesis advanced a century ago by French mathematician, Henri Poincaré, in his celebrated meditation on mathematical creativity.[38] Poincaré, as we saw in Chapter 6, gives an extended personal account of the familiar pattern of hard conscious struggle and apparently fruitless labour before leaving the problem, only to experience at an unforced moment the appearance of a beautiful new approach towards a solution. He goes further than others who have told similar stories—charging himself to

[38] Henri Poincaré (1915), Mathematical creation, in *The Foundations of Science*, G. B. Halsted, trans. Lancaster, Pennsylvania: The Science Press.

think seriously about how the non-conscious mind achieves such a miracle:

> *What is the cause then, among the thousand products of our unconscious activity, some are called to pass the threshold, while others remain below? Is it a simple chance which confers this privilege? Evidently not; among all the stimuli of our senses, for example, only the most intense fix our attention, unless it has been drawn to them by other causes. More generally the privileged unconscious phenomena, those susceptible of becoming conscious, are those which, directly or indirectly, affect most profoundly our emotional sensibility.*

This is a remarkable idea; Poincaré believes that it might be meaningful to ascribe dual categories such as logical thought and emotion, or cognition and affect, to non-conscious as well as to conscious mental processes. For him, there lies below the surface of consciousness a mirror-world of thought that recapitulates the dual structure of *aspectus* and *affectus* of the western medieval thought-world.

He is not alone among modern writers in suggesting a structure to the non-conscious world whose elements map onto aspects of our awareness. The great twentieth-century psychologist, Carl Jung, treated patients on the basis of his theory of the 'personal unconscious,'[39] in which 'complexes' of thoughts, emotions, memories, and attitudes acquired during a person's life could assemble just below the limit of conscious awareness, influencing their conscious thought-world. The structure of such submerged complexes is induced from conscious precursors. The notion of 'depth' here is important—Jung was deliberately distancing himself from the views of Freud, who had urged the notion of a 'deeply' repressed emotional world of the unconscious. Poincaré predates both of these psychologists, of course, but uncannily anticipates Jung in the way that he requires a complex of thought and affect to explain the role of non-conscious processing in mathematical creativity.

[39] C. G. Jung (1933), *Modern Man in Search of His Soul.* trans. W. S. Dell and Cary F. Baynes. Oxford: Routledge.

The reason that Poincaré is driven to a structured theory of the creative mind 'below the surface,' is simply that he knows that a purely mechanical model, in which alternative arrangements of basic mathematical structures are sequentially and blindly processed and assessed for correctness, simply will not work. His conscious knowledge of the way he solves mathematical problems amounts to trying many different approaches. Each strategy is, in turn, a sort of combination of the starting ingredients of formal structures, functions, geometry, mappings—all that he has worked with for decades. He is aware that one reason that his purely conscious struggles seem to get nowhere is that there are just far too many possible combinations, orderings of the symbolic ingredients, to try without a filter that prevents the huge waste of time implicated in checking through myriads of fruitless arrangements. The possibilities for mathematical strategies grow in a similar way to the number of possible games of chess, as the number of moves increases. Poincaré suggests that the ability of a chess master to navigate the game's exponential complexity is none other than a non-conscious version of the aesthetic preference for elegance, simplicity, and symmetry that has long been the guiding light of a mathematician's conscious world of thought. Shadow-emotional reaction to the scan of subliminal possibilities draws non-conscious thought towards those that possess the aesthetic appeal of fruitfulness, acquired through long experience. His suggestion is all the more intriguing because it blurs the distinction that we proposed earlier between the set of those emotions that respond to creation, and those which propel or energize it. Poincaré maps the second sort directly onto the first—the only difference being that they are the non-conscious mirror-emotions, the hidden shadows of our conscious feelings.

It is almost as though Poincaré had been reading Anselm or Grosseteste on the role of *affectus*, once set on the road to a proper end by mental sight, of propelling mental actions towards that which builds up and away from that which tears down. But what attested experience is there today of emotional generation of scientific creativity in the face of so much public denial? It turns out that paying attention to scientists who combine a reflective attitude to their work with a candid and admirable honesty about their personal journeys of research, opens a window to plenty of examples that provide a contemporary setting to the creatively emotional in science.

Putting out fire with fire: A case study in creative scientific affect

The worst aircraft accident in history, in terms of loss of human life, occurred not in the air, but on the runway of a busy passenger airport. On a foggy day in 1977, a 747 'jumbo' jet had come to a stop while taxiing, across the main runway of the island airport of Tenerife. At the far end of the same runway, another 747 was waiting to start its take-off run. Unaware that the first plane was blocking its way, the captain began his take-off run. By the time the stationary aircraft loomed into view it was far too late to avoid a collision – at a speed too low to fly over the obstacle, but too high to stop. In the enormous fireball that ensued, from two fully fuelled giant aircraft in collision, one of their aluminium fuselages had melted within thirty seconds. Only sixty-one of the 650 passengers and crew survived.

I was sitting listening to this horrifying story in the unlikely setting of an international meeting on polymer science, held annually in an elegant hotel in the north English Peak District. The speaker was Professor Julie Kornfield, an energetic and enthusiastic professor of chemical engineering from the California Institute of Technology in Pasadena. I have known Julie for a long time—we used to meet as research students at conferences. We have followed over the years each other's stories of research, and of our ever-changing teams of younger researchers. As well as one of the most innovative scientist–engineers in her discipline (Julie engineers at the level of molecules as well as of the apparatus that they flow through), she is also one of the most appreciative I know of the human qualities that make or mar the creation of good ideas and their development within healthy teams. The story she told to the gathering of academic and industrial scientists was full of wonderful technical detail and intricate scientific insight, but it was also a rare example of a scientific tale told full of determination, desire, passion, and despair. It furnishes us with a current example of the creative potential of emotion when woven into a matrix with acts of logical thought and deliberative cognition.

The story began with a senior colleague, Virendra Sarohia, at Caltech's neighbouring institution of NASA's Jet Propulsion Laboratory, who had been involved during the 1980s in an international research programme of polymer scientists and combustion engineers. Its goal was to develop a jet-fuel additive that would prevent explosions on the horrific scale of the twin 747 accident. The solution was understood in principle: the fuel becomes explosive when it disperses into tiny droplets on impact—it is the high surface area of all the droplets in a mist of jet-fuel that allows access to oxygen at the enormous rates needed to cause an explosion.

The challenge was therefore to identify an additive to the fuel, which at very low concentrations might prevent the natural breakup of liquid into these tiny droplets. A class of candidates was compelling to investigate: the very macromolecules, or 'polymers', that we met with in Chapter 2 in their entangled melt form, but in this case highly diluted in a solution of jet-fuel. The elastic properties of polymers hold even at the level of single chains—stretch out a polymer and let go, and as a result of its Brownian motion (yet again), it will recoil into the denser cloud of random contortions that characterize its equilibrium state. In a droplet of fluid containing many polymer chains, the concerted elasticity of all the macromolecules works to maintain its original, spherical, shape. The idea for the jet-fuel stabilizer emerges from this insight—for in the process of a large droplet breaking up into two smaller ones, the intermediate shape must contain a stretched tube of fluid linking the two halves of the original droplet. If this fluid contained sufficient long molecular chains, stretched along with it, then the two halves of a breaking drop would be brought back together again as the chains re-contracted. The dangerous cascade of division into the catastrophically explosive mist would be nipped at the very first stage.

The earlier US–UK research programme developed just such a long-chain polymer additive against a set of intimidating requirements: they must suppress of droplet formation on impact, but not inside jet engines themselves, where the injected fuel *must* vaporize in order to combust. Nor must the passage of fuel from the tanks tear the long-chain molecules apart. Finally, there were constraints on cost, delivery, and stability in storage. The project was making progress with enough of these to permit a highly publicized test—a crash of a pilotless jet into a concrete barrier on a disused runway. The international research team had planned for a flash-free impact in full view of live cameras, but something went wrong. As a result of the difficulties of steering a large aircraft remotely, the artificial concrete obstacle met not with the intended wing, but with one of the operating engines. The resultant fireball was far briefer and less damaging than it would have been with the unmodified fuel, but the sight of billowing clouds of flame, however controlled, was enough to shelve the project for lack of credibility.

Sarohia was bitterly disappointed, yet was able to pursue successful research projects in other topics for several years without serious regrets—until, that is, the terrible events of 11th September 2001 in Manhattan. For it was not the impacts themselves of the two hijacked airliners that caused the towers of the World Trade Centre to collapse, but the many minutes of searing fire unleashed by the aircraft fuel, dispersed and explosively ignited. Sarohia knew that, in principle, it would have been possible to save the towers, and thousands of lives, had the fuel not exploded. A renewed search for a safe fuel became a consuming passion for him, and one that caught Julie Kornfield's own interest. The difference was that now there was no hope of a national, let alone internationally funded programme. The remaining challenges would have to be met on the shoestring budgets that may sometimes be generated at the margins of university research groups.

Kornfield is generous to a fault of others' contributions in the way she tells the story, yet the underlying breakthrough-concept is hers. The central problem is to deliver the long-chain polymers to the fuel without the strong flows of fuel-delivery tearing the chains apart along the way. Her idea was that the polymers should consist of short and robust segments when they are being delivered to and mixed in with the fuel, then assemble *in situ* into the functional long chains through the incorporation of chemically 'sticky' points at the extremities of the segments. Such 'self-assembly' of inanimate matter sounds like a pipe-dream; indeed, I have heard even scientists unfamiliar with its details describe it as 'magic'. But a moment's reflection on the biological growth of every living thing from its embryonic form is reassuring. For the hierarchy of living cell, tissue, and organism is really a complex and evolved form of molecular self-assembly. Reproducing such marvels of complexity is of course entirely beyond us at present, but the task of creating short molecular chains that assemble into longer ones when left to their own devices, is feasible. Food-additives and shampoos are examples of technologies that make use of such polymer self-assembly.

Julie's highly creative idea brought its own challenges. The first was the chemistry behind the 'sticky ends' of the polymer chains. Fortunately, many candidates present themselves here, from oppositely charged groups of atoms that attract each other electrically, to the use of 'hydrogen bonds,' the natural affinity of incorporated hydrogen atoms with oxygen atoms on a neighbouring molecule, that ensures water stays liquid at room temperatures.

The next obstacle proved more taxing, for demanding the self-assembly of highly dilute components quickly enough to be useful requires the very rare chain ends to find each other very rapidly. The mutual search of the sticky chain ends can be propelled only by random Brownian motion. Kornfield's initial idea to accelerate the rare pairing of polymers was to multiply the number of 'stickers': creating polymer chains with many associating units (the star polymers in new guise) would increase the chance of two neighbouring polymers finding each other. The task of synthesizing candidate structures was assigned to research student Ameri David, who began working through options of association numbers, chain length, and sticker-chemistry in a systematic search for the perfect self-assembler. As so often in any creative project exploring unknown territory, however, a concept that seemed at first so promising began to fade, as none of the formulations behaved as the team desired. The increased number of arms in each polymer unit did indeed create a higher probability of joining up, but brought with them a fresh drawback: gelled particles rather than long stretchable chains.

As Feynman reminds us, finding dead-ends in science projects is common, and faced with such an outcome, a research student would typically cut their losses and request another research problem to tackle. But by this time the passion and urgency injected into the project by Sarohia and Kornfield had infected the whole team. David adopted a strategy that had then become, and remains now, extremely unusual. In current practice, scientists usually specialize in either theoretical or experimental methods. But at the risk of losing recognition as an experimentalist through spreading his effort too widely, he decided to try a theoretical approach to the problem.

The mathematics of statistical mechanics allows the modelling of self-assembly, and the calculation of quantities such as the average size of a self-assembled clusters. The price that has to be paid is that of approximation. Chemical details must be lost in a simplified mathematical picture of a polymer. The great strength of a mathematical model, however, is that the whole 'microuniverse' of possible designs can in principle be surveyed in a single formula. Months of empirical experimentation, of chemically wandering in darkness, are replaced by a mathematical space, lit up and transparent all at once. David was following a hunch that somewhere in the abstract design-space of chain length and sticker-group strength, there was a 'sweet spot' that would answer to the problem's stringent requirements. He found it lurking in an extreme corner of the mathematical solution, where the initial chains were relatively long, and the sticky groups counter-intuitively strong, so delivering very much longer-lasting bonds than those he had worked with until then.

There was one more appealing little detail that the calculations revealed, that pointed to an important property of the 'sticky end groups'. David had thoughtfully played with two kinds of groups: one that attracted itself, and a second type that worked in +/− pairs, so that a '+' would only bind to a '−' and vice versa. The first suffered from a severe problem: most of the unit chains would form rings by the simple process of their two ends sticking together. But by imagining unit chains of two types, one with both ends '+' and the other of both '−', he saw that no chain would be able to curl up on itself to form an infertile ring, but now the desirable multi-unit chains would grow. A diagram from the group's main published article is reproduced in Figure 7.1. It explains in pictures the way that the long unit chains drive the desired extended concatenation and avoid the clumped alternative assemblies.

Now the Caltech team was ready for another experimental phase, but this time armed with the calculated recipe that no one would have guessed was the avenue to explore. A serious search remained, for the theory gave orders of magnitude within which to design real systems, not a recipe-book instruction-set. The second research student, Jeremy Wei, needed to find the right chemistry for both the polymer chains and the powerfully attractive end groups. For the second time, as Kornfield's students found the courage to adopt the spirit, if not the letter, of her advice. For three more years the various systems proved unworkable, but Wei insisted that he would not abandon the goal

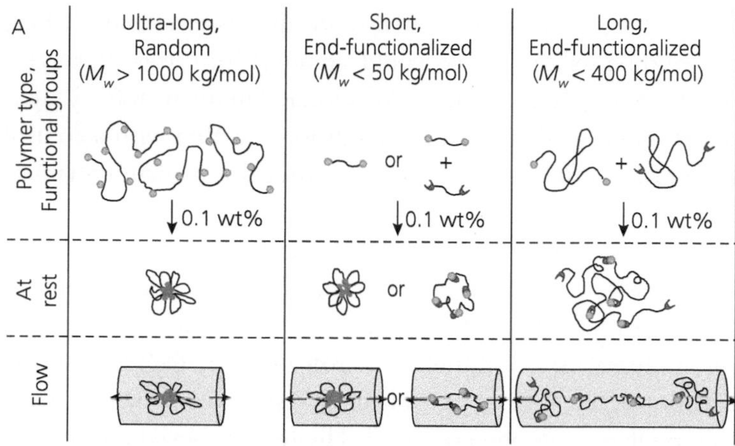

Figure 7.1 From the group's final published paper. The figure depicts, in cartoon form, the theoretical trajectory towards the final solution in the third column. Only this structure yields the desired long chains in the bottom row. From [Wei et al. (2015) Science, **350**, 72–75]. Reprinted with permission from AAAS.

of a fire-safe fuel any more than David had done. In his case, the infectious desire to realize a solution was additionally fuelled by a strong personal religious faith, which he was happy to share with the lab team—'faith is a very powerful thing,' avers Kornfield.

After the long years of research, periods of apparent hopelessness and new creative breakthroughs driven by the combination of persistence-inspired contrariness and deep personal commitment of the junior scientists, a moment of truth arrived at the point of the explosion test. The really critical experiment involved testing the polymer-enhanced fuel after a period of strong shear-mixing, similar to the disruption that would be caused during delivery of the fuel to an aircraft. The attractive stickers should break apart under the flow, but then reform as soon as the fuel arrived into a resting state.

The journey from the lab to the explosion testing apparatus is a very short one, but it must have seemed like a summons to a court-house to Kornfield, Wei, and David. There would be no doubt of the yes/no result, and no indication beforehand of which they would see. A fireball would signify failure after all; success would strangely be far less spectacular. The team took a high-speed movie of the impact and ignition, starting with a control test of fuel containing the original, simple long-chain polymers. In that case the fireball is so bright on the video recording that it saturates the camera completely in a screen of blinding white. Then the new treated fuel is introduced—99.7 per cent conventional jet-fuel containing the self-assembling polymer. The test-rig produces a spray of fuel over a row of lit gas-flames to ensure the spray ignites. Without the benefit of slow-motion, it appeared that no flame was observed at all. The team believed the ignition sources must have failed. Before repeating the experiment, however, they reviewed the video slowed down and glimpsed, over each ignition source, an occasional willow-the-wisp of darting flame—but these are immediately self-extinguished. No fireball. Hardly a flicker of flame. The quality of that moment of pure joy is evident from the way Kornfield tells the story. The international journal *Science* was at that time looking for examples of discoveries or insights that were completed in experimental realization, but had been motivated and guided by theory. This perfect example can be explored in more detail in the published account.[40]

The Caltech team's story is a remarkable view into the role of affect in scientific creativity. I do not believe that this is due to any greater generative emotion or guiding aesthetic than in other projects, but more that it is unusually 'transparent' within its story. A combination of predicament and personality on the part of both the leadership and the entire team, opens up to a far greater extent than normal the underlying emotional structure of science. In part, this is because the members needed to find ways of encouraging each other when the road ahead seemed unavoidably blocked.

[40] M. H. Wei et al. (2015), 'Megasupramolecules for safer, cleaner fuel by end association of long telechelic polymers', *Science*, **350**, 72–5.

In this case, the other force that opens a window onto the emotional as well as the cognitive thread of a creative project is the context of past human disaster and loss. Sarohia was determined, at the end of his career, to apply what he had learned in the wake of the Tenerife accident to make another 9/11-type attack impossible. In 2002, he repeatedly met with Kornfield to inspire her to work on mist-control fuel additives. The events of 9/11 were seared in the memories of both David and Wei, who were then university students on opposite sides of the world. A powerful combination of forces sustained the team: the spectre of future loss, and the hope that, through taking the narrow scientific path, it could be banished. The vision of a solution, and of a safer future that solution would enable, acts as a guiding beacon throughout the project (we are reminded of *To the Lighthouse* in Chapter 4), and 'directs the *aspectus*' of the work at each stage. At several points, a tension between unexpected ideas with compelling aesthetic speaks to the role of non-conscious reason, and even of non-reason ('if the professor's advice is getting us nowhere, then we should quietly try doing the opposite'). The entire narrative illustrates the delicate interplay of experimental and theoretical, of concrete and mathematical reasoning, and directs attention to that other creative tension—the holding together of distant ideas from whose conversation the radically new can arise.

When the *aspectus*—the vision—of this distant connection is driven and directed by the *affectus* of shared desire and determination, it becomes possible to turn imaginative insight into physical reality and then into beautiful creative innovation.

David Bohm on creativity

The moving story of the Caltech group, as well as furnishing a transparent illustration of how we might rethink the interplay of *affectus* and *aspectus* today in thinking about creativity, provides an illuminating interpretation of another, more contemporary, reflection on creative thought. David Bohm, American-born and later naturalized British physicist and philosopher, became an influential and challenging thinker throughout the second half of the twentieth century. He is responsible, and most well-known, for a remarkable mathematical alternative and radically new interpretation to the accepted version of quantum mechanics. He was able to show that one can recover a deterministic physics in place of the probabilistic theory of Heisenberg and Schrödinger, but at a price: one may remove dice from the game of modern physics but only if its laws no longer act locally. The possibility of non-local, coherent causes in physics led him to think about possible applications to the mind and brain. However controversial these ideas might be scientifically, they did drive him to reflect deeply on how the human mind manages to create the new.

Bohm's introspection and alertness to the imagination evinces a similar pattern to the slow and deep traditions of the medieval scholars. He does not echo the fashions of his time, which tended to see creativity on every painted surface, every sculpted article or designer building. Rather he seems, in his book *On Creativity*[41] to be wrestling with how rare this faculty is. He writes about a common mindset which he terms the 'mechanical,' which might at a superficial reading be interpreted as the unconscious role of the brain in animating our breathing or heartbeat. But Bohm does not mean such automatic functions by his use of the term 'mechanical,' but perfectly conscious thought—mental processes that masquerade as innovative, though they are not much more than trained reaction to stimuli. He seems to know by experience the effort it requires to haul one's train of thought out of the rut of the usual and the habitual, and to conceive of something really new. Even mental responses to the world that have 'a supreme degree of psychological significance' and conducive of the 'creative possibilities of the mind' and 'inseparable parts of one's very self' are normally only delusionally so, and actually

[41] David Bohm (1996), *On Creativity*. London and New York: Routledge Classics.

'nothing but mechanical results of past conditioning, being in fact the principal barriers to real joy and creativity.' Recall that, at two significant moments of the Caltech experience, the young scientists needed to make huge emotional and intellectual efforts that generated stark departures from the thinking around them that they could so easily have followed.

Bohm's observations resonate with another strand that threads its way through all discussion of emotion and creation—that of the exercise of will. There is innovative power in affective desire. The medieval scholars knew it; scientists and artists of all epochs have experienced it. He writes:

> *[The scientist] wishes to find in the reality in which he lives a certain oneness and totality, or wholeness, constituting a kind of harmony that is felt to be beautiful. In this respect the scientist is perhaps not basically different from the artist, the architect, the musical composer etc., who all want to create this sort of thing in their work.*

Such a desire to connect, to discover a whole from apparently discordant paths, requires, for Bohm, a personal departure from the norms of expectation 'from the very first step.' At this level he feels the same pulse that has animated this exploration—that at the depths of the human creation of the new, there is little fundamental differentiation between scientific and artistic impulse. If that were true, then the delicate interplay of affect and cognition that we have recognized, together with its connectivity to the essential non-conscious wellsprings of the new, would be apparent in artistic endeavour as well.

Picasso and Guernica—a documented journey of aspectus, affectus, and art

Documented journeys of scientific ideas, from their conception to fully articulated final form, are rare; such journaled accounts of artistic creation are rarer still. In consequence, those artists who have broken with convention and described the process in as frank and evidenced detail as in the Caltech project, have invited very well-worn paths of analysis. These are not examples of 'the road less travelled by,' but if any can claim to reward repeated examination, then that must surely be Picasso's agonized masterwork *Guernica* (Colour Plate H). The high peak of Cubist invention, Picasso himself wrote much about its genesis, and left numbered and dated the forty-five preparatory sketches for the work. A daily photographic record of its creative process also exists, thanks to the persistence of Dora Maar. The confluence of the artist's own remarkable biography, his radical invention with Braque of the Cubist form itself, and his conscious articulation of how he produced his art, motivated his selection as one of Howard Gardner's *Creative Minds*[42] of the twentieth century. Before that, the richly illustrated, exposed, and candid process of its creation inspired a celebrated book-length study by psychologist Rudolf Arnheim.[43]

Picasso was no stranger to pain felt, depicted, and inflicted. In a grossly amplified echo of Bohm's criticism of 'mechanical' thinking and the need to awake from it, Picasso's version reads:

> *A work of art must not be something that leaves a man unmoved, something he passes by with a casual glance . . . It has to make him react, feel strongly, start creating too, if only in his imagination . . . He must be jerked out of his torpor.*

If he himself could ever have been found in this state, he was shocked out of it by the hideously cruel destruction of the small Basque town of Guernica, by Franco's air force on 26th April 1937. It seems that he determined almost immediately that a large portrayal of this obscenity must constitute his response to an invitation to exhibit at the 1937 World Fair. Here, as in the transparently emotion-driven scientific project of explosion-suppression, there is the outline of a distant goal in view, of great magnitude and inviting a strongly moral motivation. The

[42] Howard Gardner (1993), *Creative Minds*. New York: Basic Books, pp. 127–72.

[43] Rudolf Arnheim (1963), *The Genesis of a Painting: Picasso's Guernica*. Berkeley: The University of California Press.

projects also share an uncertain and tentative journey of trials in the difficult terrain that must be crossed to realize their visions. In both science and art, the broad form, properties, and consequences of their creative product might be apparent from the start. But in neither case does such initial clarity of vision onto the project illuminate a clear path towards it. In Picasso's case, the series of sketches throughout the summer of 1937 display periods of continuous development, and other moments of radical shift. A vital narrative connection with the biblical 'slaughter of the innocents' introduced, at a very early stage, a distorted representation of a woman in abject grief, clutching her dead child. She glides continuously to different positions through the sequence of sketches, and finds her final framing position, at the extreme left of the canvas, by continuous development. On the other hand, a sudden modification midway through the painting's development brings the terrified horse into its central upper position. The eye is drawn back repeatedly to the agonized scream of this animal, rather than to the human figures of the work, which end up populating its periphery. This is a discontinuous and counterintuitive step. For it is principally the human suffering, and human cruelty, that the work concentrates on—that this contemplative goal is best refracted through the intervening central lens of a suffering animal was a remarkable discovery that released the final development of the work.

The artist is frank about the emotional drives of this work in particular:[44]

> *What do you think the artist is? An imbecile . . . He's at the same time a political being, consistently alive to heart-rending, fiery or happy events . . . No, painting is not done to decorate apartments. It is an instrument of war for attack and defence against the enemy.*

[44] Excerpt from an interview with Simone Téry, *Letters Françaises (Paris)*, V, 48, 24 March 1945. Translation from Alfred Barr Jnr. (1946). *Picasso: Fifty Years of his Art*. New York: Museum of Modern Art.

In *Guernica* and its documented development, we can see with unusual clarity the combined forces of cognitive artistic skill, with their deliberate designs and distortions, their informed allusion to the narrative tradition of the artist's own culture, and his consummate draftsmanship, together with a continuous and felt emotional energy. Much of this is conscious, and directs the artistic skill towards the *aspect*, the final vision held in the conscious mind. But here too, there is evidence aplenty that the affective currents behind and beneath the work manage to release discontinuous creative connections and ideas that, in hindsight, built essential pathways towards the final work.

Mathematics, science, art have been overt and open about their creations, but less communicative about their processes of creation. Quieter still has been their shared story of the foundational and subterranean, but essential emotional currents that carry them forwards. The ancient contemplated duality of *affectus* and *aspectus*, though articulated in another era, has been strangely effective in framing a narrative in which today's inner dialogues between thought and emotion, deliberation and drive, and despair and determination, can also be heard if we turn an attentive ear to them. Their own historical development shows that the apparently clear separation between emotion and cognition today looked very different in the past. The recognition that affect is at work in the creative acts of art and science, as well as in response to their creations, has opened a further path towards an understanding of how the non-conscious world of hidden imagination might be functioning alongside reason in the creative process, first illuminated by the study of poetry and theory of Chapter 5. Innovation springs from an aesthetic drive, as well as imagination. Both, in turn, respond to it. Looking at creativity across the disciplines through this double lens, artists and scientists alike talk of deeply held values of connectedness, of harmony. We are now at a point where it becomes possible to ask the question of purpose—to what end do humans create?

The End of Creation

Since sense perception, the weakest of all human powers, apprehending only corruptible individual things, survives, imagination stands, memory stands, and finally understanding, which is the noblest of human powers capable of apprehending the incorruptible, universal, first essences, stands! ζ[1]

ROBERT GROSSETESTE

The ability to bring something new and valuable into being is a wonder. At every turn we have found the process of creation to draw on the deepest human energies, most radical thought, and most powerful emotion. Hope, desire, cognition, vision, dreaming, craft, skill, expertise, community, and passion are all summoned in the task of conceiving and realizing human imagination. They weave a much more complex picture than the idea of a single creative act by single individuals with which our investigation began. That starting point posed a question about science, of whether it calls on the same, or similar, processes of imagination as do the arts or their study within the humanities. The journey through the visual, the literary, and the abstract worlds of imagination has provided an affirmative answer, but one that contains much more complexity and historical legacy than we anticipated. When we have found scientists, whether writing in former years or in conversation today, in a position to reflect deeply and to talk freely about the origins of their ideas, we have consistently found them using the same metaphorical language

ζ[1] Robert Grosseteste, *Commentary on the Posterior Analytics*, quoted in R. W. Southern (1992), Robert Grosseteste; *The Growth of an English Mind in Medieval Europe*. Oxford: Clarendon Press, p. 167.

The Poetry and Music of Science. Tom McLeish, Oxford University Press.
 DOI: 10.1093/oso/9780192845375.003.0008

as did painters, writers, and musicians. Not only that, but the notion of 'parallel tracks' of art and science, however regularly yoked together, has failed to account for their continual mutual influence.

In the busy analytical process of closely examining art, physics, music, mathematics, poetry, and fiction—instructive though that is—it is easy to forget that the primary intention of the artists and scientists was that their readers should behold their work, contemplate it, think on it, be moved by it. The juxtaposition of art and science in their creative acts has revealed a parallel work of imagination, inspiration, and intuition that scientists rarely speak of. The contemplative imperative in any comparison of art and science has generated an unexpected and surprising critique of the way we late-moderns rush to inspect artefacts at close quarters, to seek immediate impression. Like a visitor to the impressionist gallery in the Philadelphia Museum of Art, we have had to stand back from our subject at a much greater distance than we are accustomed, so that music and mathematics, or the novel and the experiment, the poem and the theory, begin to appear within the same frame. We have had to linger over examples for longer than we normally do, but the reward is the dawning light on connections and patterns between them. This is also one of the reasons that thinking and writing from the more reflective period we call the 'middle ages' has been of continual assistance. Those in a position to engage in learning, discussion, and thought in the twelfth and thirteenth centuries were comfortable with a much wider scope of scholarship than the deep but fragmented educational formations common today. The surprising way in which medieval philosophy, and its ubiquitous theological context, have proved such effective conversation-partners to the story of scientific creativity, suggests that in its now-distant thought-world, the connections between affect and cognition, desire and thought, mathematics and music that we find so hard to see, were much more evident.

One reason for this sensitivity of a theological narrative to the inner processes of natural philosophy is suggested by the theme of the previous chapter—the role of desire for and vision of an 'end', a purpose, to the creative project. Creation is, after all, a *narrative* of its own; when we stand back far enough to bring scientific and artistic accounts into the same frame, we see more than a picture, we see a *story*. Stories have plots: beginnings, ends, and complex twists in between. They are driven by narrative tensions and drawn by final goals. The duality of *affectus* and *aspectus*, that Anselm and Grosseteste saw as threading through all human action, resurfaces eight centuries later when interdisciplinary perspectives examine cognition and creation, such as in McGilchrist's *Master and his Emissary*. Rereading accounts from painting to polymer engineering, from composition to chemistry, from poetry to physics, the multiple and mutual re-excitement of thought and feeling is so striking that we wonder why the emotional and imaginative—as well as cognitive and rational—structure of creative acts surfaces so rarely.

The structure suggested by the accounts of mathematicians, scientists, and artists, and articulated by Graham Wallas in *The Art of Thought*, of four stages of creation (*ideation, incubation, illumination, verification*), could be interpreted as a type of narrative form itself, although in its original analytical context it would lack the dual emotional sequence that a real story requires. The longer time-frame that the medieval and early modern perspectives have provided suggests the missing affective elements with which this analytic structure might be filled out into a more complex and emotionally satisfying narrative scheme. When we do that, Wallas's neat sequence of stages seems to undergo more of a metamorphosis than an expansion, for in assuming the emotional tensions and resolutions of entangling affect with cognition, it begins to resemble a story in its own right. This 'story of creation,' as a summary of the tales of imagination in art and science becomes a sort of plot-type of the creative act itself.

An Ur-narrative of creative experience

There is no universal template for the narrative of creativity that maps onto every individual experience, but in so far as commonalities have emerged, it is at least possible to summarize a faithful mental and emotional story that scientists, scholars, and artists all recognize. When the continual presence of the affective is recognized alongside the cognitive, when imaginative energies are apprehended alongside the rational, and when non-conscious as well as conscious thought is accepted within the creative process, the structure that faithfully summarizes the individual examples from visual, textual, and wordless art and science assumes seven stages. Expanding on Wallas's four-stage scheme adopted pro tem in Chapter 4, I will call them *vision*, *desire*, *industry*, *constraint*, *incubation*, *illumination*, *verification*, and *arrival*:

Vision: the first proto-creative act is the overwhelming metaphor for ideation, conceiving of what might be, but is not yet. In distant, misty, and even distorted form, the vision might already glimpse an end-point: a pictorial representation of a tragic battle, a theory of elastic fluids, a symphonic celebration of the common connection through nature of all humanity, a proof of a deep theorem. It might be an intermediate step or a pathway, an experiment in form: the first piano quintet, attempting a new chemical combination, a playful exploration of a new mathematical structure or theory. The medieval thinkers knew it as *aspectus*.

Desire: an emotional drive to realize the first vision is much more entangled with the formulation of intellectual vision than a common dualism of cognition and affect would entertain. There is just too much explicit testimony from scientists and artists alike to deny that this does more than provide the necessary emotional energy to overcome difficulty and opposition. Vision (*aspectus*) and desire (*affectus*) may generate each other, just as the observant *theoria* gives creative energy to a longed-for *poiesis*.

Industry: a series of attempts to realize the first vision begins this stage, though the single term signifies a huge variety of forms, mental and physical, of arranging and rearranging the pieces of the puzzle or the ingredients of the poem, painting, theory, or experiment. It may constitute the activity of an individual, or a group or even an entire community, and its timescales vary from days (Schumann sketched his first symphony in four) to centuries (one might think of the span from Pythagoras to Fermat and Wiles).

Constraint: the experience of frustration and failure, as initial notions prove flawed or inadequate, constitutes the dark forest that a creative imagination must navigate between a distant vision and the attainment of a goal. This recalcitrant aspect of the creative experience is near universal. It signifies the complex pattern of constraints that apply to art as well as to science and scholarship. Theories are not manifestly true or false at the point of intuition, and even the loveliest frequently fails, but only after hard work maps out their consequences and limitations. Art, writing, and music are not so free from constraint that they are bound to arrive at a satisfactory realization. Sibelius destroyed his eighth and final symphony after over a decade of work on it. Yet, as even the survey of the introduction suggested, constraints should not be regarded as inimical to creation; on the contrary they guide and suggest form, reduce an unmanageably large search-space of possible ideas, and construct the pathways that eventually lead to the final work.

Incubation: this term is best used for the consciously fallow periods in which effort is relaxed, sometimes after frustrating and apparently fruitless labour. The network of pathways, methods, and constraints is left apparently at peace. Both internal structures (limited knowledge, facility, language) and external (materiality, recalcitrance of artistic materials, results of experiments) encountered at first as constraints are propelled over the horizon of consciousness. Hadamard speaks of the place of 'fringe-consciousness' in which the vision dwells while the conscious mind engages in other tasks. Here is the house of the 'shadow emotions' that operate with their veiled process of cogitation. Here, also, must lie a representation of the constraints met with consciously in the earlier stage of industry, for no fruit ever comes from periods of incubation without previous conscious work.

Illumination: an apparently spontaneous and seemingly effortless upwelling of an idea from the non-conscious mind is the recognition of some new pathway out of the maze of industry, incubation, and constraint. The musical theme that structures a piece (or even, in Mozart's case, a visual impression of the entire work), the connection between two mathematical structures never before perceived but now visible and fruitful, the perception of the sculpted form within the marble that Michelangelo wrote about, the 'clariton' of an idea in physics while peering through a misted train window—these are the moments of illumination. Appearing typically during moments of liminal or threshold experience, they arrive with thought and emotion wrapped together. Perhaps it is just at these moments at which the cognitive environment is changing, brief transitions during which the buzz of mental work involved in perceptual interpretation quiets, that the still small voice of the subconscious can make itself heard. Only when outer sources of 'imagination' become calm can the inner ones command attention. They are welcomed with affect—one does not rejoice *after* a dazzling new idea arrives within conscious thought, but *with* the revelation. 'I can remember the very spot in the road, whilst in my carriage, when to my joy the solution occurred to me,'[2] was Charles Darwin's emotionally branded moment of illumination.

Verification: the reapplication of constraint is necessary throughout the final phase of industry—for 'knowing' that one

[2] Quoted in W. I. Beveridge (1950), The Art of Scientific Investigation, revised edition. New York: W. W. Norton and Co. Inc., p. 69.

ought to be able to complete the sonata or prove the theorem is not the same as having done it. The products of illumination are not always trustworthy, and intuition is fallible. Andrew Wiles's first dramatic intuition and apparent realization that he could prove Fermat's Last Theorem turned out to be a deceptive one.

Arrival: the creative pilgrim's progress is over when all final painting-in, symbolic detail, chapter writing, calculation, checking, is done. Homecoming marks the moment when the original creative energies have found a form that respects both imagination and constraint, a pathway from the original vision to a final work. The end may, and often does, look very different from the original idea. It may even assume the form of an opposite.

I have recounted this story, or more properly narrative outline, many times to people working in scientific research, art history, theology, textual analysis, and to musicians and artists. For those who work in the 'fine arts'—those more commonly called 'creative' disciplines—there is little surprise but much familiarity. The more striking responses come from both scientists and also scholars of humanities disciplines, for whom the recognition of their own rarely articulated story may come as something of a revelation. A work of history, the analysis of the source-texts of a medieval treatise, political analysis, a philosophical argument—these are also acts of creation, bearing similar structural similarities that we have noticed between scientific programmes and artistic projects. A common and telling aspect of discussions stimulated by this 'narrative of human creativity' is the recognition that emotions of all colours play a critical role. Without the fierceness and sustaining power of desire conceived at the point of vision of a project, the counter-emotions of frustration, despair, and doubt, and even depression, would more often be overwhelming in the periods when more constraints are encountered than pathways forwards.

The 'Narrative of the Creative Act,' with its seven chapters and empirical universality, might qualify as one of the great human stories in itself. It is not, however, listed in Christopher Booker's well-known scheme of 'Seven Basic Plots,' that attempts a categorization of all possible storylines.[3] It might just be shoe-horned into his third archetype: 'The Quest,' but that would be problematic because that plot's protagonists explore existing territory rather than create worlds of their own. No 'creation narrative' is listed in even the more generous thirty-six-type scheme of French writer Georges Polti.[4] The closest approach among that phylogeny is possibly his eleventh type, the 'enigma,' but although that might represent a subset of tasks within scientific or literary creation—a mathematical step for example, or the organizing of a subplot—it touches on neither the creative passion nor the vital spark that brings the radically new into existence.

The absence of 'creation narratives' from narrative taxonomies is surprising for another reason—namely that there is a strongly related recognized literary form in ancient texts, from the *Gilgamesh* epic to the Bible. These creation stories refer, of course, to the myriad ways in which human beings have imagined that the world itself came into being, but it does not seem such a wide analogical chasm to leap between the story-forms of the creation of all that is, and those that relate acts of human creativity. If compilers of narrative forms have not noticed the ubiquitous retelling of the creation story, that is possibly because it is, among other things, the 'story about writing stories.' It is the meta-story in which writers themselves are swimming, the water that fish fail to recognize. An inherently authorial narrative, it does not easily assume the role of subject, but lies instead underneath other stories, so deeply embedded in our psychological makeup that we rarely notice when we are acting it out. The surprised delight of some scholars when they recognize the tale of their own labours

[3] Christopher Booker (2004), The Seven Basic Plots. London: Bloomsbury Press.

[4] Georges Polti (1921), The Thirty-Six Dramatic Situations. Franklin, Ohio: James Knapp Reeve.

reflected back to them, and the difficulty some scientists have of recognizing that their work possesses a narrative structure at all, are testimony to the depth at which it is buried.

Profound human stories are also likely to be old ones. Anthropologist Agustin Fuentes would agree—his book, *The Creative Spark*,[5] places creativity centre-stage as the fundamental human trait, and central human activity responsible for the evolution of *homo sapiens*'s brains and the environments we construct. For example, the multiple communal and individual innovations in behaviour, the technical and material manipulations in making the 'Acheulean' stone tools in the Africa of a million years ago, inspires him to write of the relation of brain size and complexity of behaviour:

> *The interactive process [of creativity] represents a critical part of the early human niche, and ancestors' way of making a living in the world. Neither came first. There was a mutual feedback loop between the bodies and minds of our ancestors.*

Fuentes charts the stories of creativity that became the art, science, and technology of today, but also explores transformative acts of creation in the less obvious areas of food, sex, conflict, and religion. If creativity is woven like a structural thread into the evolution of our minds and bodies, and through the very processes that keep humans alive, able to form social groups, and to reproduce, then the structural story of the creative process must dwell at similar depths. Not just the notion, but the narrative of creativity is neurologically hard-wired into our species, and has been slowly developing, redrafting itself, finding new subplots, and experimenting with imagination itself for hundreds of thousands of years.

In a final plot-twist on creation as narrative, which may help in taking the questions raised by Chapter 4 a little further, we should take account of structuralist, as well as phylogenetic, analysis of stories. Exemplified by the work of Algirdas Greimas,[6] who

[5] Agustin Fuentes (2017), *The Creative Spark*. New York: Dutton.

[6] A. J. Greimas (1966), *Sémantique structural*. Paris: Seuil.

developed diagrammatic representations of how initial and nested sequences of stories work, such an approach might start on the 'creativity story' by employing the interactive lines of agency and counter-agency:

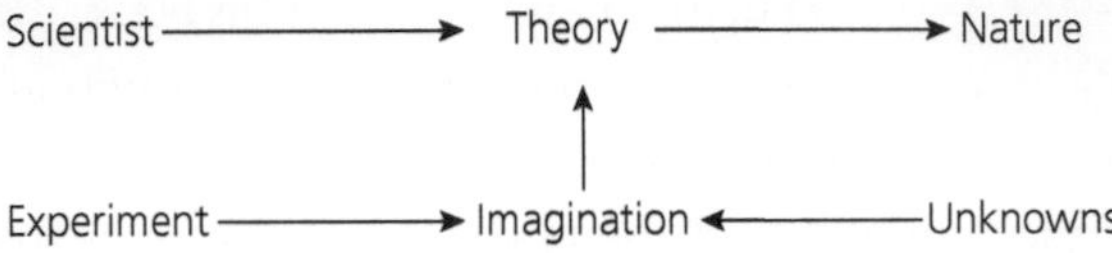

The upper line in this diagram of an 'initial sequence' reflects the original intention and desire, the lower the agency for accomplishing it. The interest as *story* is that in Greimas's scheme, agents receive opposing forces, either of which may triumph or subvert. Initial failure of the original intention in the case of failure of imagination to conceive of a workable representation then sets off a second, nested 'topical sequence' in which the agent (in this case of the story of a creative act, imagination itself) becomes the object of intent and desire. The next chapter would tell tales of industry, constraint, and incubation before the imagination can return to its work of theorizing nature, but in a transformed mode. The point here is not to desiccate a perfectly human and romantic story into dusty analysis, but as theologian N. T. Wright has urged,[7]

> *The way that stories possess the power they do, by which they actually change how people think, feel, and behave, and hence change the way that the world actually is, can be seen more clearly by means of an analysis of their essential components.*

In other words, stories themselves, when acted out, can be the agents of creation. It is a common practice among academic communities to make visiting days from one university to another. A visiting speaker gives an afternoon seminar, after holding conversations about work in progress during the morning. A most

[7] N. T. Wright (1992), *The New Testament and the People of God*. London: SPCK, p. 69.

common form of introduction in these conversations is simply, 'Let me tell you the story . . .'.

I write this chapter having just returned from such a visit to London's British Museum, and in partnership with a wonderful temporary exhibition there. Curated by Jill Cook, *Living with Gods* is a journey through a hundred or so artefacts from religious practice, ancient and modern, eastern and western, monotheistic and aboriginal.[8] Before participating in our public panel-conversation, *Science of Belief*, Jill guided our group through the exhibits. The very first was one of the exhibition's most prized—the 40,000-year-old ivory sculpture known as the 'Lion Man' (see Colour Plate I). Found in a north-facing cave (so not a dwelling place) in Swabia, together with a number of pierced animal teeth, the small[9] apparently votive object is beautifully worked. With the legs and lower body of a human, and head of the now-extinct European mountain lion, the chimera form is compelling and clear even now. After painstaking restoration, its detailed features announce a powerful command and attention. To explore the possible techniques and effort that the manufacture of the figure would have required, the British Museum asked a professional carver to take a similar sized block of hardwood and copy the figure as best he could. After 700 hours of work the overall form was achieved, but still required finishing and polishing. In an ice-age European climate, with average temperatures 15°C colder than today, a community of humans valued the time and skill of at least one of their members enough to create this strange and alluring piece of art.

The creative act occurred at eight times the temporal span into the past of the very oldest historical record of any ancient civilization we know of. To conjecture a 'religious' motivation for the Lion Man is speculation, but most visitors describe a strong resonance with their sensibilities for the numinous as

[8] Some of the exhibits are discussed by Jill Cook in a British Museum blog: https://blog.britishmuseum.org/living-with-gods-highlight-objects/

[9] The figurine of the 'Lion Man' is about 20 cm from head to foot.

they walk around the object. For me the close encounter was deeply affecting and unforgettable. The chimera form arises again, much later, in Egyptian deities, for example, so it is not without reason to suppose that it signifies an early attempt to signify the transcendent. In any case, as a piece of creative art it is extremely successful, and not in the slightest crude or childish.

The evening discussion around the exhibition set the traditions of science and religion alongside each other. Although my colleague, neuroscientist Colin Blakemore, and I have differing views on the relationship between them, it proved constructive to set aside the common 'conflict' paradigm for a moment, and to explore instead a process of mutual interpretation. Naturally, one can do this from a purely anthropological perspective, as well as from a position of religious belief. If the story of creativity, and even creativity itself, spring from the deepest and oldest sources that sustain our humanity, then that is another reason to let the equally deep-lying web of theological narratives interpret the basic narrative of creation. We need to ask what it is that drives us to unleash our imaginative energies in science and in art, and what those creative acts accomplish in the experience of being human. Religious traditions, as we have noticed, already record textual 'creation narratives' of the cosmic kind, but they also encode the stories of *homo sapiens*'s most basic values, hopes, and abilities, and how those capacities have evolved with us. These must include those that drive us to imagine and to create. As Fuentes writes:[10]

> *We have to identify the kinds of structures, behaviours, and cognitive processes that might have enhanced the role that human symbol creation and use, and the human imagination, had in the initial appearances of our religious experience, belief, ritual, and their associated institutions, in our archaeological past.*

Whether or not religious narratives refer to realities beyond the natural world is here not the issue—whether they do or do not,

[10] A. Fuentes, *The Creative Spark*, p. 215.

they most certainly tell us about ourselves. Starting with our medieval philosophical travelling companions, we lend an ear to what those stories are saying.

Telling stories of creativity and creation through theological lenses

One of the reasons that medieval philosophers have been able to shine the brightest lights even on the experiences we have related from much later centuries is that they did not assume the compartmentalized minds that we are educated to develop today. Thinking about the creative process of science in that light is an approach that Cambridge theologian, Sarah Coakley, might call a scientific form of her *théologie totale*[11]—an academic pursuit integrated seamlessly into a pattern of belief, worship, and meditation (Coakley has boldly claimed that some theological insights are accessible only through immersion in the practice of belief). This is *not* to suggest that one should become a theologian, let alone necessarily a believing theologian, to understand human creativity, but rather that there are special concepts, categories, and methodologies from theological traditions without which the picture would remain incomplete. Such a 'theological imperative' arose before—when the stories of poetry and theoretical science were set alongside each other, theological ideas and categories arose without deliberate summons. Comparing early fiction and experiment drew unexpectedly on theologies of the fall and restoration. At that point the key ideas were creation in image, and recovery of understanding; here the theological resonance is found in the exploration of ends and purposes.

[11] Sarah Coakley (2013), *God, Sexuality and the Self*. Cambridge: Cambridge University Press.

There are three further windows onto creativity through which theological approaches promise to provide special insight, in addition to their witness to evolving human culture. The first is the delight in formulating creation narratives and in locating the topic of origins as a central theme. Second is a natural accommodation to 'grand-narratives,' which may be academically unfashionable today, but which don't get much grander than the story of human creativity. Third is the frequently recurring theme of purpose; again, teleology is a quiet category within the humanities today, but goals and ends have motivated and inspired imagination from renaissance art to relativity. A meta-question that must arise from a comparative study of human creativity is the purpose of art and science themselves. Fuentes points to the powerful anthropological role that creativity plays as the driver of the extraordinary evolutionary success of our species. But there are other important ways of articulating purpose.

We have already encountered the eleventh century thinker, Anselm of Canterbury, in his lucid introduction of the duality between *affectus* and *aspectus*. He might also have furnished us at that point, or even before, with another early and delightful description of the 'aha moment' that marks an experience of illumination. It comes from the prologue to his major work, the *Proslogion*:[12]

> *After I had published, at the solicitous entreaties of certain brethren, a brief work (the Monologium) as an example of meditation on the grounds of faith, in the person of one who investigates, in a course of silent reasoning with himself, matters of which he is ignorant; considering that this book was knit together by the linking of many arguments, I began to ask myself whether there might be found a single argument which would require no other for its proof than itself alone; and alone would suffice to demonstrate that God truly exists, and that there is a supreme good*

[12] Anselm, *Proslogion* in Jasper Hopkins and Herbert Richardson (trans.) (2000) Complete Philosophical and Theological Treatises of Anselm Of Canterbury. Minneapolis: The Arthur J. Banning Press.

> *requiring nothing else, which all other things require for their existence and well-being; and whatever we believe regarding the divine Being. Although I often and earnestly directed my thought to this end, and at some times that which I sought seemed to be just within my reach, while again it wholly evaded my mental vision, at last in despair I was about to cease, as if from the search for a thing which could not be found. But when I wished to exclude this thought altogether, lest, by busying my mind to no purpose, it should keep me from other thoughts, in which I might be successful; then more and more, though I was unwilling and shunned it, it began to force itself upon me, with a kind of importunity. So, one day, when I was exceedingly wearied with resisting its importunity, in the very conflict of my thoughts, the proof of which I had despaired offered itself, so that I eagerly embraced the thoughts which I was strenuously repelling.*
>
> *Thinking, therefore, that what I rejoiced to have found, would, if put in writing, be welcome to some readers, of this very matter, and of some others, I have written the following treatise.*

Whatever one might think today of the logical, or theological, status of Anselm's 'ontological proof' of the existence of God,[13] there is no doubt of its ingenuity in assembling the rational apparatus of his day into a discussion of universals, and of the remarkable way that it touches upon modern notions of set-theory and self-reference. His account is also a 'textbook' example of the seven stages of creation advanced in this chapter. He gives the background from which his vision (of a single argument) arose, and with it an 'earnest' desire. Industry, which brought it almost 'within mental reach' and constraint, which removed it from 'mental vision', immediately follow, together with the emotional 'despair' that marks the complete loss of sight towards the goal. There is a tangible sense of loss. Incubation in the form of attempts to 'exclude the thought altogether' precedes a final illumination. In the form of a non-conscious process of imagination, so that to the conscious mind it appears as a gift ('the proof of which I had despaired offered itself') Anselm's experience elicits the emotions of eager rapture. The more sober process of writing

[13] For a fairly comprehensive account of the argument and its reception, see I. Logan (2009), *Reading Anselm's Proslogion, The History of Anselm's Argument and Its Significance Today*. Farnham: Ashgate.

the argument out leads his readers into the text itself. *Proslogion* stands as an important milestone in the formation of a tradition of open thought and disputation, opening up a prolonged debate on the tension between received doctrine, and new ideas that were the product of reason. One consequence of this new dialectic was the formation of the first European universities, and in due time the rise of early modern science.

This larger and longer narrative not only provided the foundations for innovations in logic and philosophy, but also the motivation towards a fully developed natural philosophy by thinkers such as Robert Grosseteste and Roger Bacon in the thirteenth century. The reason these thinkers give for seeking an understanding of nature has strong echoes in the modern period. It begins in a unique medieval moment of concurrence of Greek, Latin, Arab, and Hebrew thought. As we reviewed in Chapter 3, the products of the 'translation movement' of the twelfth century—including the bulk of Aristotle's works—began to receive attention from scholars in the Latin West for the first time since the late Roman Empire. Preserved and commented on in Greek within the Byzantine empire, and in Arabic throughout the early Muslim world, a treasure-chest of works on optics, physics, medicine, and mathematics provoked radical new thinking in the schools of centres such as Toledo, Paris, Oxford, and Hereford. Before looking beyond their scientific imagination to the end they have in view, there is one particularly illuminating example that forms an arch of scientific creativity from the ancient world to the modern, and whose apex traverses the intellectually explosive medieval centuries. This is the ancient problem of understanding the rainbow.

All the colours of the rainbow

For an example of sophisticated thinking in natural philosophy generated by the reuniting of middle-eastern and western traditions in the medieval period, we return to Grosseteste's short treatise on colour (his *De colore*) of around 1224, encountered in

Chapter 3. On close reading, it proves to be a highly perceptive, imaginative, and strongly mathematical text. He sets up three pairs of opposite qualities possessed by light that permeates translucent materials, and counts possible combinations between them. A recent scientific and philological analysis[14] has shown conclusively that the argument of *De colore* proposes a three-dimensional theory for an abstract space of colour. This is remarkable not only for its sophisticated development of Aristotle's account (in which all the observable colours could be ranked along a one-dimensional line between the poles of white and black), but also because Grosseteste's idea seems to correspond very closely to our modern theory of colour.

The perception of colours does indeed occupy a three-dimensional space. This is because normal human retinas contain three different types of pigmented light-dependent cells (the long (L), medium (M), and short (S) wavelength 'cone' cells). Under exposure from any source of light, each cell sends a series of spiked electrical pulses towards the optic nerve—the spike-rate codes for a number that records the intensity of incident light within the wavelength range to which it is sensitive. In consequence, information on the spectral content of light from any source is received by the visual cortex in the brain in the form of a triplet of three numbers: the signals from the L, M, and S cones. For the same reason, the perception of nearly any colour can be simulated by the combination of three 'primary' colours on a screen or paper. These are the hues we call red (R), green (G), and blue (B). Computer screens quantify colours using such an 'RGB' triplet. The existence of a three distinct colour-sensitive cells in the human eye was first brilliantly deduced by Thomas Young in 1802 from these perceptual properties of colour mixing alone.

The three colour terms of Grosseteste's *De colore* are not red, green, and blue, but the more abstract terms *multa* (greatness),

[14] Dinkova-Brun, Greti, et al. (2013), *Dimensions of Colour: Robert Grosseteste's De Colore; Edition, Translation and Interdisciplinary Analysis*. Durham: Durham Medieval and Renaissance Texts.

clara (clarity), and *pura* (purity), together with their opposites. The treatise is careful to ascribe only the first two to the inherent properties of light, but the quality *pura*, and its opposite *impura*, to the material through which the light passes. The significance of that move appears in the final statement of the work, inviting doubtful readers to reassure themselves of its claims by passing different lights through a variety of materials and so reconstructing 'all types of colour.' No wonder some writers have been tempted to ascribe to Grosseteste the creation of a full-blown tradition of experimental science. This is, however, not the carefully constrained and hypothesis-driven experimental method that we considered in Chapter 4, alongside the early novel, however demonstrably it might be on a path towards it. Grosseteste's account implies an acute sense of observation, but not the constrained abstraction and simplification of the experiment. It also reminds us that the scientific imagination developed over many centuries.

Nevertheless, that other essential prerequisite of experimental science, the ability to think by analogy and across scales, is clearly within his grasp. For although there is no clue in *De colore* of how we might map its triplets of (greatness, clarity, purity) onto our (red, green, blue), one of his later works gives us a clue to reconstruct just such a correspondence. For by 1225 he had written another colour-related treatise, this time addressing the rainbow—the *De iride*—also within Chapter 3. Most of this remarkable work deals with the geometric optics of refraction, writ large on great clouds of mist in the sky that Grosseteste thinks act as giant lenses. That itself is an impressive, if ultimately fruitless, notion. But within the treatise's very last section, the triplet colour coordinate system of *De colore* is revisited, but now written on the meteorological sky of sun and rain. There lies the explanation of the three dimensions of colour—because rainbow colours depend on three parameters: (i) the colours at different *angles of sight* within one rainbow, (ii) the different *types of cloud* or mist that can make a rainbow, and

(iii) the quality of impinging light from the sun, depending upon its altitude, redder at sunset and yellower when the sun is higher in the sky. This is an elegant example of a motif within scientific creativity that we have met repeatedly—the leap of thinking by analogy across scales. Connection by the same physics operating at widely separate scales is the driving force of Newtonian gravity: that the attraction of objects near the Earth's surface might have common cause with the path of the moon in the sky. But here we see the same connections between widely separated scales three centuries earlier. A three-dimensional scheme for variation of colour in objects of the scale that can be placed on a table is directly related to the colours that appear when sunshine and rain coexist in vast volumes in the atmosphere.

The science of the rainbow makes a marvellous connection of a different type: between the oldest scientific thinking and the newest. No eye ever tires of resting on splendid examples of this most colourful of atmospheric phenomenon. The desire to understand it is as old as we have records of natural philosophy. Aristotle thought about it in his *Meteorologica*—suggesting that it arose through reflection of sunlight from clouds. In the early eleventh century, Ibn Al-Haytham, in Baghdad, took Aristotle's ideas further, noting that a concave mirror can generate an apparent arc of reflected light from a point source, in his *On the Rainbow and Halo.*

One of Grosseteste's achievements was to show that Aristotle (and by implication, Al-Haytham) could not be right about this, since the apparent image of a rainbow is not in the place that any reflection of the sun would be. This is his motivation for suggesting, for the first time in the history of science, that refraction, rather than reflection, was the underlying optical process responsible for the rainbow. In this he was correct, although he did not identify the individual droplets of rain as the important geometrical refracting objects, favouring, as we have seen, the entire cloud that the raindrops make up as the critical structure.

The final leap of insight into the optical origin of the rainbow required one more century of incubation. Theodoric of Freiburg was a German Dominican who, between 1304 and 1310, developed Grosseteste's ideas of refraction, but cross-fertilized with the thinking of other late-thirteenth-century scholars, Albertus Magnus of Cologne and Roger Bacon of Oxford. Albertus had suggested that individual drops of rainwater were as important in generating refraction as the clouds themselves.[15] Bacon had made the key move to realize that the appearance of colours at particular angles from the sun depended on the directions from which those colours came from when entering the eye, rather than requiring a projected image of a semicircle onto a screen of cloud.[16] This is the reason that a rainbow seems to travel along with a moving observer—it is not an 'object' in the sky but a set of directions of light rays. Theodoric, and independently his Arab contemporary Kamal al-din al-Farisi,[17] showed that the paths of light into and out of a raindrop, and reflecting internally from its back surface, could account for the rainbow's colours and the angles at which they were seen. The material requirement is simply that blue light be refracted through a greater angle than red on passing through an interface between air and water.[18] Compare a diagram from his treatise with a modern version in Colour Plate J.

Theodoric won his essential insight from a practical approach strongly suggestive of Grosseteste's advice in the postscript to *De colore*—he filled glass spheres with water, and carefully traced the paths of light from the sun as they passed through them. He performed the same experiments on hexagonal crystals for comparison. Walking around his glass spheres he saw that the colour

[15] Albertus Magnus (1681), *De Coronis et Iride quae Apparent in Nubibus*, in his *Opera Omnia*, Petri Jammy, ed. Lugduni.

[16] Roger Bacon (1897), The Opus Majus. J. H. Bridges, ed. Oxford.

[17] See, for example, E. Wiedermann (1910), *SPMSE*, 42, 15 ff.

[18] Theodoric of Freiburg (1914), *De Iride et Radialibus Impressionibus*. J. Würschmidt, ed. Münster.

of light seen from the drop depended on the placement of his eye, so realizing that, in the rainbow, the different colours perceived by an observer came from different raindrops—those that were placed at the correct angle of sight for each specific colour.

A characteristic experience in science is that when one first starts working with a substantially correct theory for a phenomenon, new insights and explanations appear almost unrequested. So it was with Theodoric, for when an explanation for the rainbow appeared from the rays he traced on his detailed diagrams, reflecting once from the reverse of each raindrop, it was natural to ask what the result would be for rays reflected twice. The answer emerged from further diagrams, one reflection more complex than the one reproduced in Plate J. Theodoric's geometric representations pointed to a second, fainter rainbow visible outside the primary bow, and due to the extra mirror reflection inside each raindrop, with the order of colours reversed. He had a theory that automatically accounted for the secondary rainbow, visible when the incident sunlight is particularly bright and the background sufficiently dark (see Colour Plate J).

The story of the rainbow has further chapters: vital nuances to the geometrical optics of Theodoric were treated by Descartes,[19] who identified why each colour emerges from the droplet at a dominant, but not unique angle. The nineteenth-century English mathematician, George Airy, applied the wave theory of light to the phenomenon for the first time,[20] accounting for the rapid sequence of 'supernumerary bows' seen just inside the primary bow under favourable conditions. It took the twentieth-century quantum mechanics of matter and light to explain *why* light of different wavelengths is refracted by water to different degrees. At that level the imagination has to descend in scale once more, to the electron clouds within water molecules, to conceive

[19] *Oeuvres de Descartes* (ed. Charles Adam and Paul Tannery, 12 vols., Paris, 1897–1913), 6, 327.

[20] See D. Hammer, 'Airy's Theory of the Rainbow', *J. Franklin Inst.*, **156**(5), 335–336 (1903).

and to paint with mathematics the way that they deform under the passing light wave. As the negatively charged electrons sway from side to side away from their positively charged atomic nuclei they add little contributions of re-radiated light of their own. The electron clouds respond a little more readily to the slightly faster oscillations of blue light, than to red, and so change the effective direction of travel a little more. The beautiful bow is re-woven in the scientific imagination, from its great arcs of colour to the tiniest threads of electron dynamics within its molecular optics.

The rainbow's story illuminates not only the similarities and consonances between the scientific and artistic imagination, but also their vital differences and complementarities. The rainbow itself has proved an exceptionally challenging subject for artists. The English landscape artist, John Constable, for example, made a careful study of the optical geometry of rainbows, including double bows.[21] As a result, his are some of the most convincing portrayals in paint. However, he was prepared to stretch physical constraints when the artistic demands prevailed, such as in the later addition of a rainbow to *Salisbury Cathedral from the Meadows* out of its geometric position relative to the sun, to mark the death of his friend, Archbishop John Fisher.[22]

A contemporary example of rainbow-inspired art arose from Grosseteste's treatise, *On the Rainbow*, itself. Scientists within the interdisciplinary *Ordered Universe* project had followed up his intriguing suggestion that all possible colours are contained within the set of all possible rainbows, under variation of cloud-type and sunlight.[23] In a beautiful example of 'neoclassical imagination'

[21] See Andrew Wilton (1980), *Constable's 'English Landscape Scenery'*. London: British Museum Publications.

[22] For a fascinating account, see the Tate research publication by John E. Thornes, A Reassessment of the Solar Geometry of Constable's Salisbury Rainbow, https://www.tate.org.uk/research/publications/in-focus/salisbury-cathedral-constable/reassessing-the-rainbow.

[23] Hannah E. Smithson, et al. (2014), 'Color-coordinate system from a 13th-century account of rainbows', Journal of the Optical Society of America, 31, A341-9. An account written for a broad readership can be found in H. E. Smithson, et al. (2014), 'All the colours of the rainbow', *Nature Physics*, **10**, 540–2.

operative in science, the thirteenth-century natural philosophy inspired a twenty-first-century computation of rainbow colours from all possible raindrop sizes and solar illuminations. Their projection into the standard RGB space of human perceptual colour is shown in Colour Plate J(d). The result possesses a structure of unexpected beauty: a nested set of spirals within the abstract colour-cube. The publication caught the attention of glass artist Colin Rennie from the UK's National Glass Centre in Sunderland. He began to work on a representation of the colour spirals in glass, but found that a straightforward implementation would require a confusing mesh of supports, wires, and other engineering. He writes:

> *As with scientific hypothesis, if there are more and more supporting elements required to hold up a theory, the likelihood is that the theory is wrong. So, I abandoned the idea, taking it apart completely and examining in a thought experiment which elements were needed in order to retain the integrity of the idea and its relevance to the research. The cube form was crucial, this defined a space, and the axis, but the cube on its corner was more truthful to Grosseteste imaginings. Colour elements were also essential, but the spiral essence could be distilled. The fluidity and elegance of the initial form was what had attracted my attention in the first place so I took that as a start, asking myself the question, 'why?' Why was this form exciting, and what relevance did it have to both glass, colour and medieval history? My thinking led me to the idea that from the observations in the field, colours or flowers, plants, the sky, the rainbow, were elusive, transitory and could be hard to pin down, as if they were moving in the ethereal conceptual space, the notion of the firmament was brought to mind. I decided to explode the spiral, to make it shatter into a myriad of moving twisting interlocking interrelating blends of colours.*

Rennie's rotated and 'shattered' forms start from the science, exercise both new freedoms and work under new constraints, but authentically reflect the contours of the 'creation narrative' once more. The tension of order and chaos enters, and a palpable emotional journey through the creative process responds to scientific insight and creative imperative. The rainbow, in both artistic and scientific representation, serves as a beautiful visual metaphor for the imaginative process in science itself. Perhaps

Grosseteste had his thoughts on the puzzle of the rainbow in mind when he explained, in his commentary on Aristotle's account of scientific method, the meaning of that 'extra sense' of *sollertia*.[24] As he talks about the mind's eye descending into the structure of a translucent body, we can almost see rays of light, carrying properties—that he has named but as yet only partially understands—into material and translucent bodies. He imagines them refracting at surfaces, creating an unseen geometric web of light that interacts with the material properties, emerging as the colour of glass, or of a rainbow. The striking words in his definition are 'a thing naturally linked to itself'—he touches on a primary experience, of the mind's imagination establishing a connectivity with the inner structure of nature. The persistence of his vision, that goes on 'again penetrating this connectivity,' seems to anticipate the further kaleidoscopic dives of scale into droplets, the material of water, electron clouds, and atoms that later centuries would continue.

From the mid-point of this centuries-long 'relay' narrative, as Hadamard would describe it, of a slow growth in understanding of the rainbow, comes a poetic articulation of the structure by which the 'vision of the mind' reaches out into the world. The motivation that powers such extended labour must lie at an equal depth. The end to which the hard labour of science is directed has also surfaced in differently articulated forms, ancient and modern. It seems indeed to lie at a level within the human psyche at which the commonalities shared between art and science become indistinguishable.

The end of creativity

Why do we reach out with the 'penetrative power of the vision of the mind' beneath the superficial appearance of nature, delineating and disentangling, with this precious human ability, the hidden structure of the world? Even the great medieval thinkers

[24] See the section 'The ancient aesthetic of active seeing' in Chapter 3.

were taciturn on the topic of teleology. Most of their natural philosophy is framed within, but did not explicitly refer to, their theological worldviews. But there are occasions when they reveal the layer of twinned *affectus* and *aspectus* that describes their deepest goals. For natural philosophers in the West, from late antiquity to the early modern era, the reason that human beings set their minds to comprehending the natural world is that a state of vision and understanding is both their origin and their destiny. To read and to reimagine the world is to engage with a task of recovering an original knowledge, at an almost divine level of insight, lost when humankind rejected an outward-facing orientation of contemplation and care, in favour of an inward turn to self-interest.

Our palpable ignorance of nature, and our discomfort within it are, for both medieval and early modern natural philosophers, articulated as consequences of the Christian notion of the Fall. Ignorance is neither a natural nor an intended state, but a moral failing. The biblical grand narrative does not finish at the point of such tragedy, of course, but subsequently follows the long tale of calling, redemption, and the eventual vision of a new creation. Of the many aspects of the theological creation-fall-redemption-re-creation shape of history, the possibility of a regained deep knowledge of nature begins with the gifts of human senses and minds. The rest is a combination of grace and hard labour. This is the context for the quotation at the start of this chapter. Sense perception, imagination, memory, and understanding are the rungs of a scientific Jacob's ladder that starts with human perception, works with imagination, and starts a journey back towards human apprehension of the fundamental essences of the world. It is intriguing that medieval philosophy identifies *memory* also as a faculty to be reconstituted, especially in the light of recent research into dementia, memory-loss, and personhood.[25] It is a capacious vision, not simply of deep insight when we

[25] See, for example, T. Fuchs, 'Embodiment and personal identity in dementia.' *Med Health Care and Philos* **23**, 665–676 (2020).

turn to look in a particular direction, but a simultaneous grasp and contemplation of all of nature, mentally held in permanent view.

Whether this way of talking about the human relationship to nature sits comfortably with us today (it may for some, most certainly will not for others) it is an enduring and powerful story that finds different language and forms in different epochs and traditions. At the most basic level it identifies a feature of the human condition that is at once so strange, yet so commonplace that it goes unnoticed—our continual discomfort with the shape we find the world in. It is what Kant would call the difference between *is* and *ought*. We find the world to be not as it should, morally or materially. Time, too, seems to us 'out of joint,' passing too quickly or too slowly but never conveniently.

A simpler version of this rationale for science appears at the dawn of the European middle ages in the writings of the great Northumbrian cleric and scholar we know as the 'Venerable Bede'. Encountered most often today for his early eighth-century *Ecclesiastical History of the English Speaking Peoples*, the first post-Roman history written in England, Bede also wrote a short course in the science of his day—his *De Rerum Natura*. This short but lovely book (first translated from Latin into English in only 2010, though it was copied more frequently than the *History* for five centuries) contains Bede's own material as well as glosses on the earlier natural histories by Pliny (in the first century) and Isidore of Seville (in the sixth). As its editor and translator Faith Wallis points out,[26] Bede strips out much of the moralizing of its first Christian editor, Isidore. Rather than suggest that nature is the source of metaphorical moral and theological messages, he urges that Christian scholars should learn and teach natural philosophy primarily so that people should not be afraid. When storm or earthquake or gale descends, science helps them understand that this is the way that the world works, and not fear in it any deliberate agent of terror. The connection between natural science and the

[26] Bede (2010), *On the Nature of Things and On Times*. Calvin B. Kendall and Faith Wallis. Liverpool: Liverpool University Press.

banishing of fear found pithy resonance in the great twentieth-century double Nobel Laureate Marie Skłodowska-Curie:[27]

> *Nothing in life is to be feared, it is only to be understood. Now is the time to understand more, so that we may fear less.*

Fear is the ultimate tell-tale of a damaged relationship. Relational pathology begins with ignorance, a missing of clues and misinterpretation of words, but if the strangeness of the other is not reconciled, ignorance develops into a mutual propensity for harm, and the permanent fear of it. There could be no starker reminder that humans, and the world in which they have evolved, are in a relationship of this type, prone to mutual fear and potential harm, than the declaration around the turn of the third millennium of the 'Anthropocene' age. In Bede's time, people could legitimately fear the potential that natural forces had to harm them; in our own, we have come to recognize, and to fear, the actual harm humankind is wreaking upon nature, first glimpsed two centuries ago by Humboldt.

Only rarely do the great natural philosophers of past ages draw back the curtain on their motivations. In the same way that Grosseteste's theology for doing science appears in just one or two places of commentary, there is just one point, but a telling one, where Theodoric of Freiburg points to the teleological text behind his project to understand the rainbow. He explains, in the treatise *On Light and its Origins*, that his experiments and optical calculations are a response to the question in the biblical *Book of Job* (chapter 38 v. 24) "'In what way is the light scattered, and the heat distributed over the earth?"[28]—This difficult question the Lord proposed to holy Job.'[29] Grosseteste also treats *Job*, especially this same lengthy wisdom poem of chapters 38–42 that

[27] Quoted by Melvin A. Benarde (1973), *Our Precarious Habitat*. New York: W. W. Norton, p. v.

[28] The Hebrew of the book as a whole, and of this poem, the Lord's Answer is notoriously difficult—other translations of this same verse render it, for example as, 'What is the way to the place where lightning is dispersed, or the place where the east winds are scattered over the earth' (NIV).

[29] E. Krebs (1906), *Meister Dietrich. Sein Leben, seine Werke, seine Wissenschafte*. BGPM v. 5–6, Münster.

constituted the height of Hebrew nature poetry in the survey of our Chapter 5—God's 'Voice from the Whirlwind.' The reading of *Job* in that discussion was metaphorical of the insight of scientific imagination, and of the timeless imaginative power of the appropriate question. In Theodoric, it steps stage-front as direct scientific inspiration. A century before his work, the text had become a source of inspiration and wisdom underpinning the contemplation of nature, in Grosseteste's most extensive work, the *Hexameron*—a commentary on the six days of creation of the book of Genesis. Early modern theological writing connected with the early intimations of experimental science and the new astronomy of the sixteenth century also identify the reception of *Job* as a strongly positive influence. Calvin's pupil Lambert Daneau, for example, was one of the early Reformation theologians who sought, in his *Physica Christiana* not to refute natural philosophy by scriptural truth, but to provide it with a more solid support.[30]

The *Book of Job* has been a constant inspiration for thinkers of all traditions and eras since. Contemporary neo-Kantian philosopher Susan Neiman has recently even urged that *Job* be held alongside Plato as a foundational text for western thought.[31] Here Neiman sums up the tension between order and chaos in both the natural and moral worlds of *Job*:

> *As Kant would later put it, two things fill the mind with awe and wonder the more often and more steadily we look upon them: the starry heavens above me and the moral law within me. They are both awesome and wonderful, but entirely separate—the one stands for a cosmos described by the Voice from the Whirlwind, a cosmos so vast and impersonal that it strikes down our self-conceit and makes us feel, as Job put it, that we are but dust. Yet the moral law within me, which Job so beautifully upholds in his darkest hours—he may wish he had never been born, but he never once wishes he had behaved anything less than righteously—that moral*

[30] See, for example, Peter Harrison (2007), *The Fall of Man and the Foundations of Science*. Cambridge: Cambridge University Press pp. 108ff.

[31] Susan Neiman (2016), *The Rationality of the World: A Philosophical Reading of the Book of Job*, ABC net, http://www.abc.net.au/religion/arti-cles/2016/10/19/4559097.htm.

> *law reveals our power to step in and change a piece of the world if it seems to be gone wrong.*

The *Book of Job*, of all ancient literature, succeeds in articulating in timeless and plangent depth the difference between what human beings consider the world ought to be, and how they find it. Its response, in poetic dialogue of beautifully structured form, but of brutally honest content, has also shocked and offended many of its readers. One of its enduring puzzles is that, when God finally answers long-suffering yet righteous Job's complaints 'from the whirlwind,' his Answer seems to bypass the moral dimensions of Job's predicament, directing him instead with over 160 questions about the natural world (of which Theodoric's is just one):

> *Have you entered the storehouses of the snow?*
> *Or have you seen the arsenals of the hail,*
> *Where is the realm where heat is created, which the sirocco spreads across the earth?*
> *Who cuts a channel for the torrent of rain, a path for the thunderbolt?*
> *Can you bind the cluster of the Pleiades, or loose Orion's belt?*
> *Can you bring out Mazzaroth in its season, or guide Aldebaran with its train?*
> *Do you determine the laws of the heaven?*
> *Can you establish its rule upon earth?*

The text takes Job and his readers into the fields and phenomena we now call meteorology, astronomy, earth sciences, then later zoology. At every turn, it probes the aleatory and wild side of nature. More recent scholarship has perceived, in the catalogue of nature-questions from the lips of the Creator, not a petulant put-down of an ignorant complainer, but, more faithfully to the ancient Jewish tradition of pedagogy through questions, an invitation to think more deeply about the structure of nature in its full dynamism and variety. I have elsewhere developed this ancient tributary to the rivers of ideas that became natural philosophy,[32] but here, with Neiman, it is almost enough to point

[32] Tom McLeish (2014), *Faith and Wisdom in Science*. Oxford: Oxford University Press.

out its existential analysis of the human condition within suffering, and its immediate reference to human engagement with the natural world as a response.

'Almost enough,' because there is another strand to the wisdom-writing in *Job* that connects with questions of creativity in science and art, and with the modes of the visual, written, and wordless imagination. The book contains one earlier poetic insertion into the otherwise regular alternation of speeches between Job and his friends. When their arguments reach a bitter impasse of accusation and denial, a new voice (known as the 'Hymn to Wisdom') starts to sing a new song (chapter 28):

> *Surely there is a mine for silver, and a place where gold is refined.*
> *Iron is taken from the soil, rock that will be poured out as copper.*
> *An end is put to darkness, and to the furthest bound they seek the ore in gloom and deep darkness.*
> *A foreign race cuts the shafts; forgotten by travellers, far away from humans they dangle and sway.*
> *That earth from which food comes forth is underneath changed as if by fire.*
> *Its rocks are the source of lapis, with its flecks of gold.*
> *There is a path no bird of prey knows, unseen by the eye of falcons.*
> *The proud beasts have not trodden it, no lion has prowled it.*
> *The men set their hands against the flinty rock, and overturn mountains at their roots.*
> *They split open channels in the rocks, and their eye lights on any precious object.*
> *They explore the sources of rivers, bringing to light what has been hidden.*

The metaphor is unmistakable. It is the same as Grosseteste's definition of *sollertia*—the same as Newton's 'seeing further than others' on the seashore of the world, as Darwin's 'power of seeing' into the world. Human eyes are given, according to the Hymn to Wisdom, the unique capacity to see below the surface of the world, and so to behold its inner structures, its veins of precious metals and of heat-transformed precious stones. Not the hawk with her hunter's vision, continues the text, nor may the prowling lion share such an inner vision, achieved as it is by human technical ability to create illuminated subterranean pathways of sight where none existed before. As a metaphor for the reach of the mind's eye into nature, it is matchless. At the close of the poem, the writer makes an astonishing move in ascribing the same form of transparent insight into the world, even in terms of quantification of its forces and fields, to the Creator's own deployment of wisdom:

> *But God understands the way to it; it is he who knows its place.*
> *For he looked to the ends of the earth, and beheld everything under the heavens,*
> *So as to assign a weight to the wind, and determine the waters by measure, when he made a decree for the rain and a path for the thunderbolt—then he saw and appraised it, established it and fathomed it.*

Some translators of Job 28 have baulked at the obvious similarity in super-visual abilities of the miners at the start of the hymn, and of God at its close. Yet there is no escaping the claim that a deep and inner insight into the world is one, within the worldview of the writer, that human beings share with their Creator, and that such an ability is of critical relevance to their inadequate, puzzling, and painful relationship with nature. There are far fewer students of Old Testament wisdom among scholars of science now than there were in medieval or early modern centuries. But those that do read them find still a strong resonance that helps them understand their scientific motivation, and even to suggest healthier ways of framing science than the instrumentalist narratives of today's public science policy. Mathematician Ennio Di Giorgi spoke of a related ancient wisdom-book at a 1996 congress of philosophers, scientists, and theologians gathered to reflect on the theme of wonder in the natural sciences:[33]

> *A reminder of the oldest roots of wisdom might seem out of place as an answer to the problems posed by the developments of science and modern technology, but I believe that if we want, if not to resolve such problems, at least to take a correct approach to them, we must put them in a very broad perspective which embraces the most concrete and lowliest realities as well as the highest and most abstract ones.*

[33] As quoted in Marco Bersanelli and Mario Gargantini, trans. John Bowden (2009), *From Galileo to Gell-Mann*. Conshohoken: Templeton Press.

> *It seems to me that this perspective is that of the book of Proverbs, which speaks at length of the most common human conditions and finally of the life of the smallest and most common animals, and in which Wisdom herself says of herself that she was with the Lord at the beginning, before the creation of the world, which delighted in this creation and loves to stand with the sons of men (cf. Proverbs 8:22-31)*

The passage Di Giorgi quotes is a delightful description of 'Wisdom' (*Sophia* in Greek, *hokhma* in Hebrew) as a little girl playing within the early epochs of the created world, as its order emerges in the form of land and sea, the depths, and the heavens. Like the poems in *Job*, it values a wisdom that blends a deep contemplation of the natural world with an active and practical engagement.

In a fascinating development that describes the experience of multiple perspectives in scientific and artistic negotiation of subject and object, Di Giorgi brings left- and right-hemisphere perspectives into focus when he describes the consequence of wisdom in science:

> *The humility and commitment to daily work must be combined with an attitude of respect and attention to every branch of knowledge since in life everyone succeeds in informing themselves only on a limited number of subjects, but can and must love all of wisdom in the broadest sense of the word.*

The path of a wise engagement with the world seems to describe an elliptic orbit, from distant points of 'panoptic' compass and integration, to the close approach of detailed study and work. In other terms, this is once more the creative process that must reciprocate between incubation and industry, between landscape and canvas, integration and analysis.

Embedded in Di Giorgi's words, but also echoing through the longer comparative stories of creativity in picture, word, and number, is also solid practical advice. Little of the radically new emerges from a narrow obsession or labour within established boundaries. There is value in broad, 'interdisciplinary' excursions, not only for their own sake, or for the benefit of recuperation, valuable though these are, but for the new patterns and connections that they offer for specific creative demands. Although the most distant connections require the deepest and longest incubation, even at the still-mysterious depths of the non-conscious, the long wait for their surfacing is worth the patience. For an early modern articulation of the same vision, we can turn to Newton's contemporary, mathematician and theologian, Isaac Barrow, who wrote in his collection of sermons:[34]

> *He can hardly be a good scholar who is not a general one, for one part of learning doth confer light upon another.*

A long tradition within the same worldview that encompasses the wisdom texts of *Job* and *Proverbs* is that, in some sense, humankind is created *in Imago Dei*—in the image of God—an idea whose hermeneutics have spawned an extensive literature, but which surely captures at its heart the fundamental urge and ability to *create*. Contemporary theologian, Philip Hefner, is probably the leading voice in the development of this approach to the challenges of being 'citizens in the commonwealth of the natural world.' His *The Human Factor, Evolution, Culture and Religion*[35] locates the human ability to create, or in collaboration with their creator, to 'co-create,' at the nexus of the evolving freedom of the world. The act is simultaneously constrained, or conditioned, by the past, but embodies freedom to explore potential in the future. Creation assumes the moral value of choice in doing so. Hefner has a late-modern take on the disjointed relationship of humankind with the world, pointing out the dangers of a runaway technology of our own making that we are no longer able to control.

The story of a creator out of control of creation transports the discussion back once more to *Job*, for the Hymn to Wisdom and the Voice from the Whirlwind hold the balance of chaos and order in constant tension. The *Book of Job*'s context of pain, and the shocking implication from both of its great poems that humans may share in the perceptive and imaginative vision of

[34] *The theological works of Isaac Barrow*, ed. A. Napier, 9 vols. (1859) Cambridge: Cambridge University Press.

[35] Philip Hefner (1993), *The Human Factor. Evolution, Culture and Religion*. Minneapolis: Fortress Press.

God, hints at another aspect to *Imago Dei*—that we share not only the ability to create but also a related propensity to suffer. The pain of separation, of disjointedness with the world, is the import of Job's anguish. The immersed—one might say incarnational—experience of questioning engagement with nature that Job experiences in the whirlwind is by no means the end of his healing, but it does signal its beginning.

Surprisingly, the theological lens through which the reflective thinkers of the twelfth and thirteenth centuries looked on the questions of human predicament, creativity, and purpose, projects those insights forwards into our own age, while at the same time drawing on much older traditions of wisdom. It is no coincidence that 'wisdom' appears explicitly (in the Greek *sophia*) in the older names for science—'natural philosophy' declares by its etymology a love of wisdom, after all. But contemporary talk about what science or art achieves is unlikely to be couched (Hefner is an exception) in explicitly theological terms. We need to listen to other ways of articulating the experience of absence in the relationship between humans and the natural world that resonate without repeating.

Modern minding of the gap

The strange persistence of a chasm to be bridged between humans and the world is discussed in other contexts and language than the theological. In the late-modern world, thinkers have used other ideas that reflect the same ancient notion. Hannah Arendt begins her book *The Human Condition* with an observation on a journalist's response to the new possibility of space exploration—a 'first step from men's imprisonment to the earth.'[36] She concludes that the history of modernism has been a turning away from the world that has increased its inhospitality, so that we are suffering

[36] Hannah Arendt (1998 [1958]), *The Human Condition*, 2nd edition, Chicago: Chicago University Press.

from 'world alienation.' It would be ironic if science were, at least in part, a path towards making the world hospitable once more, if its own framing had served to widen the gulf between us. Yet there is an inherent distancing in the very act of declaring the world an object, ourselves as observing subjects. In his third *Critique*, Kant explains the self-imposed limits of reason by saying that:[37]

> *Between the realm of the natural concept, as the sensible, and the realm of the concept of freedom, as the supersensible, there is a great gulf fixed, so that it is not possible to pass from the former to the latter by means of the theoretical employment of reason.*

Kant's suggestion is that aesthetic judgement may constitute a bridge over the chasm, although critics have pointed out that any judgement, reasoned or aesthetic, fails to overcome an intransigent divide between reasoning subject and sensed object. Kant, and the multiple other sources of this study which have pointed to the role of emotion in the creative act of imaging the world, talks to another of Arendt's concerns in *The Human Condition*, even more relevant in our century than hers. For her, one avenue of human creativity, namely the development of the electronic computer, threatened not to bridge the gulf from the human to the material world, but render it meaningless by dehumanizing:[38]

> *If we compare the modern world with that of the past, the loss of human experience .. is extraordinarily striking. It is not only and not even primarily contemplation which has become an entirely meaningless experience. Thought itself, when it became 'reckoning with consequences', became a function of the brain, with the result that electronic instruments are found to fulfil these functions much better than we ever could. . . . The trouble with modern theories of behaviourism is not that they are wrong but that they could become true*

It is sometimes difficult to believe that Arendt was writing in 1958. Three-quarters of a century on, with artificial intelligence, deep-learning, computer-aided and even computer-generated

[37] Immanuel Kant (1952) [1790], *Critique of Judgement*, trans. J. C. Meredith. Oxford: Oxford University Press, p. 11.

[38] Hannah Arendt, *The Human Condition*, p. 321

art and music a reality,[39] her warnings deserve close scrutiny. Mathematician Marcus du Sautoy, whose musical collaboration concluded Chapter 6, has considered the promise and problems of machine creativity from several angles in a recent book, *The Creativity Code*.[40] He is less fearful than Arendt for a paradoxical duality of reasons, first that computational discoveries are to be celebrated (the entirely new play in Go invented by the deep-learning algorithm 'Alpha-Go'), second that 'anything to match human creativity is still way beyond the reach even of these amazing new tools.' For true human-like creativity, he claims, would require true human consciousness. Strange then that so much of the exploration of this book has encountered the *subconscious* as the wellspring of creativity. Perhaps the real challenge of artificial intelligence from the perspective of creativity is to imagine how it could support the multiple threads that weave a human creative process: the dual submerged and sunlit worlds of thought, their intercommunication within a narrative of purpose, and their braiding together of cognition and emotion. Finally, could a machine, learning its relationship with the world, engage with the irrational elements of its encounter?

In seemingly very different guise, the same experience of unreconciled chasm between human subject and an objective world is the source of French existentialist Albert Camus' celebrated development of Søren Kierkegaard's notion of the 'absurd.' Camus means by the term not the comic, but the felt incongruence between the form of desired human relationship with the world and the one experienced:[41]

> *Man stands face to face with the irrational. He feels within him his longing for happiness and for reason. The absurd is born of this confrontation between the human need and the unreasonable silence of the world.*

[39] See Andrew Hugill, 'Transboundaries: Moving across the art/science divide', *Interdisciplinary Science Reviews*, **45**(1), 29–34 (2020).

[40] Marcus du Sautoy (2019), *The Creativity Code*. London: HarperCollins.

[41] Albert Camus (1955), *The Myth of Sisyphus and Other Essays*, trans. Justin O'Brien. London: Hamish Hamilton.

Camus' celebrated solution—'one must always struggle against the absurd'—looks futile out of context, but within the achievements of his masterly novels such as *La Peste*, the struggle becomes creative and participative, even if ultimately bleak.

Oxford theologian, Paul Fiddes takes Kant's hint at the role of the aesthetic together with its late-modern critics to suggest that the role of creativity is critical in the face of the human predicament: 'Human beings can and must cultivate a wisdom of responsible decision-making, using their substantial (if not absolute) freedom to be creative.'[42] If science's pure hypothesis-testing and rational exercise of the mind represents that segment of its orbit in which the world is objectivized, then its more hidden side of creative, imaginative ideation is the blend of close-approach and contextual connectivity that opens up a very different relationship. Creation is immersive rather than objective, it is connective rather than analytical, yet both movements of the dance are necessary to complete it. In remarkable commonality with Di Giorgi, and from an entirely different disciplinary perspective, Fiddes also advocates a close rereading of ancient Wisdom tradition in the task of restoring a broken sense of meaning in the humanities.

If Kant and Descartes are the philosophers of the objective cycle of science, then the phenomenologists, Husserl and Levinas, represent those who give voice to the immersive phase. These thinkers echo medieval entanglements of *aspectus* and *affectus* in their insistence on desire generating the appearance of objects in consciousness, and delight in regarding them when they do. For Levinas, we live immersed in 'a sensed and wanted world.'[43] For this reason, he mounts a criticism of the visual stance and is suspicious of visual metaphor altogether. We encountered in Chapter 3 its power to conceive and assemble, even to 'see' with

[42] Paul Fiddes (2014), *Seeing the World and Knowing God*. Oxford: Oxford University Press.

[43] Emmanuel Levinas (1995), *Theory of Intuition in Husserl's Philosophy*. Evanston: Northwestern Press.

deep understanding, and to re-portray, in both artistic and scientific forms, an internal and comprehendible representation of the world. But Levinas points out that the act of seeing can never escape entirely from an objectifying and distancing, even an imperialistic stance. He prefers the vocal and auditory, the immersive senses by which humans are both immersed in the world, and also surprised by it:[44]

> *While in vision a form espouses a content and soothes it, sound is like the sensible quality overflowing its limits, the incapacity of form to hold its content.*

The wordless, imaginative aspects of music and mathematics share an experience of immersion within their forms and ideas, on the part of musicians and mathematicians. These are the modes in which creators are 'drawn in' to their art and their science; within the visual they stand back. Levinas' opposition of vision and hearing bears close parallel to a puzzling critique of science by Picasso's cubist collaborator Georges Braque. When Braque claimed that, 'Art is meant to disturb, Science reassures' one wonders to what extent he had allowed the shocking content of relativity or quantum mechanics to pervade his consciousness. But if he was instead aware only of the more publicly displayed, distancing, objectifying processes within science, rather than the inward, creative and immersive, intimate and exposed aspect of all creative engagement with materiality, then his remark becomes a pointer to the role that both science and art must play in a real relationship with the world. Art and science must both reassure and trouble, call on extensions of both seeing and hearing, must both distance and immerse. A long view of even the visual metaphor for imagination, one that translates the old insight of the extramissive as well as the intromissive aspects of seeing, recognizes that the establishment of cognitive connection with the world is not a static gaze from a remove, but a dance between alternate states of immersion and inspection, of close reading and panoramic contemplation.

[44] Emmanuel Levinas (1993), *Outside the Subject*, trans. M. B. Smith. London: Athlone Press.

Both art and science have emerged from our study with elliptical orbits of the world, with perigee of self-loss and exposure, apogee of distance and regard.

The drawing out of this more complex late-modern description of human relationship to nature, with its duality of immersion and alienation, bears a strong similarity to the medieval duality of the ladder from sense to understanding, and the predicament of dulled sensibility. In very different language, the ancient and modern accounts both frame the human condition in terms of a story—a predicament – and articulate purpose to resolve it. This commonality also suggests an intriguing parallel to the 'social neuroscientific' thesis of Iain McGilchrist's *Master and His Emissary*. The close reading, analytical, immersive, and auditory aspects of our relationship with nature resonate with his summary of left-hemisphere thought, the contemplative, connective, and visual with the right. Correctly critiquing the shallow claims that science is a left-brain-dominant and art a right-dominant activity, McGilchrist points out that both approaches on their own suffer from a failure to balance and to integrate a necessary dynamic between their two poles. So, of art, the critical equilibrium is, 'the balance between the facticity of the medium and the something that is seen through the medium.'[45] We recall the shared double focus of the impressionists, and its scientific resonance. He notes that this same notion of generating *transparency* into the world through the creative act is the view of phenomenalist philosopher, Maurice Merleau-Ponty, who wrote of the purpose of art itself that, 'we see not *art* but according to art.'[46]

The relational purpose of art, its role as mediator between artists, their community, and a distorted world seen more transparently because of it, surfaces repeatedly in modern

[45] Iain McGilchrist (2009), *The Master and His Emissary*. Yale University Press, p. 373.

[46] See Duane H. Davies (2016), Merleau-Ponty and the Art of Perception. Albany: State University of New York Press.

commentary. Philosopher and mathematician A. N. Whitehead wrote:[47]

> *Great Art is the arrangement of the environment so as to provide for the soul vivid but transient values so that something new must be discovered . . . the permanent realization of values extending beyond its former self.*

Though the act of art-making itself is transient, there is a lasting quality of discovery in Whitehead's words. His description parallels the central move of the experimental method, but shows that to go beyond observation to active rearrangement is the revealing act. The artistic project becomes a moral and a necessary one. It connects, in a form of projective geometry, the self with the world beyond. Writer Julian Barnes makes the same point by completely inverting a usual way of talking about the artistic creative process:[48]

> *It is the art which illuminates, which gives the artist both his being and his significance, rather than the other way around.*

Like the precious moment of conception at the threshold of a new insight into molecular structure or the variation of hue in a rainbow, the internal creation of the new, when it is faithful to the external world, lights up the human condition. Jean-Paul Sartre explains why this experience lies at the core of being human:[49]

> *The final aim of art is to reclaim the world by revealing it as it is, but as if it had its source in human liberty.*

Sartre is very close to describing the experience of scientific imagination at its most poetic and deeply felt. In 'as it is, but. . .' he captures the meeting of form and freedom, of creativity and constraint, that must govern the exercise of freedom to explore and reimagine the world, but faithful to the form in which it is found.

[47] A. N. Whitehead (1926), quoted in Iain McGilchrist, *Master and His Emissary*, p. 280.

[48] Julian Barnes (2015), *Keeping an Eye Open*. London: Jonathan Cape, p. 237.

[49] Quoted in Frank Kermode, *The Sense of an Ending*.

Yet he writes of art, not science. At this point we must revisit the moving claim by literary scholar George Steiner that puzzled in the introductory chapter:

> *Only art can go some way towards making accessible, towards waking into some measure of communicability, the sheer inhuman otherness of matter.*

After a long journey, Steiner's words seem strange no longer, for to bridge the discomfort of distance from a strange and inhuman world, to refuse to remain static and suspended at a height, abstracted from it, but to immerse ourselves also within it, and to re-create at both poles of this orbit, is not the task of art alone, but of science as well. We have found that open avenues of re-engagement with nature are not best delineated by the division of 'art' or 'science,' but by the senses they recruit in real and metaphorical terms. Seen from within the process of a creative duality, the 'Two Cultures' picture thins, fragments, and ultimately dissolves. As anthropologist and naturalist, Loren Eiseley, wrote at the conclusion of a contemplative essay that, like Fuentes, reflected on an ancient stone tool:[50]

> *It is because these two types of creation—the artistic and the scientific—have sprung from the same being and have their points of contact even in division that . . .in a sense the 'two cultures' are an illusion, that they are a product of unreasoning fear, professionalism and misunderstanding.*

Art and science share the same three springs of imagination. The visual image offers perspective, insight, illumination. The written and spoken word bring the possibility of *mimesis* through the textual, the experimental, and the narrative form for the story of creativity itself. The wordless depths of number, the musical and mathematical, draw on the ancient insights of the liberal arts at the limits of comprehension. These are the trinity of disciplines and of modes of creation that transport our present longings for

[50] Loren Eiseley (1978), *The Illusion of the Two Cultures in The Star Thrower*. New York: Harcourt Books.

a fruitful and a peaceful home in the world, towards a future in which we are less ignorant, wiser in our relationship, but no less caught up in wonder at it.[51]

The world is connected and complex; to divide our communities of learning so deeply that we fail to recognize the commonality of creation between them is to detach ourselves even further from it, to widen rather than heal the gulf felt by Job, Kant, Arendt, Camus, and Steiner, and which generates science itself, as Grosseteste and Bacon knew. If our communities and centres of learning do not soon rediscover this, they will fail to provide the citizens of the twenty-first century with the complete and connected minds of thought and affect that they and their communities will need to flourish. The present quest to understand creativity has found itself directed towards a search for wisdom, and to the heart of what it means to be human, and to heal. If we can recognize the intense visual imagination that art and science share, the creative impulses and history of literary narrative and experimentation that is one of the greatest stories ever told, and relearn and share the ancient wisdom of wordless measure and music; if we can impart a little more of their glory to the next generation than we received from the last, and woven together in a more apparent pattern, then we might glimpse the wisdom of which the human mind is capable, and about which an ancient visionary sang:[52]

People set their hand against the flinty rock, and lay bare the roots of the mountains.
They tunnel through the rock; their eyes see all its treasures.
They explore the sources of the rivers bringing to light what has been hidden.
But where can wisdom be found?
And where is the place of understanding?

[51] The Trinitarian motif in literary and artistic creativity has been movingly explored by Dorothy Sayers (1941), *The Mind of the Maker*, 1st edition. London: Methuen.

[52] Book of Job 28, vv10–13.

Bibliography

Arendt, Hannah (1998) *The Human Condition*, 2nd edition. Chicago: Chicago University Press

Arnheim, Rudolf (1963) *The Genesis of a Painting: Picasso's Guernica*. Berkeley: The University of California Press

Auerbach, Erich (1953) *Mimesis: The Representation of Reality in Western Literature*. trans. Willard R. Trask. Princeton: Princeton University Press

Ball, Philip (2001) *Stories of the Invisible: A Guided Tour of Molecules*. Oxford: OUP

Baofu, Peter (2008) *The Future of Post-Human Unconsciousness*. Newcastle: Cambridge Scholars Publishing

Barnes, Julian (2015) *Keeping an Eye Open: Essays on Art*. London: Jonathan Cape

Barzun, Jacques (1964) *Science: The Glorious Entertainment*. London: Harper and Row

Bede (2010) *On the Nature of Things* and *On Times*. Calvin B. Kendall and Faith Wallis, eds. Liverpool: Liverpool University Press

Beer, Gillian (1983) *Darwin's Plots*. Cambridge: Cambridge University Press

Begbie, Jeremy (2014) *Theology, Music and Time*. Cambridge: Cambridge University Press

Bernard, Claude (1957) *Introduction to the Study of Experimental Medicine*. New York: Dover Publications

Bersanelli, Marco and Mario Gargantini (2009) *From Galileo to Gell-Mann*. trans. John Bowden. Conshohoken: Templeton Press

Beveridge, William I. B. (1950) *The Art of Scientific Investigation*. New York: W. W. Norton and Co. Ltd.

Bohm, David (1996) *On Creativity*. London and New York: Routledge

Booker, Christopher (2004) *The Seven Basic Plots*. London: Bloomsbury Press

Boyer, C. B. (1959) *The Rainbow: From Myth to Mathematics*. New York: Yoseloff

Boyle, Robert (1772) *The Works of the Honorable Robert Boyle*, A new edition. W. Johnson et al. eds. London: Printed by Miles Flesher for Richard Davis, bookseller in Oxford

Brooks, Michael (2011) *The Secret Anarchy of Science*. London: Profile Books

Brueggemann, Walter (2001 [1978]) *The Prophetic Imagination*. Minneapolis: Fortress Press

Camus, Albert (1955) *The Myth of Sisyphus and Other Essays*. trans. Justin O'Brien. London: Hamish Hamilton

Carey, John (2006) *What Good are the Arts?* Oxford: Oxford University Press

Margaret Cavendish (1953), *A World Made by Atomes*. Electronic edition available at Emory Women Writers' Project. Atlanta: Emory University. https://womenwriters.digitalscholarship.emory.edu/toc.php?id=atomic (accessed 18.02.2021)

Chandrasekhar, Subrahmanyan (1987) *Truth and Beauty: Aesthetics and Motivations in Science*. Chicago: University of Chicago Press

Christianson, Gale E. (2005) *Isaac Newton*. Oxford: Oxford University Press

Coakley, Sarah (2013) *God, Sexuality and the Self*. Cambridge: Cambridge University Press

Coleridge, Samuel Taylor (1817) *Biographia Litteraria*, ed. J. Engell and W. Jackson Bate (Princeton 1983)

Cornford, Francis (1991) *From Religion to Philosophy: A Study in the Origins of Western Speculation*. Princeton: Princeton University Press

Crawford, Robert (ed.) (2006) *Contemporary Poetry and Contemporary Science*. Oxford: Oxford University Press

Csikszentmihalyi, Mihaly (1996) *Creativity: The Psychology of Discovery and Invention*. New York: Harper Perennial

Da Vinci, Leonardo (1952) *Notebooks*, ed. Irma A. Richter. Oxford: Oxford University Press

Daston, Lorraine and Katherine Park (2001) *Wonders and the Order of Nature 1150–1750*. New York: Zone Books

Daverio, John (1997) *Robert Schumann, Herald of a 'New Poetic Age'*. Oxford: Oxford University Press

Daverio, John (2002) *Crossing Paths: Schubert, Schumann, and Brahms*. Oxford: Oxford University Press

Demasio, Antonin (2006) *Descartes' Error: Emotion, Reason and the Human Brain*. New York: Random House

Dinkova-Brun, Greti, et al. (2013) *Dimensions of Colour: Robert Grosseteste's De Colore; Edition, Translation and Interdisciplinary Analysis*. Durham: Durham Medieval and Renaissance Texts

Du Châtelet, Madame la Marquise *Principes mathématiques de la philosophie naturelle par feue* (1st edition, 1756; 2nd edition, 1759)

Du Sautoy, Marcus (2019), *The Creativity Code*. London: Harper Collins

Ede, Sian (2005) *Art and Science*. New York: Tauris, Palgrave MacMillan

Einstein, Albert (1954) *Ideas and Opinions*. New York: Bonanza Books

Einstein, Albert and Leopold Infeld (1938) *The Evolution of Physics*. London: Cambridge University Press

Eiseley, Loren (1978) *The Illusion of the Two Cultures in 'The Star Thrower'*. New York: Harcourt Books

Eliot, Thomas Sternes (1951), *Poetry and Drama*. Scholar's Select

Elson, Rebecca (2018), *A Responsibility to Awe*. Manchester: Carcanet Press Ltd.

Emerson Ralph Waldo (2010) *Works*, vols. 8, 66. Boston: Harvard University Press

Epstein, Mikhail (2012) *The Transformative Humanities—A Manifesto*. New York: Bloomsbury

Fara, Patricia (2021), *Life After Gravity*. Oxford: Oxford University Press

Fay, Laurel (1999) *Shostakovich: A Life*. Oxford University Press

Feynman, Richard P. (1992) *Surely You're Joking Mr. Feynman?* London: Random House

Fiddes, Paul (2014) *Seeing the World and Knowing God*. Oxford: Oxford University Press

Frascina, F. and C. Harrison (eds.) (1982) *Modern Art and Modernism*. London: Harper-Row in association with the Open University

Frenkel, Edward (2013) *Love and Math: The Heart of Hidden Reality*. Basic Books

Fuentes, Agustin (2017) *The Creative Spark*. New York: Dutton

Fuller, Steve, Mark de May, and Steve Woolgar eds. (1989) *The Cognitive Turn: Social and Psychological Perspectives on Science*. Dordrecht: Springer

Funkenstein, Amos (1986) *Theology and the Scientific Imagination—From the Middle Ages to the Seventeenth Century*. Princeton University Press, Princeton, New Jersey

Gál, Hans (1979) *Schumann Orchestral Music*. London: BBC Publications

Gardner, Howard (1993) *Creating Minds—An Anatomy of Creativity*. New York: Basic Books

Gasper, Giles E. M., Cecilia Panti, Tom C. B. McLeish, and Hannah E. Smithson, eds. (2019) *The Scientific Works of Robert Grosseteste Vol. 1 Knowing and Speaking: Robert Grosseteste's De artibus liberalibus 'On the Liberal Arts' and De generatione sonorum 'On the Generation of Sounds'*. Oxford: Oxford University Press, trans. therein Sigbjørn O. Sønnesyn

Ghiselin, Brewster (ed.) (1985) *The Creative Process*. Berkeley: University of California Press

Gombrich, Ernst (2002) *Art and Illusion A Study in the Psychology of Pictorial Representation*. London: Phaidon Press

Gould, Stephen J. (2003) *The Hedgehog, The Fox and the Magister's Pox*. Vintage, London

Grant, Edward (1996) *The Foundations of Modern Science in the Middle Ages*.Cambridge: Cambridge University Press

Greimas, A. J. (1966) *Sémantique structural*. Paris: Seuil

Guite, Malcolm (2012) in *Faith, Hope and Poetry*. Oxford: Ashgate

Hacking, Ian (2006) *The Emergence of Probability*, 2nd edition. New York: Cambridge University Press

Hadamard, Jacques S. (1945) *A Mathematician's Mind, Testimonial for An Essay on the Psychology of Invention in the Mathematical Field*. Princeton: Princeton University Press

Hallam, S., Cross, I. and Thaut, M., eds. (2009) *The Oxford Handbook of Music Psychology*. Oxford: Oxford University Press

Harding, Rosamond E. M. (1940) *An Anatomy of Inspiration*. Oxford: Frank Cass & Co.

Harrison, Peter (2007) *The Fall of Man and the Foundations of Science*. Cambridge: Cambridge University Press

Harrison, Peter (2015) *The Territories of Science and Religion*. Chicago: Chicago University Press

Hart, David Bentley (2003) *The Beauty of the Infinite—the Aesthetics of Christian Truth*. Grand Rapids: Eerdmans

Hart, David Bentley (2017) in *A Splendid Wickedness and Other Essays*. Grand Rapids: Eerdmans

Hefner, Philip (1993) *The Human Factor. Evolution, Culture and Religion*. Minneapolis: Fortress Press

Heisenberg Werner (1972) *Physics and Beyond: Encounters and Conversations*, trans. Arnold J. Pomerans. New York: Harper Torchbooks

Holmes, Richard (2008) *The Age of Wonder*. London: Harper

von Humboldt, Alexander (1808) *Ansichten an die Natur*, Stephen T. Jackson and Laura Dassow Walls (eds.), trans. Mark W. Person as *Views of Nature* (2014). Chicago: University of Chicago Press

von Humboldt, Alexander (1858) *Cosmos: A Sketch of the Physical Description of the Universe*, trans. E. C. Otte, (1997). Maryland: Johns Hopkins Press

Hume, David (1960) *A Treatise of Human Nature*. L. A. Selby-Bigge, ed. Oxford: Oxford University Press

Hume, David (2007) *An Enquiry Concerning Human Understanding*. P. Milligan, ed. Oxford: Oxford University Press

Hunt, Robert (1850 [1848]), *The Poetry of Science: Or the Studies of the Physical Phenomena of Nature*. Boston: Gould, Kendall, and Lincoln

Iamblichus (1920) *De Vita Pythagorica (On the Pythagorean Life)*, c. 300 AD *Iamblichus, Life of Pythagoras*, trans. Kenneth Sylvan Guthrie. Alpine, NJ: Platonist Press (1919)

Illingworth, Sam (2019) *A Sonnet to Science*. Manchester: Manchester University Press

Jacobsson, Martin (2002) *Aurelius Augustinus De musica liber VI A Critical Edition with a Translation and an Introduction*. Stockholm: Almqvist & Wiksell International

James, Henry (1934) *The Art of the Novel*. Chicago: University of Chicago Press (2011)

Jones, David (2012) *The Aha! Moment*. Baltimore: Johns Hopkins Press

Jung, Carl G. (1933) *Modern Man in Search of his Soul*, trans. W. S. Dell and Cary F. Baynes. Oxford: Routledge

Kant, Immanuel (1996) *Critique of Judgement, Critique of Pure Reason*, trans.Werner Pluhar. Indianapolis: Hackett

Kaufman, James C. and Robert J. Steinberg (2010) *The Cambridge Handbook of Creativity*. Cambridge: Cambridge University Press

Kermode, Frank (2000) *The Sense of an Ending: Studies in the Theory of Fiction*, Oxford: Oxford University Press

Killeen, Kevin, ed. (2014) *Thomas Brown, Selected Writings*. Oxford: Oxford University Press

Killeen, Kevin (2021), 'Poetry and Natural Philosophy: The errant soul in John Davies, John Donne and Phineas Fletcher' in *Oxford Handbook of Renaissance Poetry*, Jason Scott-Warren and Andrew Zurcher (eds). Oxford: OUP

Kuhn, Thomas (1966) *The Structure of Scientific Revolutions*. Chicago: Chicago University Press

Levinas, Emmanuel (1993) *Outside the Subject*. trans. M. B. Smith. London: Athlone Press

Levinas, Emmanuel (1995) *Theory of Intuition in Husserl's Philosophy*. Evanston: Northwestern Press

Lewis, C. S. (1960) *The Four Loves*. London: Geoffrey Bless

Lewis, C. S. (1994) *The Discarded Image*. Cambridge: CUP Canto Edition

Lightman, Alan (2005) *A Sense of the Mysterious*. New York: Vintage Books

Logan, I. (2009) *Reading Anselm's* Proslogion *The History of Anselm's Argument and Its Significance Today*. Farnham, UK and Burlington, VT: Ashgate

MacDonald, George (1893), *A Dish of Orts* (Lexington, 2015)

McCarty, Willard (ed.) (2020) 'Tom McLeish, The Poetry and Music of Science: Précis, reviews and a response,' *Interdisciplinary Science Reviews*, **45**(1), 1–70

McEvoy, James (2000) *Robert Grosseteste*. Oxford: OUP

McGilchrist, Iain (2009) *The Master and his Emissary*. New Haven: Yale University Press

McLeish, Tom (2014) *Faith and Wisdom in Science*. Oxford: Oxford University Press

McLeish, Tom (2020) *Soft Matter—A Very Short Introduction*. Oxford: Oxford University Press

Medawar, Peter (1984) *An Essay on Scians* in *The Limits of Science*. Oxford: Oxford University Press

Medawar, Peter (1984) *Pluto's Republic*. Oxford: Oxford University Press

Midgley, Mary (2001), *Science and Poetry*. Abingdon: Routledge

Motion, Andrew (1993) *Philip Larkin: A Writer's Life*. London: Faber

Murdoch, Iris (1998) *Existentialists and Mystics*. New York: Penguin Books

Peelen, Mary (2019) *Quantum Heresies*. Glenview: Glass Lyre Press

Planck, Max (1932) *Where is Science Going?* New York: W. W. Norton and Co.

Poincaré, Henri (1915) Mathematical Creation, in *The Foundations of Science*, trans. G. B. Halsted. Lancaster, Pennsylvania: The Science Press

Polanyi, Michael (1962) *Personal Knowledge*. London: Routledge

Polti, Georges (1921) *The Thirty-Six Dramatic Situations*. Franklin, Ohio: James Knapp Reeve

Popper, Karl (1976 [2002]) *Unended Quest: An Intellectual Autobiography*. London and New York: Routledge

Popper, Karl (2002 [1934]) *The Logic of Scientific Discovery*. London and NewYork: Routledge

Porteous, Katrina (2019) *Edge*. Hexham: Bloodaxe Books

Principe, Lawrence (2013) *The Scientific Revolution: A Very Short Introduction* Oxford: Oxford University Press

Riordan, Maurice and Jon Turney (eds.) *A Quark for Mister Mark*. London: Faber and Faber (2000)

Rothenberg David (2012) *Survival of the Beautiful: Art, Science and Evolution*.Bloomsbury, London

Ruskin, John (1905) *The Eagle's Nest*. George Allen, London

Saunders, Corrine and Jane Macnaughten, eds. (2015) *The Recovery of Beauty*. London: Palgrave MacMillan

Sayers, Dorothy L. (1941) *The Mind of the Maker*, 1st edition. London: Methuen

Scarre, Chris and Graeme Lawson (2006) *Archaeoacoustics*. McDonald Institute for Archaeological Research and Oxbow Books

Schaefer, Donovan O. (2017) *The Wild Experiment: Emotion, Reason, and the Limits of Science*, in Gillian Straine, ed. *Are There Limits to Science?* Newcastle: Cambridge Scholars Publishing

Schumann, Robert (1965) *Schumann on Music; A Selection from the Writings*, trans. and ed. Henry Pleasants. New York: Dover

Scruton, Roger (1998) *On Hunting*. London: Random House

Shapin, Steven (1996) *The Scientific Revolution*. Chicago: University of Chicago Press

Shavina, L. V., ed. (2003) *The International Handbook on Innovation*. New York: Elsevier Science

Shimamura, Arthur P. and Stephen E. Palmer, eds. (2014) *Aesthetic Science*.Oxford: Oxford University Press

Singh, Simon (1997) *Fermat's Last Theorem*. London: Forth Estate Ltd.

Snow, C. P. (1959 [1998]) *The Two Cultures*. Cambridge: Cambridge University Press

Southern, R. W. ed. and trans. (1979) *Life of Saint Anselm, Archbishop of Canterbury*, by Eadmer. Oxford: Clarendon Press

Southern, R. W. (1986) *Robert Grosseteste: The Growth of an English Mind in Medieval Europe*. Oxford: Clarendon Press

Sprat, Thomas (1667), History of the Royal Society of London, for the Improving of Natural Knowledge. London: The Royal Society

Steinberg, Robert J., ed. (1988) *The Nature of Creativity: Contemporary Psychological Perspectives*. Cambridge: Cambridge University Press

Steinberg, Robert J. and James C. Kaufman (2018) *The Nature of Human Creativity*. Cambridge: Cambridge University Press

Steiner, George (1989) *Real Presences*. London: Faber and Faber

Storr, Anthony (1993) *The Dynamics of Creation*. New York: Ballantine Books

Strang, V., T. Edensor and J. Puckering, eds. (2018) *From the Lighthouse, A Collection of Interdisciplinary Essays*. Abingdon: Routledge

Strathern, Marilyn (2005) *Kinship, Law and the Unexpected.* New York: Cambridge University Press

Summers, David (2003) *Real Spaces: World Art History and the Rise of Western Modernism.* London: Phaidon Press

Theodoric of Freiburg (1914), *De Iride et Radialibus Impressionibus*. J. Würschmidt, ed. Münster

Wallas, Graham (1926 [2016]) *The Art of Thought*. Tunbridge Wells: Solis Press

Walls, Laura Dassow (2009) *The Passage to Cosmos: Alexander von Humboldt and the Shaping of America*. Chicago: University of Chicago Press

Whewell, William (1837) *History of the Inductive Sciences*. London: John W. Parker

White, Michael (1997) *Isaac Newton: The Last Sorcerer*. London: Harper Collins

Wilczek, Frank (2015) *A Beautiful Question*. London: Allen Lane, Random House

Wilton, Andrew (1980) *Constable's 'English Landscape Scenery'*. London: British Museum Publications

Wolterstorff, Nicholas (1980) *Art in Action—Towards a Christian Aesthetic*. Grand Rapids: Eerdmans

Wright, N. T. (1992) *The New Testament and the People of God*. London: SPCK

Wulf, Andrea (2015) *The Invention of Nature*. New York: Alfred Knopf

Zola, Emile (1964) The experimental novel, in *The Naturalist Novel*, Maxwell Geismar, ed. Ste. Anne de Bellevue: Harvest House Ltd.

Subject index